Graduate Texts in Mathematics 132

**Springer**
*New York*
*Berlin*
*Heidelberg*
*Barcelona*
*Hong Kong*
*London*
*Milan*
*Paris*
*Singapore*
*Tokyo*

# Graduate Texts in Mathematics



*(continued after index)*

Alan F. Beardon

# Iteration of Rational Functions

## *Complex Analytic Dynamical Systems*

**With 35 Illustrations**

Springer

Alan F. Beardon
Department of Pure Mathematics
and Mathematical Statistics
University of Cambridge
Cambridge CB2 1SB
UK



Library of Congress Cataloging-in-Publication Data
Beardon, Alan F.
Iteration of rational functions/Alan Beardon.
p. cm. — (Graduate texts in mathematics; 132)
Includes bibliographical references and indexes.
ISBN 0-387-95151-2
1. Iterative Methods (Mathematics) 2. Mappings (Mathematics)
I. Title. II. Series.
QA297.8.B43 1991
511'.4—dc20 91-24082

Printed on acid-free paper.

*First softcover printing, 2000.*


Typset by Asco Trade Typsetting Ltd., Hong Kong.
Printed and bound by Edwards Brothers, Inc., Ann Arbor, MI.
Printed in the United States of America.

9 8 7 6 5 4 3 2 1

ISBN 0-387-95151-2 SPIN 10783032

Springer-Verlag New York Berlin Heidelberg
*A member of BertelsmannSpringer Science+Business Media GmbH*

To Francis, Jessica, Yvonne and Luke

# Preface

This is not a book for the experts, nor is it written by one; it is a modest attempt to lay down the basic foundations of the theory of iteration of rational maps in a clear, precise, complete and rigorous way. The author hopes that those who wish to learn something about the subject will be able to do so from this book in a relatively painless way, and that it will serve as a starting point from which many recent, and much deeper, works can be tackled with some confidence.

The book begins, and ends, with a chapter consisting entirely of examples. In the first chapter, the examples are quite straightforward and are discussed from first principles without the advantage of any theoretical developments. Many readers will want to omit this chapter, but its purpose is two-fold. First, this subject is of interest to a large number of people not all of whom are mathematicians, and it is hoped that some of these readers will appreciate the more gentle start offered by this chapter; and second, in this chapter I illustrated most of the basic results of the theory in specific examples. The last chapter also consists entirely of examples but, by contrast, a claim about a particular example here demands as much formal verification as does the proof of a theorem. The primary purpose of these examples is, of course, to illustrate the theory developed earlier, but in addition to this, they have been chosen to show the variety of possibilities that can occur, and some at least go beyond those for which the computer-generated illustrations are now so familiar. For the convenience of readers, I have included an index of examples at the end of the text.

I have included a brief section at the beginning which describes some of the elementary topics that I shall assume the reader is familiar with. Other (more advanced) material is assumed at several other places in the text, but there, some explanation is merged with the general discussion. Each chapter starts with a summary outlining the main objectives in that chapter. There are, of course, occasions when I need to use more advanced results from other parts of mathematics, and where I have thought that a brief discussion of these would materially assist the reader I have included such a discussion in the text. Where I felt that it would not, I have relegated further discussion to an appendix to that chapter. Finally, and perhaps inevitably, I accept that some important items are omitted (most notably, the existence of Herman rings), but this is not in any sense meant to be a complete account of the subject.

It has been my objective to provide as much detail as seems appropriate for an average graduate student to understand the argument completely and without too much effort, and the criterion for the inclusion of detail has been whether or not I thought that it would assist the reader. In several places there is some minor repetition of material; this is simply an acknowledgement that most readers do not read (and authors do not write) books in the same order as their pages are numbered and so, on occasions, it is helpful to some readers to have this repetition. The greatest difficulty seemed to be in placing the material in a coherent order, and to avoid constantly changing from one topic to another as seems to happen so often in other accounts of the subject: I believe that I have been reasonably successful in this but, ultimately, it is for the reader to judge. I believe that important mathematical points should be stressed (even when they are mathematically trivial), and I have written this book in the belief that the onus lies with authors, not readers, to provide the details.

There are references given in the text, but I have not attempted to include references to all results, nor to trace the results back to the original source: indeed, given some of the informal, expository (and sometimes incomplete) accounts of the subject that exist, this would have sometimes been difficult, although, of course, almost all of the results originate with Fatou and Julia. There are no original illustrations in the text; the existing pictures are more than adequate for my purposes and I am grateful for those who have allowed me to use their illustrations.

In writing this text, I have had to learn the subject myself, and I have relied heavily on the help, encouragement and advice of many people. Noel Baker generously supplied me with notes for a course he gave, and as well as reading the manuscript, has responded willingly to a stream of questions (not all sensible) from me. Keith Carne has also read the manuscript, and has listened patiently and responded to the ideas and difficulties I have had, and his interest and support in this project has been most valuable. David Herron, Bruce Palka, Cliff Earle, Kari Hag, Pekka Koskela and Shanshuang Yang participated in a seminar which worked through a large portion of the manuscript and their comments and suggestions have led to a significant improvement in the text. Norbert Steinmetz provided one of the ideas in Chapter 7, and Fred

Gehring, as before, has been a great support. To all these people, and others who have helped in various ways, I offer my thanks. Of course, I take full responsibility for any errors that remain.

Cambridge, England
November 1990

Alan F. Beardon

# Contents

# Prerequisites

This section contains notation, terminology and some of the results that are taken for granted in the text. First, the notation. The real line, the complex plane and the extended complex plane are denoted by $\mathbb{R}$, $\mathbb{C}$ and $\mathbb{C}_\infty$ respectively, and throughout the text, $\Delta$ denotes the unit disc in $\mathbb{C}$. For any set $A$, the closure, the boundary and the interior of $A$ (all with respect to some underlying space $X$ which will be clear from the context) are $\bar{A}$, $\partial A$ and $\mathrm{Int}(A)$, or $A^0$, respectively. For sets $A$ and $B$, $A - B$ denotes the set difference (rather than $A \backslash B$ which I find visually unattractive); thus

$$A - B = \{x \in A : x \notin B\},$$

and the complement of $A$ in $X$ is $X - A$.

The symbol $\mapsto$ defines a function $f$ (for example, $x \mapsto x^2$) as well as, of course, $f(x) = x^2$. Often, visual clarity is improved if brackets are omitted, so I use $f(x)$ and $fx$ interchangeably. Likewise, if the composition $x \mapsto f(g(x))$ is defined, it is denoted by $fg$. These liberties allow one to inject a particular emphasis into a formula; for example, $f(gx)$ is to be thought of as the $f$-image of $g(x)$, while $fg(x)$ (the same point) is the $fg$-image of $x$. The composition of $f$ with itself $n$ times is the $n$-th iterate $f^n$ of $f$, and $f^0 = I$, the identity map. As usual, both notations $f'$, $f''$, and $f^{(n)}$ are used for the derivatives of $f$.

A small amount of complex analysis is taken for granted, roughly speaking that which would be covered in a first (and conventional) course in the subject. For example, we shall assume familiarity with the Maximum Modulus Theorem, Schwarz's Lemma and Rouché's Theorem. All of these results can be found in, for example, [3]. We say that $f$ is a $d$-fold map of $V$ onto $W$ if, for every $w$ in $W$, the equation $f(z) = w$ has exactly $d$ solutions in $V$ (counting multiple solutions by their multiplicity); for example, a polynomial of degree $d$ is a $d$-fold map of $\mathbb{C}$ onto itself. If $d = 1$ we say the map is univalent, and at various points in the text we shall use Hurwitz's Theorem (that if a sequence univalent analytic maps $f_n$ converge uniformly to $f$ on a domain $D$, then $f$ is either constant or univalent in $D$). This too can be found in [3].

Finally, we shall assume familiarity with the very basic ideas of metric spaces, namely those up to, say, uniform continuity, compactness and connectedness. We stress, however, that the material in Chapter 1 needs none of these ideas, and that some attempt has been made to match progression through the text with an assumption of increasing mathematical maturity.

CHAPTER 1

# Examples

In this chapter we introduce some of the main ideas in iteration theory by discussing a variety of simple examples. The discussions involve only elementary mathematics, and our sole objective is to illustrate and stress those features that will be met in a general context later.

## §1.1. Introduction

This book is about the repeated application, or *iteration*, of a rational function,

$$R(z) = \frac{a_0 + a_1 z + \cdots + a_n z^n}{b_0 + b_1 z + \cdots + b_m z^m},$$

of a complex variable $z$. Specifically, we select a starting point $z_0$ in the complex plane $\mathbb{C}$ and then apply $R$ repeatedly constructing, in turn, the points

$$z_0, z_1 = R(z_0), z_2 = R(z_1), \ldots.$$

In general, we denote the composition of two functions $f$ and $g$ by juxtaposition so $fg$ is the function $z \mapsto f(g(z))$, and we allow ourselves to write either $fg(z)$ or $f(gz)$ depending on which of these we wish to emphasize. With this notation, $z_n = R^n(z_0)$, and by convention, $R^0 = I$, where $I$ is the identity map.

Many questions now present themselves; for example, does the sequence $z_n$ converge, or, better still, for which values of the initial point $z_0$ does the sequence $z_n$ converge? If the sequence $z_n$ does not converge, can we say anything else about its behaviour and, in any case, how robust are the answers to these questions under a small change in the initial point $z_0$? Instead of looking at the future progress of $z_0$, we can also look at its history as represented, say, by the sequence

$$\ldots, z_{-2}, z_{-1}, z_0,$$

where again $z_{n+1} = R(z_n)$. In general, for a given $z_0$ there will be several different possibilities for $z_{-1}$, even more for $z_{-2}$, and so on, so here there is a case for considering the totality of such sequences arising from a given point $z_0$.

We can gain a little insight immediately by making some elementary observations about fixed points. A point $\zeta$ is a *fixed point* of $R$ if $R(\zeta) = \zeta$, and it is clear that such points must have a special role to play in the theory. Suppose now that for some choice of $z_0$, the sequence $z_n$ converges to $w$. Then (because $R$ is continuous at $w$)

$$w = \lim_{n\to\infty} z_{n+1} = \lim_{n\to\infty} R(z_n) = R\left(\lim_{n\to\infty} z_n\right) = R(w),$$

so $w$ is a fixed point of $R$: thus *if* $z_n \to w$, *then* $R(w) = w$. For example, if

$$R(z) = z^2 - 4z + 6, \tag{1.1.1}$$

then, regardless of the choice of $z_0$, if the sequence $z_n$ converges it can only converge to 2, 3 or $\infty$ (we will discuss $\infty$ later). As

$$R(z) - 2 = (z - 2)^2,$$

the reader can now find those $z_0$ for which $z_n \to 2$.

If the fixed point $\zeta$ of $R$ lies in $\mathbb{C}$, then the derivative $R'(\zeta)$ is defined and we say that $\zeta$ is:

(1) an *attracting fixed point* if $|R'(\zeta)| < 1$;
(2) a *repelling fixed point* if $|R'(\zeta)| > 1$; and
(3) an *indifferent fixed point* if $|R'(\zeta)| = 1$.

This classification will be discussed again in much greater detail in Chapter 6, but it will be helpful to make some preliminary remarks now. If $z$ is close to the fixed point $\zeta$, then, *approximately*,

$$|R(z) - \zeta| = |R(z) - R(\zeta)| = |R'(\zeta)|.|z - \zeta|,$$

so points close to an attracting fixed point move even closer to it when we apply $R$, while points close to a repelling fixed point tend to move away from it. In particular, if $z_0$ lies sufficiently close to an attracting fixed point $\zeta$, then $z_n \to \zeta$ as $n \to \infty$. On the other hand, if $z$ is close to (but not equal to) a repelling fixed point $\zeta$, initially it is repelled away from $\zeta$, *but it may return to the vicinity of* $\zeta$ (or even to $\zeta$ itself) at a later stage. In fact, the only way that $z_n$ can converge to a repelling fixed point $\zeta$, is to have $z_n = \zeta$ for $n \geq n_0$, say. To see this, we suppose that $z_n \to \zeta$, where $z_n \neq \zeta$ for any $n$, and seek a contradiction. Certainly, the fact that the $z_n$ converge to, but are distinct from, $\zeta$ implies that for infinitely many $n$,

$$|z_{n+1} - \zeta| < |z_n - \zeta|.$$

However, we can choose a number $k$ such that $|R'(\zeta)| > k > 1$, and a neighbourhood $N$ of $\zeta$ such that for $z$ in $N$,

$$|R(z) - \zeta| = |R(z) - R(\zeta)| > k|z - \zeta|,$$

and then putting $z = z_n$, we obtain a contradiction. For example, the function (1.1.1) has attracting fixed points 2 and $\infty$, and a repelling fixed point 3. Thus for this function, if $z_n \to w$, then either $w$ is 2 or $\infty$, or $z_n = 3$ for some $n$ (this happens, for example, when $z_0 = 1$).

We shall now describe, in the simplest possible way, the central idea in iteration theory. Given two starting points $z_0$ and $w_0$, we construct the sequences $z_n$ and $w_n$ as above, and ask the question *if $w_0$ is sufficiently close to $z_0$, do the two sequences $z_n$ and $w_n$ exhibit roughly the same behaviour as $n$ tends to $\infty$?* In practical terms, if we wish to investigate the sequence $z_n$, where $z_0 = \sqrt{2}$, will it help if we take $w_0$ to be 1·414, and then investigate the sequence $w_n$? The answer naturally depends on the choice of $z_0$, so we divide the complex plane into the set $F$ consisting of those $z_0$ for which the answer is "*yes*" (for example, when the sequences $z_n$ and $w_n$ converge to the same point), and the set $J$ consisting of those $z_0$ for which the answer is "*no*" (for example, when the sequences $z_n$ and $w_n$ oscillate violently and independently). The action of the $R^n$ on the set $F$ is to preserve the proximity of points (the formal notion is equicontinuity), while the repeated action of $R$ on $J$ is to drive the points apart. *The division of the plane into the sets $F$ and $J$ is fundamental and ever present, and the reader should keep it clearly in mind.*

The seminal ideas, and the foundations, of this subject originated with the French mathematicians Pierre Fatou and Gaston Julia around 1918, and more recently (due largely to computer graphics which were not available to Fatou and Julia) the subject is enjoying a resurgence of interest at all levels of mathematics. The subject can be studied theoretically and empirically, and both approaches are valuable, or even essential; however, the purpose of this book is to provide a careful and precise treatment of the fundamental theoretical ideas and we shall have very little to say about computer graphics (although, naturally, we use computer-generated pictures to illustrate our results).

In this introductory chapter, we shall be content to consider a variety of explicit examples, all of which will be discussed using fairly primitive techniques. *The sole purpose of these examples is to aquaint the reader with some of the central ideas in an informal and concrete setting* and readers who are familiar with the subject can safely omit this chapter.

## §1.2. Iteration of Möbius Transformations

A Möbius transformation is a rational map of the form

$$R(z) = \frac{az + b}{cz + d}, \qquad ad - bc \neq 0,$$

where we have the usual convention that

$$R(\infty) = a/c, \qquad R(-d/c) = \infty,$$

if $c \neq 0$, while $R(\infty) = \infty$ when $c = 0$. Möbius transformations are among the small class of functions whose iterates can be computed explicitly and, as a result, we can easily see that for the vast majority of these functions, the sequence $z_n$ converges almost regardless of the choice of the starting point $z_0$.

We begin by considering an example, namely

$$R(z) = \frac{3z - 2}{2z - 1}. \tag{1.2.1}$$

Then (by induction)

$$R^n(z) = \frac{(2n + 1)z - 2n}{2nz - (2n - 1)} = 1 + \frac{z - 1}{2nz - (2n - 1)}, \tag{1.2.2}$$

and so we find that for all $z$,

$$R^n(z) \to 1 \tag{1.2.3}$$

as $n \to \infty$. The reader should be warned, however, that despite this convergence, the point 1 is *not* an attracting fixed point of $R$. Indeed, $R'(1) = 1$, and a simple calculation shows that starting at $z_0 = 1 - \varepsilon$, say, where $\varepsilon$ is small and positive, the iterates $z_n$ start to move *away from* the point 1. Nevertheless, (1.2.3) shows that the points $z_n$ eventually return to a neighbourhood of, and indeed converge to, the point 1.

As one might expect, there is a general result underlying this discussion, and we now consider this in detail. Any Möbius $R$ either has a single (repeated) fixed point, or it has two distinct fixed points, and we consider each case in turn.

*Case* 1: *$R$ has a single fixed point.*

Suppose first that $R$ has $\infty$ as its only fixed point. Then $R(z) = z + \beta$, say, where $\beta \neq 0$, so

$$R^n(z) = z + n\beta,$$

and for every $z$, $R^n(z) \to \infty$ as $n \to \infty$.

Now suppose that $R$ has some point $\zeta$ in $\mathbb{C}$ as its only fixed point. In this case, let $g(z) = 1/(z - \zeta)$ (a Möbius map taking $\zeta$ to $\infty$), and define $S$ by

$$S(z) = gRg^{-1}(z).$$

As $S$ fixes $z$ if and only $z = \infty$, it follows that $S$ is a translation, and hence that $S^n(z) \to \infty$ as $n \to \infty$. Also,

$$\begin{aligned} S^n(z) &= (gRg^{-1})(gRg^{-1})\cdots(gRg^{-1})(z) \\ &= gR^n g^{-1}(z), \end{aligned} \tag{1.2.4}$$

and so replacing $z$ by $g(z)$, and applying $g^{-1}$, we find that

$$R^n(z) \to g^{-1}(\infty) = \zeta.$$

We have now shown that *if $R$ is a Möbius transformation with a unique fixed point $\zeta$, then, for every $z$, $R^n(z) \to \zeta$*. Note that this discussion contains a method for actually computing the explicit form of $R^n$, for $R^n(z) = g^{-1}S^n g(z)$ and $g$ and $S^n$ are known explicitly.

*Case 2: R has exactly two (distinct) fixed points.*

Suppose first that $R$ fixes 0 and $\infty$. Then $R(z) = kz$ and $R^n(z) = k^n z$. Clearly, $R^n$ fixes 0 and $\infty$, while for all other $z$,

$$R^n(z) \to 0 \quad \text{if } |k| < 1;$$

$$|R^n(z)| = |z| \quad \text{if } |k| = 1;$$

$$R^n(z) \to \infty \quad \text{if } |k| > 1.$$

Further, when $|k| = 1$, we have either:

(i) *$k$ is an $n$-th root of unity* and $R^n$ is the identity; or
(ii) *$k$ is not a root of unity*, and the points $R^n(z)$ are dense on the circle with centre the origin and radius $|z|$.

Now suppose that $R$ has exactly two fixed points $\zeta_1$ and $\zeta_2$ where $\zeta_1 \neq \zeta_2$. First, we construct a Möbius transformation $g$ which maps $\zeta_1$ to 0, and $\zeta_2$ to $\infty$: for example, if $\zeta_1$ and $\zeta_2$ are both finite, we can take

$$g(z) = \frac{z - \zeta_1}{z - \zeta_2}.$$

Now let $S = gRg^{-1}$: then $S$ fixes 0 and $\infty$ and so our previous remarks apply to $S$. As $g$ maps circles (including straight lines) to circles, we find that *if $R$ has two fixed points, then either the $R^n(z)$ converge to one of the fixed points of $R$, or they move cyclically through a finite set of points, or they form a dense subset of some circle.* As before, this gives a method of finding $R^n$ explicitly.

We have just seen that the iterates of a rational function of degree one, namely the Möbius maps, behave in a very simple manner. *These functions stand alone in their simplicity, and for the rest of the text we shall only be concerned with rational functions of degree at least two.*

EXERCISE 1.2

1. Show that if $R$ is given by (1.2.1), and if $x = 1 - \varepsilon$, where $\varepsilon$ is small and positive, then $R(x) < x < 1$.

2. Verify (1.2.2) by using (1.2.4) in the form $R^n = g^{-1}S^n g$.

3. Show that the map $R(z) = 2z/(z + 1)$ fixes 0 and 1, and compute $R'(0)$ and $R'(1)$. Show also that for every $n$,

$$R^n(z) = \frac{2^n z}{(2^n - 1)z + 1}.$$

Evaluate $\lim R^n(z)$ for each $z$ (including 0, 1 and $\infty$), and also $\lim (R^{-1})^n(z)$, that is, the limit of the iterates of the Möbius map $R^{-1}$.

4. Let $R(z) = z/(2z+1)$. Find an explicit formula for $R^n(z)$ and examine $\lim R^n(z)$ for each $z$.
5. Show that if
$$R(z) = \frac{(1+i)z + (1-i)}{(1-i)z + (1+i)},$$
then for all $z$, $R^4(z) = z$. [Try to avoid computing $R^4$.]

## §1.3. Iteration of $z \mapsto z^2$

It is natural to continue our investigation by examining the simplest rational function of degree two, namely
$$R(z) = z^2,$$
and we shall see that even for this, the situation is much more complicated than for Möbius maps. We shall continue to include $\infty$ in our discussion in an intuitive way, so the fixed points of $R$ are 0 (attracting), 1 (repelling), and $\infty$ (attracting), and $z_n \to 0$ when $|z_0| < 1$, while $z_n \to \infty$ when $|z_0| > 1$.

It is evident that the interesting dynamics of $R$ (that is, the action of the iterates $R^n$) occurs on the unit circle
$$C = \{z: |z| = 1\},$$
and we now focus our attention on this. First, the circle $C$ has the striking property that it is both forward and backward invariant under $R$ (that is, each point of $C$ has its entire history and future lying on $C$): in fact, for this $R$, $J = C$, and the invariance of $J$ is one of the basic results in the general theory.

The behaviour of the points $R^n(e^{i\theta})$ on the unit circle $C$ is rather complicated and depends on the number-theoretic properties of $\theta$. Writing $z = e^{i\theta}$, we have
$$R^n(z) = \exp(2^n i\theta),$$
so if $z$ is of the form
$$\exp(2\pi i r/2^m) \tag{1.3.1}$$
for some integers $r$ and $m$, then $R^m(z) = 1$ and consequently,
$$R^n(z) = 1 \qquad \text{when} \quad n \geq m.$$
Note that the set of points described in (1.3.1) is dense in $C$ and starting at any of these points the iterates reach the fixed point 1 after a finite number of steps and remain there thereafter. By contrast, if we start at any point $z_0$ on $C$, but not of the form (1.3.1), then the sequence $z_n$ of iterates cannot converge to any point. Indeed, if $z_n \to w$, then $w$ must be a fixed point of $R$ on $C$ and so $w = 1$. But this is a repelling fixed point of $R$, so (as in §1.1), we must have $z_n = 1$ for some $n$. This, however, shows that $z_0$ is of the form (1.3.1) contrary to our assumption.

The points of the form (1.3.1) are dense in $C$, as also are the points on $C$ not of the form (1.3.1), and we can now begin to see the complicated and "chaotic" behaviour of the iterates $R^n(z)$ on the unit circle $C$. Indeed, *every arc of $C$ contains infinitely many points which eventually reach* 1 *and stay there, and also infinitely many points which move on around the unit circle in a relentless fashion, never converging to any point.* It is clear from this that given any point $z_0$ on $C$, we can choose a point $w_0$, on $C$ and arbitrarily close to $z_0$, for which the iterates $z_n$ and $w_n$ exhibit quite different behaviour: thus the set $J$ contains the circle $C$. It is also clear that if $z_0$ and $w_0$ lie in the unit disc

$$\Delta = \{z: |z| < 1\},$$

then the points $z_n$ and $w_n$ exhibit a similiar behaviour (for they both converge to zero). The same is true if $z_0$ and $w_0$ lie in $\{z: |z| > 1\}$, for then both sequences converge to $\infty$ (any suspicion that they may nevertheless diverge from each other stems from the reader's implicit use of the Euclidean distance, and this will be replaced in Chapter 2 by other distances which cope satisfactorily with $\infty$). These facts show that in this example, *$J$ is the unit circle.*

There are many other noteworthy features of this example and we shall consider just a few more. First, let $I$ be any arc of positive length on the unit circle $C$. Now the map $R: z \mapsto z^2$ doubles the polar angle of points on $C$, hence if $I$ subtends an angle $\theta$ at the origin, then the arc $R(I)$ subtends an angle $2\theta$. It follows immediately that for all sufficiently large $n$, the arc $R^n(I)$ covers, and hence *is*, the unit circle $C$. This holds regardless of how small $I$ is: the action of the $R^n$ on each (arbitrary small) arc of $C$ is to "explode" it to such an extent that it eventually covers $C$. This too is a particular case of a general result about the set $J$.

Next, consider the periodic points of $R$, that is, the fixed points of some iterate $R^n$. As

$$R^n(z) = z^{2^n},$$

the fixed points of $R^n$ on $C$ are the $(2^n - 1)$-th roots of unity and, taking these into account for all values of $n$, we find that *the periodic points are dense in $C$.* Further, it is clear that if

$$\zeta = \exp\left(\frac{2\pi i}{2^n - 1}\right),$$

then $\zeta$ is fixed by $R^n$, but not by $R^m$ for any $m$ in $\{1, 2, \ldots, n - 1\}$: thus *for each natural number $k$, there are periodic points on $C$ with exact period $k$.*

Other features of this example can be easily handled by using the idea of a binary expansion (see Exercise 1.3.5). Given $z$ on $C$, write

$$z = \exp(2\pi i\theta),$$

where $\theta$ satisfies $0 \leq \theta < 1$. The action of $R$ is to map $z$ to $\exp(2\pi i 2\theta)$ and we can ignore the integral part of $2\theta$ as exp is periodic with period $2\pi i$. It follows, then, that we can understand the action of $R$ on $C$ if we understand the action of the map

$$\theta \mapsto 2\theta \pmod 1$$

on the interval [0, 1). This map is most easily understood by writing $\theta$ in binary expansion, say

$$\theta = 0.a_1a_2a_3\cdots, \tag{1.3.2}$$

for then, modulo 1,

$$2\theta = 0.a_2a_3a_4\cdots,$$

and the effect is to ignore the first coefficient and move the other coefficients one place to the left. It is now possible to construct a point $z_0$ in $C$ with the property that the sequence $z_n$ is dense in $C$: we simply list all *finite* sequences containing only 0's and 1's and place them next to each other to form a sequence $a_1, a_2, \ldots$ of 0's and 1's. By applying the map $\theta \mapsto 2\theta \pmod 1$ a suitable number of times, the sequence $a_1, a_2, \ldots$ is transformed into a sequence which begins with any predetermined initial sequence. Because of this, the angle $\theta$ corresponding to the sequence $a_1, a_2, \ldots$ gives rise to a point $z_0 = e^{i\theta}$ on $C$ for which the corresponding sequence $(z_n)$ is dense in $C$. In a similar way, we can construct a point $z_0$ in $C$ with the property that the set of $z_n$ is infinite, but not dense, in $C$ (see Exercise 1.3.6).

It should be clear to the reader that in essence, everything we have said about $z^2$ is also true for $z \mapsto \alpha z^d$, where $d \geq 2$ and $|\alpha| = 1$, and for these maps, $J = C$. We shall now show that the converse is true.

**Theorem 1.3.1.** *Suppose that $P$ is a polynomial of degree $d$, where $d \geq 2$, and that the unit circle $C$ is both forward and backward invariant. Then $P(z) = \alpha z^d$, where $|\alpha| = 1$.*

PROOF. Let $\Delta = \{|z| < 1\}$ and $\Delta^* = \{|z| > 1\} \cup \{\infty\}$. As $C$ is both forward and backward invariant, it follows that $P(\Delta)$ is either $\Delta$ or $\Delta^*$, and as $P$ is bounded on $\Delta$, $P(\Delta) = \Delta$. As $C$ is invariant, it now follows that $|P(z)| \to 1$ as $|z| \to 1$. These facts imply that

$$P(z) = g_1(z)\cdots g_d(z), \tag{1.3.3}$$

where each $g_j$ is a Möbius map of $\Delta$ onto itself (see Exercise 1.3.7). From this, and the fact that $P$ has no poles in $\mathbb{C}$, we deduce that all of the zeros of $P$ are at the origin, and hence that $P$ is of the required form. □

EXERCISE 1.3

1. Discuss the iteration of $R: z \mapsto z^{-2}$. Show, for example, that $J$ is the unit circle. Does $R$ have any attracting fixed points?

2. Let $R(z) = z^2$ and suppose that $W$ is an open subset of the complex plane which meets the unit circle $C$, but which does not contain the origin. Show that

$$\bigcup_{n=1}^{\infty} R^n(W) = \mathbb{C} - \{0\}.$$

3. Show that $z$ satisfies (1.3.1) if and only if for some $n$, $R^n(z) = 1$.
4. Let $R(z) = z^2$, and suppose that $z_0$ is any point other than 0 or $\infty$. Show that given a point $\zeta$ on $C$, it is possible to choose points $z_{-1}, z_{-2}, \ldots$, where $R(z_n) = z_{n+1}$, such that the sequence $z_{-n}$ converges to $\zeta$. [This is a special case of a general result.]
5. The binary expansion $0.a_1a_2\cdots$ of a number $\theta$ in $[0, 1)$ is determined by
$$\theta = \sum_{n=1}^{\infty} a_n 2^{-n},$$
where each $a_j$ is 0 or 1. Show that if $\theta$ has two such expansions $(a_n)$ and $(b_n)$ so that
$$\sum_{n=1}^{\infty} a_n 2^{-n} = \sum_{n=1}^{\infty} b_n 2^{-n},$$
then either $a_n = b_n$ for all $n$, or for some $N$, $a_N = 1$, $a_n = 0$ when $n > N$, and
$$b_n = \begin{cases} a_n & \text{for } n < N; \\ 1 - a_n & \text{for } n \geq N. \end{cases}$$
6. Let $\theta$ be given by the binary expansion
$$0.1010010001\cdots,$$
where there are exactly $n$ 0's following the $n$-th occurrence of 1, and let $z_0 = \exp(2\pi i\theta)$. Show that the set $\{z_1, z_2, \ldots\}$ is infinite, but that the sequence $(z_n)$ is not dense in $C$.
7. Show that in the proof of Theorem 1.3.1, $P$ satisfies (1.3.3). [Consider the quotient $P(z)/B(z)$ where $B$ is a finite product of this type whose zeros coincide with those of $P$ in $\Delta$, and apply the Minimum Modulus Theorem.]
8. By modifying the proof of Theorem 1.3.1, discuss those rational functions for which $J$ is the unit circle.

## §1.4. Tchebychev Polynomials

It is an elementary fact that $\cos(kz)$ can be expressed as a polynomial $T_k$ in $\cos z$, so
$$\cos(kz) = T_k(\cos z); \tag{1.4.1}$$
for example, $T_2(z) = 2z^2 - 1$. The polynomials $T_k$ are the famous Tchebychev polynomials and, because of the functional relationship (1.4.1), it is fairly easy to analyse the behaviour of their iterates. Indeed, (1.4.1) implies that for all $n$,
$$(T_k)^n(\cos z) = \cos(k^n z),$$
and so we can investigate the iterates of $T_k$ simply by studying the iterates of $z \mapsto kz$ and then applying the function cos.

For notational simplicity, let us fix a value of $k$, $k \geq 2$, and denote $T_k$ by $T$. First, we examine the action of the iterates of $T$ on the real interval $[-1, 1]$.

Each point in $[-1, 1]$ is of the form $\cos\theta$ for some real $\theta$ and as

$$T(\cos\theta) = \cos(k\theta),$$

we find that $T$ maps $[-1, 1]$ into itself (that is, $[-1, 1]$ is forward invariant).

Now let $V$ be any sub-interval of $[-1, 1]$ of positive length and construct a sub-interval $U$ of $[0, \pi]$ such that $\cos(U) = V$. As

$$T^n(\cos x) = \cos(k^n x),$$

we find (by letting $x$ range over $U$) that

$$T^n(V) = \cos(k^n U).$$

Now $U$ has positive length so, for sufficiently large $n$, $k^n U$ is an interval of length at least $2\pi$ and then

$$T^n(V) = \cos(k^n U) = [-1, 1].$$

We have shown that if $V$ is a non-empty sub-interval of $[-1, 1]$, then, for sufficiently large $n$, $T^n$ maps $V$ onto $[-1, 1]$. This is reminiscent of the action of the iterates of $z^2$ on the unit circle (see §1.3), and it certainly shows that the iterates $T^n$ cannot preserve the proximity of points in $[-1, 1]$: thus in this case, $J$ contains $[-1, 1]$.

We shall now show that on the complement $\Omega$ of $[-1, 1]$, $T^n(z) \to \infty$ as $n \to \infty$. It follows that the iterates preserve proximity there (once we have discussed $\infty$ properly) so in this example, the sets $J$ and $F$ are $[-1, 1]$ and $\Omega$ respectively.

There are two ways to show that $T^n \to \infty$ on $\Omega$. First, we take any point $w$ in $\Omega$ and write $w = \cos z$ for some complex $z$, where $z = x + iy$. Note that as $w \notin [-1, 1]$, $y \neq 0$. Next,

$$\begin{aligned}
|\cos(x + iy)|^2 &= |\cos x \cosh y + i \sin x \sinh y|^2 \\
&= \cos^2 x \cosh^2 y + \sin^2 x \sinh^2 y \\
&= \cos^2 x(1 + \sinh^2 y) + \sin^2 x \sinh^2 y \\
&\geq \sinh^2 y.
\end{aligned}$$

We deduce that

$$|T^n(w)| = |\cos(k^n z)| \geq |\sinh(k^n y)|$$

which tends to $\infty$ as $n \to \infty$, as promised.

The second method uses the important idea of conjugacy of maps. First, we define $R(z) = z^k$, and

$$\varphi(z) = \frac{z + 1/z}{2}.$$

Then if $z = e^{i\theta}$, we have

$$\varphi R(z) = \cos(k\theta) = T\varphi(z)$$

(because $T = T_k$), and so, by analytic continuation,

$$\varphi R(z) = T\varphi(z) \tag{1.4.2}$$

for all $z$. Observe that this shows that as $R$ maps $\Delta^*$ into itself, so $T$ maps $\Omega$ into itself, and hence that $[-1, 1]$ is backward invariant under $T$. Now $\varphi$ is a conformal bijection of the exterior $\Delta^*$ of the closed unit disc onto $\Omega$, so from (1.4.2), for $z$ in $\Omega$,

$$T(z) = \varphi R \varphi^{-1}(z). \tag{1.4.3}$$

We express this by saying that $T$ is *conjugate in* $\Omega$ to $R$, and as this implies that

$$T^n(z) = \varphi R^n \varphi^{-1}(z),$$

it is easy to see that $T^n(z) \to \infty$ $(= \varphi(\infty))$ on $\Omega$. Note that although (1.4.2) holds on the unit circle $C$, (1.4.3) does not because $\varphi$ is not injective there: we call (1.4.2) a *semi-conjugacy* on $C$, and using this we can see that many of the properties of the iterates $z^k$ on $C$ are transmitted to the iterates of $T_k$ on $[-1, 1]$.

The link between $T_k$ and $z^k$ given in (1.4.3) suggests that there should be a result for $T_k$ corresponding to Theorem 1.3.1. We prove

**Theorem 1.4.1.** *Suppose that $T$ is a polynomial of degree $k$, where $k \geq 2$. Then the interval $[-1, 1]$ is both forward and backward invariant under $T$ if and only if $T$ is $T_k$ or $-T_k$.*

PROOF. We know that $[-1, 1]$ is forward and backward invariant under $T_k$ and this implies that $[-1, 1]$ is forward and backward invariant under $-T_k$. Now suppose that $[-1, 1]$ is invariant under an arbitrary polynomial $T$ of degree $k$, and let $\Delta$ be the open unit disc. The map $\varphi$ is a conformal bijection of $\Delta$ onto $\Omega$ and as $\Omega$ is forward and backward invariant under $T$, the conjugate map $\varphi^{-1}T\varphi$ is an $k$-fold analytic map of $\Delta$ onto itself. Further, it is clear that $|\varphi^{-1}T\varphi(z)| \to 1$ as $|z| \to 1$ and this implies that $\varphi^{-1}T\varphi$ is a finite product with $d$ factors as in (1.3.3). Also, if $w$ is a zero of $\varphi^{-1}T\varphi$, then

$$T\varphi(w) = \varphi(0) = \infty,$$

so $\varphi(w) = \infty$ and hence $w = 0$. We deduce that

$$\varphi^{-1}T\varphi(z) = az^d, \tag{1.4.4}$$

where $|a| = 1$, and hence that

$$T\left(\frac{z + z^{-1}}{2}\right) = \frac{az^d + (az^d)^{-1}}{2}.$$

Equating coefficients of $z^d$ and $z^{-d}$ on both sides of this equation, we find that $a = 1$ or $-1$. If $a = 1$, then (1.4.3) (in which, $T$ is $T_d$) and (1.4.4) show that

$T = T_d$. If $a = -1$, then

$$-T_d\varphi(z) = -\varphi(z^d) = \varphi(-z^d) = T\varphi(z),$$

and so $T = -T_d$. □

We end this section with some general remarks about these ideas. There is an interesting class of examples whose iterates can be analysed in a broadly similar way, and to see what these are, we extract the essential features from the discussion above. Let $G$ be a group of Euclidean isometries which gives rise to some tesselation of the Euclidean plane. A function $f$ is said to be *automorphic* with respect to $G$ if $f$ is meromorphic on $\mathbb{C}$ and if

$$f(g(z)) = f(z)$$

for every $g$ in $G$ and every $z$ in $\mathbb{C}$. For example, if $f(z) = \cos z$, then $G$ is the group generated by the two maps

$$z \mapsto z + 2\pi, \qquad z \mapsto -z:$$

for a second example, if $f(z) = \sin z$, then $G$ is generated by $z \mapsto z + 2\pi$. Borrowing material from the more advanced theory of automorphic functions (which will not be discussed here), we find that there are other examples of the general phenomenon suggested by (1.4.1), namely the case when an automorphic function $f$ satisfies an algebraic relation of the form

$$f(kz) = R(f(z)) \tag{1.4.5}$$

for some rational function $R$ and some complex number $k$. This yields

$$f(k^n z) = R^n(f(z))$$

and so starting at $f(z_0)$, the iterates $R^n(fz_0)$ can be examined by considering the *single* function $f$ at the sequence of points

$$z_0, kz_0, k^2z_0, k^3z_0, \ldots.$$

The Tchebychev polynomials are just one class of examples of this type and, in fact, all solutions of (1.4.5) are known (see [82]). We shall consider another more complicated example of this type in Chapter 4. Further information on the action of the Tchebychev polynomials on $[-1, 1]$ can be found in [67], p. 54, or in [99].

## Exercise 1.4

1. Use the formula for $\sin(3x)$ to examine the iterates of

$$T(z) = 3z - 4z^3.$$

Now consider the formula for $\cos(3x)$.

2. Using the relation (1.4.2), show that $(T_k)^n\varphi = \varphi R^n$. Deduce that the periodic points of $T_k$ are dense in $[-1, 1]$.

3. Derive the backward invariance of $[-1, 1]$ from its forward invariance and the fact that (as in the text) $T^n \to \infty$ on $\Omega$.
4. Show that for all $m$ and $n$, $T_n$ commutes with $T_m$.

## §1.5. Iteration of $z \mapsto z^2 - 1$

In this section we shall simply describe, without proof, the dynamics of the iterates of the polynomial

$$P(z) = z^2 - 1.$$

The complexity of the division of $\mathbb{C}$ into the sets $F$ and $J$ for $P$ is illustrated in Figure 1.5.1 and it must be emphasized that this example is typical of rational functions: the cases when $J$ is a smooth curve (such as an interval or a circle as in §1.3 and §1.4) are rare indeed. No proofs are offered in this section, but by the end of the text the reader will be in a position to verify all of the claims made here.

In Figure 1.5.1, the set $J$ is the collection of black curves, and the set $F$ consists of the white regions (including the outer region which we denote by $F_\infty$). There are infinitely many components of $F$, and each is simply connected (it has no holes). The two marked points are $-1$ and $0$.

If $z$ is in $F_\infty$, then $P^n(z) \to \infty$ as $n \to \infty$: this is easy to establish whenever $|z|$ is sufficiently large (see Exercise 1.5.3), and the same holds for all other $z$ in $F_\infty$ (from general results in complex analysis). Next, the two regions $F_0$ (containing 0) and $F_{-1}$ (containing $-1$) are interchanged by $P$, that is,

$$P(F_0) = F_{-1}, \qquad P(F_{-1}) = F_0:$$

this is a consequence of the fact that $P(0) = -1$ and $P(-1) = 0$.

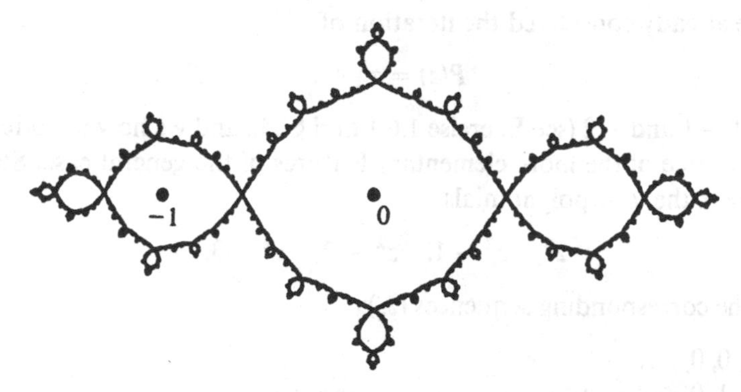


Figure 1.5.1. $z \mapsto z^2 - 1$. Fractal image reprinted with permission from *The Beauty of Fractals* by H.-O. Peitgen and P.H. Richter, 1986, Springer-Verlag, Heidelberg, New York.

If $\Omega$ is any component of $F$, then so is $P(\Omega)$ and, roughly speaking, the components of $F$ are moved one onto another by the action of $P$ (this is a general fact). The region $F_\infty$ has the special property that it is both forward and backward invariant under $P$ (essentially because $\infty$ is). For any other component $\Omega$ of $F$, the sequence $P^n(\Omega)$ of regions is attracted towards the pair $F_0, F_{-1}$ in the sense that the sequence of components $P^n(\Omega)$, $n \geq 1$, eventually becomes

$$\ldots, F_{-1}, F_0, F_{-1}, F_0, \ldots.$$

EXERCISE 1.5

1. Show that the two fixed points of $P$ in $\mathbb{C}$ are $(1 + \sqrt{5})/2$ and $(1 - \sqrt{5})/2$, and that both of these are repelling fixed points. By considering the graph of $P(x)$ for real $x$ (together with the graph $y = x$), show that $P(x) > x$ when $x > (1 + \sqrt{5})/2$, and deduce that $P^n(x) \to \infty$ there. As a repelling fixed point of $P$ must lie in $J$ (this will be proved later), this shows that the extreme right-hand point of $J$ (see Figure 1.5.1) is $(1 + \sqrt{5})/2$. Show also that $(1 - \sqrt{5})/2$ is the common boundary point of $F_{-1}$ and $F_0$.
2. Show that the left-hand end-point $\xi$ of $F_{-1}$ is $\sqrt{(1 + \sqrt{5})/2}$. [*Hint*: Either draw the graph of $P^2(x)$ for real $x$, or assume (as will be shown later) that $P$ is a bijection of $F_{-1}$ onto $F_0$ and show that $P(\xi) = -(1 - \sqrt{5})/2$.]
3. Show that if $|z| > (1 + \sqrt{5})/2$, then $P^n(z) \to \infty$. [*Hint*: Show that $|P(z)| \geq k|z|$ if $k > 1$ and $|z| \geq k/2 + (k^2/4 + 1)^{1/2}$.]
4. Show that $P'(-1) = -2$ and $(P^2)'(-1) = 0$, and interpret this geometrically for the sequence $z_0, z_1, \ldots$ when $z_0$ is close to $-1$.

## §1.6. Iteration of $z \mapsto z^2 + c$

We have already considered the iteration of

$$P(z) = z^2 + c$$

for $c = 0, -1$ and $-2$ (see Exercise 1.6.1 and §1.4), and we now introduce the reader to some of the more elementary features of the general case. Starting with $z_0 = 0$, the four polynomials

$$z^2, \quad z^2 - 1, \quad z^2 - 2, \quad z^2 - 3,$$

induce the corresponding sequences $(z_n)$:

(i) $0, 0, 0, 0, \ldots$;
(ii) $0, -1, 0, -1, \ldots$;
(iii) $0, -2, 2, 2, \ldots$;
(iv) $0, -3, 6, 33, \ldots$.

These sequences exhibit quite different behaviours: the first is constant, the second has period two, the third ultimately becomes constant, while the fourth converges to $\infty$. We now invite the reader to imagine what is happening to the sequence

$$z_0\,(=0), z_1, z_2, z_3, \ldots \tag{1.6.1}$$

computed for the function $P: z \mapsto z^2 + c$ as $c$ moves *continuously* from 0 to $-3$: somehow, the sequence (1.6.1) changes in character through the examples (i)–(iv) and we wish to explain these changes. The explanation lies in recognizing different regions of values of $c$ in each of which the iterates of $P$ have common features, can then investigating the transition from one state to another as $c$ moves from one region to another.

To illustrate this, let us identify the set of values $c$ for which $z^2 + c$ has an attracting fixed point in $\mathbb{C}$. Now $z^2 + c$ has two fixed points, say $\alpha$ and $\beta$, in $\mathbb{C}$, and as these are solutions of

$$z^2 - z + c = 0, \tag{1.6.2}$$

they satisfy

$$\alpha + \beta = 1, \qquad \alpha\beta = c.$$

This shows that

$$P'(\alpha) + P'(\beta) = 2,$$

and this in turn shows that not both $\alpha$ and $\beta$ can be attracting (for if they are, $|P'(\alpha)| < 1$ and $|P'(\beta)| < 1$). It follows that $P$ can have *at most one* attracting fixed point, and the condition that one of the fixed points, say $\alpha$, is attracting is

$$2|\alpha| = |P'(\alpha)| < 1.$$

However, from (1.6.2),

$$c = \alpha - \alpha^2$$

so the set of $c$ we are seeking is just the image of the disc $\{\alpha: |\alpha| < \frac{1}{2}\}$ under the map $z \mapsto z - z^2$. As this map is $fgh$, where

$$h(z) = z - \tfrac{1}{2}, \qquad g(z) = z^2, \qquad f(z) = \tfrac{1}{4} - z,$$

the set of $c$ is the cardioid illustrated in Figure 1.6.1.

In a similar way we can locate the set of $c$ for which $z^2 + c$ has an attracting 2-cycle: this is so when $c = -1$, for example, for then 0 and 1 are attracting fixed points of the second iterate of $z^2 - 1$. Now the fixed points of the second iterate $P^2$ are the solutions of

$$P^2(z) - z = 0.$$

As $\alpha$ and $\beta$ are fixed by $P$, they are also fixed by $P^2$: thus $P(z) - z$ divides $P^2(z) - z$ and we can write

$$\begin{aligned} P^2(z) - z &= (z^2 - z + c)(z^2 + z + 1 + c) \\ &= (z - \alpha)(z - \beta)(z - u)(z - v). \end{aligned}$$

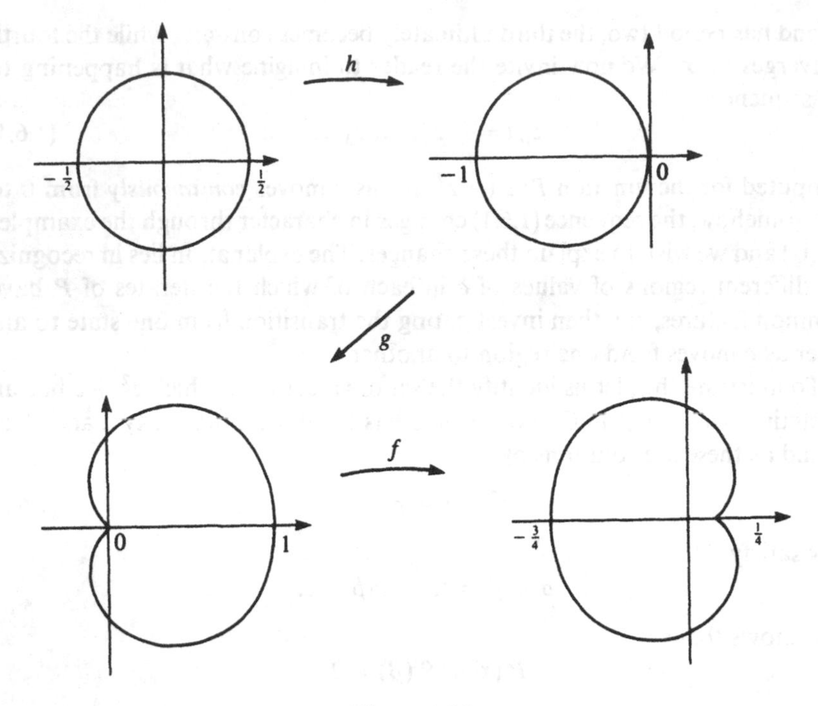


Figure 1.6.1

We are seeking conditions on $c$ which imply that $\{u, v\}$ is an attracting 2-cycle; that is,

$$P(u) = v, \qquad P(v) = u, \qquad u \neq v,$$

together with

$$|(P^2)'(u)| < 1, \qquad |(P^2)'(v)| < 1,$$

(so $u$ and $v$ are attracting fixed points of $P^2$). Now by the Chain Rule,

$$\begin{aligned}(P^2)'(u) &= P'(P(u))P'(u)\\ &= P'(v)P'(u)\\ &= 4uv\\ &= 4(1 + c),\end{aligned}$$

so the set of $c$ we are seeking is the disc $\{c: |1 + c| < \frac{1}{4}\}$ (see Exercise 1.6.2). This disc is illustrated, together with the cardioid obtained previously, in Figure 1.6.2.

The point of tangency of the disc and the cardioid is $-\frac{3}{4}$ and we shall now describe the change in the dynamics as $c$ moves from the cardioid to the disc through the value $-\frac{3}{4}$. When $c$ is in the cardioid, $P$ has an attracting fixed point $\alpha$ and a repelling 2-cycle $\{u, v\}$ (that is, $u$ and $v$ are repelling fixed points

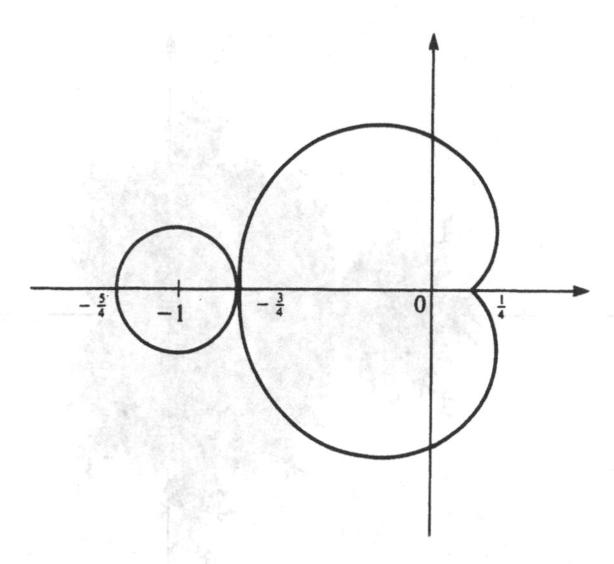


Figure 1.6.2

of $P^2$). As $c \to -\frac{3}{4}$, the points $\alpha$, $u$ and $v$ converge to a common value $-\frac{1}{2}$, and when $c = -\frac{3}{4}$, $\alpha$, $u$ and $v$ coincide (so $P$ has no "genuine" 2-cycles) at an indifferent fixed point of $P$. As $c$ now moves into the disc, these points separate again, this time with $\alpha$ being a repelling fixed point and $\{u, v\}$ being an attracting 2-cycle: thus during this process the attracting nature of $\alpha$ has been transferred to the pair $\{u, v\}$. This process is repeated, $\{u, v\}$ losing its attracting nature to a 4-cycle, then to an 8-cycle, and so on: this is known as *period doubling* and the reader can easily verify it on a computer (by examining the limit of $P^n(0)$ for various values of $c$).

Unfortunately, this analysis does not seem to carry us any further; for example, in searching for $N$-cycles, we can divide $P^N(z) - z$ by $P(z) - z$ but the resulting polynomial has degree $2^N - 2$ (and this is at least 6 when $N \geq 3$). Because of this, we resort to an experimental approach (with computer graphics) and for reasons that will be described in Chapter 9, it is appropriate to consider the set

$$\mathcal{M} = \{c\colon P^n(0) \not\to \infty\}:$$

see Figure 1.6.3.

A rough version of $\mathcal{M}$ appeared in a paper by Brooks and Matelski given in 1978, [29], where it was used in the context of discrete groups, and a little later Mandelbrot produced very detailed pictures of the set $\mathcal{M}$ which now bears his name. In fact, the map $z^2 + c$ is the same (up to a change of coordinates) as the *logistic map* $z \mapsto \lambda z(1 - z)$, and for real values at least, this has a very long history. For further information, we refer the reader to (for example), [16], [35], [67], [68], [69], [73] and [80].

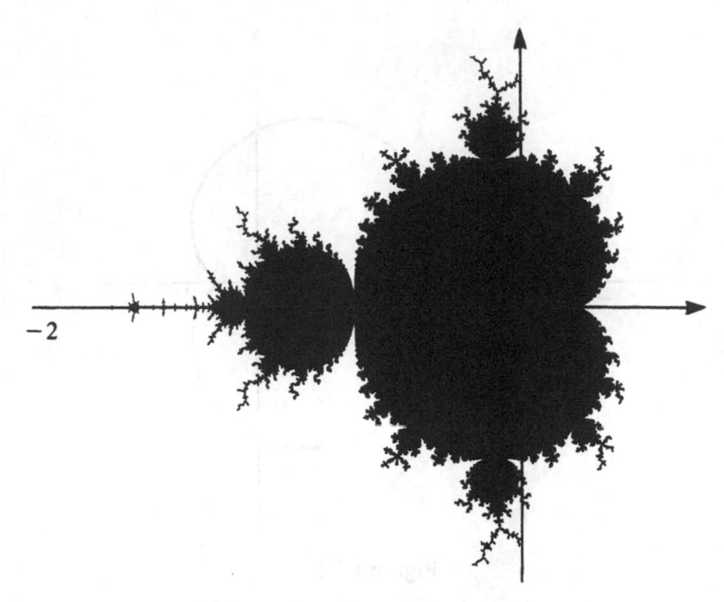


Figure 1.6.3. The Mandelbrot set.

EXERCISE 1.6

1. Show that $P(z) = z^2 - 2$ and $Q(z) = 2z^2 - 1$ are conjugate (that is, $P = \varphi Q \varphi^{-1}$, for some $\varphi(z)$ of the form $az + b$).

2. Suppose that $|1 + c| < \frac{1}{4}$. Using the notation in the text, show that $\alpha$, $\beta$, $u$ and $v$ are distinct. Show also that because $P^2$ fixes $u$, it also fixes $P(u)$, and deduce that $P(u) = v$, $P(v) = u$, and $u \neq v$.

3. Show that the map $P: z \mapsto z^2 + c$ is conjugate to the map $Q: z \mapsto \lambda z(1 - z)$ for a suitable $\lambda$ and find the precise relationship between $c$ and $\lambda$.

4. Let $P(z) = z^2 + c$. Show that:
   (i) $P^n(z) \to 0$ if $|z| < \frac{1}{2} + \sqrt{\frac{1}{4} - |c|}$;
   (ii) $P^n(z) \to \infty$ if $|z| > \frac{1}{2} + \sqrt{\frac{1}{4} + |c|}$.
   This shows that when $|c|$ is small, the set $J$ lies in a thin annulus $\{z: 1 - \varepsilon_1 < |z| < 1 + \varepsilon_2\}$ and so is approximately the unit circle in $\mathbb{C}$.

5. Use a computer to investigate the limiting behaviour of $P^n(0)$ in the cases $c = 0.2$, $c = -0.8$, $c = -1.3$ and $c = -\frac{7}{4}$. Experiment with other values of $c$.

6. In the following, $P(z) = z^2 + c$, $c_n = P^n(0)$ and

$$\alpha = \frac{1 - \sqrt{(1 - 4c)}}{2}, \qquad \beta = \frac{1 + \sqrt{(1 - 4c)}}{2},$$

   (the fixed points of $P$). Prove:
   (i) if $c < -2$, then $c_n \geq 2 + n|c + 2|$ when $n \geq 2$;
   (ii) if $-2 \leq c < 0$, then $P$ maps $[-\beta, \beta]$ into itself;

(iii) if $0 \le c \le \frac{1}{4}$, then $P$ maps $[0, \alpha]$ into itself; and
(iv) if $c > \frac{1}{4}$, then $c_n \ge n(c - \frac{1}{4})$.

Use (i)–(iv) to show that the intersection of the Mandelbrot set $\mathcal{M}$ with real axis is $[-2, \frac{1}{4}]$.

7. Show that $i \in \mathcal{M}$.

8. Prove that every quadratic polynomial has a 3-cycle. [Show that we may take $P$ in the form $az + z^2$. Then $P$ fixes 0 and $\alpha$, say, and if $P$ has no 3-cycle, then $P^3(z) - z$ has all its zeros at 0 and $\alpha$, thus $P^3(z) = z + z^m (z - \alpha)^{8-m}$ for some $m$.]

## §1.7. Iteration of $z \mapsto z + 1/z$

We now examine the iterates of

$$R(z) = z + 1/z.$$

It is clear that $R$ has no fixed points in the complex plane, and as $R(z) \to \infty$ when $z \to \infty$, and also when $z \to 0$, it is natural to define

$$R(\infty) = \infty = R(0);$$

thus $\infty$ is the only fixed point of $R$. If $x > 0$, then $x < R(x)$, so

$$0 < x < R(x) < R^2(x) < \cdots,$$

and hence $R^n(x) \to \infty$ (for any finite limit would have to be a fixed point of $R$). The behaviour of $R$ on the imaginary axis is quite different: if $y > 1$, then

$$R(iy) = iy + 1/iy = i(y - 1/y),$$

so points of the form $iy$, $y > 1$, move away from $\infty$ (and towards the origin) under the action of $R$. It appears, then, that $\infty$ acts as an attracting fixed point for some choices of $z_0$, and as a repelling fixed point for others: in fact, $\infty$ is an indifferent fixed point (see Exercise 1.7.1).

Now consider any $z_0$ with $z_0 = x_0 + iy_0$, where $x_0 > 1$, and for the iterates $z_n$, write $z_n = x_n + iy_n$. Then

$$x_{n+1} = x_n(1 + |z_n|^{-2}), \tag{1.7.1}$$

and so

$$1 < x_0 < x_1 < x_2 < \cdots < x_n < \cdots.$$

This implies that for all $n$, $r_n > 1$; hence

$$|y_{n+1}| = |y_n(1 - |z_n|^{-2})| < |y_n|,$$

and we see that $z_n \to \infty$ in the horizontal strip

$$\{x + iy: x > 1, |y| \le |y_0|\}:$$

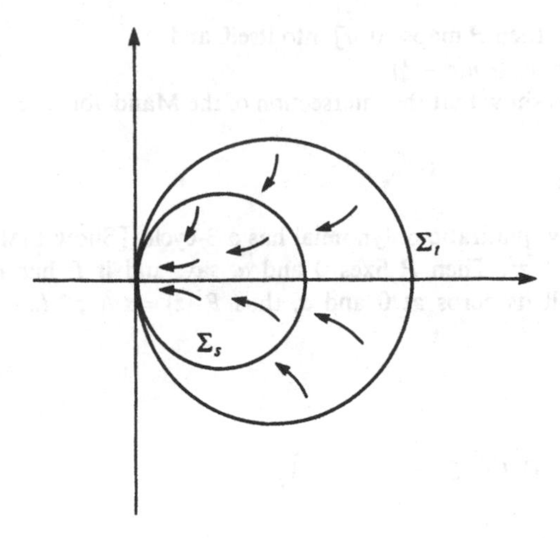


Figure 1.7.1

later, we shall see that this sort of behaviour is typical near indifferent fixed points. It is also clear from (1.7.1) that $R$ maps each set

$$\Sigma_t = \{x + iy: x > t\}, \qquad t > 0,$$

into itself.

It is worthwhile to express these results in terms of the conjugate function

$$S(z) = gRg^{-1}(z) = \frac{z}{1 + z^2},$$

where $g(z) = 1/z$. As $R$ maps $\Sigma_t$ into itself, we have

$$Sg(\Sigma_t) = gR(\Sigma_t) \subset g(\Sigma_t),$$

and as

$$g(\Sigma_t) = \left\{z: \left|z - \frac{1}{2t}\right| < \frac{1}{2t}\right\},$$

we find that $S$ acts in the manner indicated in Figure 1.7.1.

Exercise 1.7

1. Show that the origin is an indifferent fixed point of $S$.
2. Show that if $z$ lies in some disc $g(\Sigma_t)$, then

$$S^n(z) \to 0, \qquad \arg S^n(z) \to 0,$$

and interpret this geometrically.

3. Show that $z \mapsto z + 1/z$ is conjugate to the map $fg$, where

$$f(z) = \frac{3z + 1}{z + 3}, \qquad g(z) = z^2.$$

Show also that the map $fg$ preserves both the unit circle and the unit disc. Show that 1 is an attracting fixed point of $f$, a repelling fixed point of $g$, and an indifferent fixed point of $fg$.

## §1.8. Iteration of $z \mapsto 2z - 1/z$

Let

$$R(z) = 2z - 1/z.$$

As before, we adjoin $\infty$ to the plane and define both $R(0)$ and $R(\infty)$ to be $\infty$, so the fixed points of $R$ are $1$, $-1$ and $\infty$. The fixed points 1 and $-1$ are repelling, and as $|z| > 2$ implies that

$$|R(z)| > 2|z| - 1/|z| > 3|z|/2,$$

we find that $|R^n(z)| > (3/2)^n|z|$, so $\infty$ is an attracting fixed point of $R$.

We shall now focus our attention on the action of $R$ on the interval $I = [-1, 1]$. As $R$ is a strictly increasing map of each of the intervals

$$I_0 = [-1, -\tfrac{1}{2}], \qquad I_1 = [\tfrac{1}{2}, 1],$$

onto $I$, we can define two branches, say $g_0$ and $g_1$, of $R^{-1}$ on $I$ by $g_0(I) = I_0$ and $g_1(I) = I_1$. Further, as $R'(x) > 2$ on $I_0$ and $I_1$, it follows that

$$|g_0'(x)| < \tfrac{1}{2}, \qquad |g_1'(x)| < \tfrac{1}{2},$$

on $I$, so if we apply either $g_j$ to an interval lying in $[-1, 1]$, the length of the interval decreases by a factor of at least 2.

Now let $K_n$ be the union of the $2^n$ mutually disjoint closed intervals $\varphi_1 \cdots \varphi_n(I)$ where each $\varphi_j$ is either $g_0$ or $g_1$: for example, $K_2$ consists of the four intervals

$$g_0^2(I), g_0g_1(I), g_1g_0(I), g_1^2(I).$$

As $\varphi_{n+1}(I) \subset I$, we see that

$$\varphi_1 \cdots \varphi_n\varphi_{n+1}(I) \subset \varphi_1 \cdots \varphi_n(I)$$

and so $K_{n+1} \subset K_n$, and, as each $K_n$ is non-empty and compact, it follows that $K$ is also, where

$$K = \bigcap_{n=0}^{\infty} K_n.$$

In fact, $K$ is Cantor set: it is uncountable, and it cannot contain an interval for such an interval would lie in $K_n$ for every $n$, and each interval in $K_n$ has length at most $2^{-n}$. It is easy to see that

$$R(K_{n+1}) = K_n, \qquad R^{-1}(K_n) = K_{n+1}, \tag{1.8.1}$$

and from this we see that $K$ is both forward and backward invariant under $R$.

Next, the iterates $R^n$ fail to preserve proximity on $K$: indeed, each interval

$\varphi_1 \cdots \varphi_n(I)$ is mapped monotonically by $R^n$ onto $I$, so, for example, the end-points of $\varphi_1 \cdots \varphi_n(I)$ (which are at most a distance $2^{-n}$ apart) are mapped by $R^n$ onto $-1$ and 1. In this example, $J = K$.

Finally, let us write

$$g_0 g_0 g_1(I) = I(0, 0, 1),$$

and similarly for other sequences of $g_j$. Then each point $x$ in $K$ corresponds to a unique sequence $(x_1, x_2, \ldots)$ of 0's and 1's which is defined by

$$\{x\} = \bigcap_{n=1}^{\infty} I(x_1, \ldots, x_n), \tag{1.8.2}$$

and with this representation, we have

$$R(I(x_1, \ldots, x_n)) = I(x_2, \ldots, x_n).$$

It follows that

$$\{R(x)\} = \bigcap_{n=1}^{\infty} I(x_2, \ldots, x_n),$$

and this shows that $K$ contains many periodic points of $R$: for example, if

$$(x_1, x_2, \ldots) = (1, 0, 1, 1, 0, 1, 1, 0, 1, \ldots),$$

the corresponding $x$ given by (1.8.2) is fixed by $R^3$.

Exercise 1.8

1. Verify (1.8.1), and deduce that $x$ is in $K$ if and only if $R(x)$ is.
2. Show that if $|z| > 1$, then $|R(z)| > |z|$, and deduce that $R^n(z) \to \infty$. Show also that if $x$ is real and not in $K$, then $R^n(x) \to \infty$.
3. Show that the periodic points of $R$ are dense in $K$ (explicitly, given any open interval $I$ that meets $K$, there are points $z$ in $I$ which are fixed by some $R^n$).
4. Find positive values of $\alpha$ such that an analysis similar to that in the text holds for the map $z \mapsto \alpha z - 1/z$.
5. For which $\alpha$ and $\beta$ are the maps $\alpha z - 1/z$ and $\beta z - 1/z$ conjugate?
6. Show that $2z - 1/z$ is conjugate to $z/(2 - z^2)$.

## §1.9. Newton's Approximation

In 1669, Newton discussed the equation $x^3 - 2x - 5 = 0$, Starting with an approximation $x_0 = 2$ to the real root $\zeta$, he wrote $x = 2 + y$ and so obtained the equation

$$y^3 + 6y^2 + 10y - 1 = 0.$$

Neglecting the non-linear terms, he then found that $y = 1/10$ and so took

$x_1 = 2{\cdot}1$ as his next approximation to $\zeta$. The method was systematically discussed by Joseph Raphson in 1690, and by using the derivative (which Newton did not), we obtain the now familiar Newton–Raphson iterative scheme

$$x_{n+1} = x_n - \frac{f(x_n)}{f'(x_n)}$$

for finding the solutions of the equation $f(x) = 0$. Given $f$, this means iterating the new function

$$F(x) = x - \frac{f(x)}{f'(x)}, \tag{1.9.1}$$

and so this method falls within the scope of this text.

In 1879, Cayley ("*throwing aside the restriction as to reality*") proposed using the method to find complex roots of complex functions, and referred to it as the Newton–Fourier method [30]. He asked for conditions under which the sequence $z_n$ converges to a root $\zeta$ ("*selected at pleasure*"), and solved the problem for the general quadratic polynomial. In the same year, he commented in a one-page note [31] that "*the case of the cubic equation appears to present considerable difficulty*"; and eleven years later, and apparently without having made further progress, he again raised the issue: "*J'espere appliquer cette theorie au cas d'une equation cubic, mais les calculs sont beaucoup plus difficiles*", [32]. With the benefit of the computer-generated pictures that we now have, we can see that Cayley's failure to make any real progress at that time was inevitable: the problem is one of finding the Julia set $J$ for $F$, and this is illustrated in part for the polynomial $z^3 - 1$ in Figure 1.9.1.

The key to understanding Newton's method lies in two observations: first, (1.9.1) shows that the zeros of $f$ correspond to fixed points of $F$ and, second, that successive iterates under $F$ have a tendency to converge towards an attracting fixed point of $F$. Thus if $\zeta$ is a zero of $f$ which gives rise to an *attracting* fixed point of $F$, then the iterates

$$z_n = F^n(z_0)$$

necessarily converge to $\zeta$ provided at least that the initial guess $z_0$ is sufficiently accurate. This much is straightforward, but the problem of deciding if $z_n \to \zeta$ when the initial guess $z_0$ is far from $\zeta$ is much harder.

Our immediate task is to show that a zero of $f$ is an attracting fixed point of $F$. We may assume that

$$f(z) = (z - \zeta)^m g(z),$$

where $m \geq 1$ and $g(\zeta) \neq 0$: then

$$\frac{f(z)}{f'(z)} = \frac{(z - \zeta)g(z)}{(z - \zeta)g'(z) + mg(z)},$$

and, as the denominator on the right is non-zero at $\zeta$, we can differentiate

Figure 1.9.1. $z \mapsto z^3 - 1$. Fractal image reprinted with permission from *The Beauty of Fractals* by H.-O. Peitgen and P.H. Richter, 1986, Springer-Verlag, Heidelberg, New York.

both sides of (1.9.1) and obtain

$$F'(\zeta) = \frac{m-1}{m}.$$

This shows that $0 < F'(\zeta) < 1$ and hence that every zero of $f$ corresponds to an attracting fixed point of $F$. Note that the convergence of the $z_n$ to $\zeta$ is fastest when $m = 1$ (this is when $f$ has a simple zero at $\zeta$).

This argument guarantees the convergence for a sufficiently accurate initial guess, but this was not Cayley's question; Cayley was interested in finding *all* starting points which yield convergence to the given root. If we apply Newton's method to the quadratic polynomial

$$f(z) = (z - \alpha)(z - \beta),$$

where $\alpha$ and $\beta$ are distinct, we are led to consider the iteration of

$$F(z) = \frac{z^2 - \alpha\beta}{2z - (\alpha + \beta)}.$$

This shows explicitly that $F$ fixes both $\alpha$ and $\beta$ (the zeros of $f$) and it seems reasonable to hope (on grounds of symmetry, say) that the $z_n$ will converge to $\alpha$ provided only that we start the process at a point $z_0$ which is nearer to $\alpha$ than to $\beta$ (and similarly for $\beta$). We shall now verify that this is so.

The set of points nearer to $\alpha$ than to $\beta$ is conveniently described as $\{|w| < 1\}$ where

$$w = \frac{z - \alpha}{z - \beta},$$

and because of this, it is best to work with the variable $w$ rather than $z$. Note that the two interesting values of $z$, namely $\alpha$ and $\beta$, correspond to the values 0 and $\infty$ for $w$. Converting the problem into the variable $w$ means that we should consider the conjugate function $R$ given by

$$R(z) = gFg^{-1}(z),$$

where $g(z) = (z - \alpha)/(z - \beta)$. A computation (see Exercise 1.9.1) shows that $R(z) = z^2$, and this is all we need. Indeed, if $z_0$ is closer to $\alpha$ than to $\beta$, then $|g(z_0)| < 1$ and so as $n \to \infty$,

$$R^n(gz_0) \to 0.$$

But this means that as $n \to \infty$, we have

$$F^n(z_0) = g^{-1}R^n g(z_0) \to g^{-1}(0) = \alpha$$

as required. Of course, a similar argument holds for $\beta$.

Finally we remark that more is known about this problem: for example, if $P$ is a polynomial with real distinct roots, then the sequence $z_n$ converges to one of the roots of $P$ for *essentially every choice of* $z_0$ (technically, this means that the set of $z_0$ for which the convergence fails has zero area). For more details, see [34], [39], [67] (pp. 91–92), [79] and [80].

Exercise 1.9

1. Using the notation in the text, show that $gF(z) = [g(z)]^2$, and deduce that $R(z) = z^2$. Show also that if $z_0$ is equidistant from $\alpha$ and $\beta$, then $F^n(z_0)$ will not converge to either $\alpha$ or $\beta$.

2. Let $f(z) = z^3 - 1$, and let $z_n$ be constructed using Newton's method. Find an explicit value of $r$ such that if $|z_0 - 1| < r$, then $z_n \to 1$.

## §1.10. General Remarks

The preceding examples have, we hope, given the reader some insight into iteration theory. Having said this, we should stress that some of these examples are *not* typical of the general case. For the typical rational function $R$, the complex plans (with $\infty$ attached) divides into the two sets $F$ and $J$. The sequence of forward and backward iterates preserve this subdivision: if $z$ lies in $F$ then so does its entire history and future under applications of $R$, and the same is true of $J$. The iterates preserve proximity on $F$ (and usually converge) whereas their action on $J$ is in every sense chaotic. In the general case, the geometry of $J$ is beautiful, delicate and extremely complicated, a typical example being illustrated in Figure 1.5.1. The sets $J$ are *Julia sets* and there are now many stunning pictures widely available in the literature: see, for example, [80], [81] and [100]. For general references, we refer the reader to [16], [35], [42], [43], [57], [69], [73] and [96].

The remainder of the book is devoted to a serious and rigorous study of these ideas, but before we can proceed with this, we must undertake a careful study of rational maps (Chapter 2) and normal families of analytic functions (Chapter 3).

CHAPTER 2

# Rational Maps

We present here some of the elementary properties of rational maps. A point $\infty$ is adjoined to the complex plane and a rational map of degree $d$ is considered as a $d$-fold map of the extended complex plane onto itself. The spherical and chordal metrics are introduced, the important notions of conjugacy, valency, fixed points and critical points are discussed, and the Riemann–Hurwitz relation is proved.

## §2.1. The Extended Complex Plane

We begin by taking an abstract point, which we denote by $\infty$, and adjoining it to the complex plane $\mathbb{C}$: the *extended complex plane* is then simply the union

$$\mathbb{C}_\infty = \mathbb{C} \cup \{\infty\}.$$

To obtain a metric on $\mathbb{C}_\infty$ we identify $\mathbb{C}$ with the horizontal plane

$$\{(x_1, x_2, x_3) \in \mathbb{R}^3 : x_3 = 0\}$$

in $\mathbb{R}^3$ and proceed to construct the usual model for $\mathbb{C}_\infty$ as a sphere. Let $S$ be the sphere in $\mathbb{R}^3$ with unit radius and centre at the origin, and denote the point $(0, 0, 1)$ (the top point of $S$) by $\zeta$. We now project each point $z$ in $\mathbb{C}$ linearly towards (or away from) $\zeta$ until it meets $S$ at a point $z^*$ distinct from $\zeta$: the map $\pi: z \mapsto z^*$ is called the *stereographic projection* of $\mathbb{C}$ into $S$. Clearly, if $|z|$ is large, then $z^*$ is near to $\zeta$, and with this in mind, we define the projection $\pi(\infty)$ of $\infty$ to be $\zeta$. With this definition, $\pi$ is a bijective map from $\mathbb{C}_\infty$ to $S$, and this explains why $\mathbb{C}_\infty$ is also called the *complex* (or *Riemann*) *sphere*: see Figure 2.1.1.

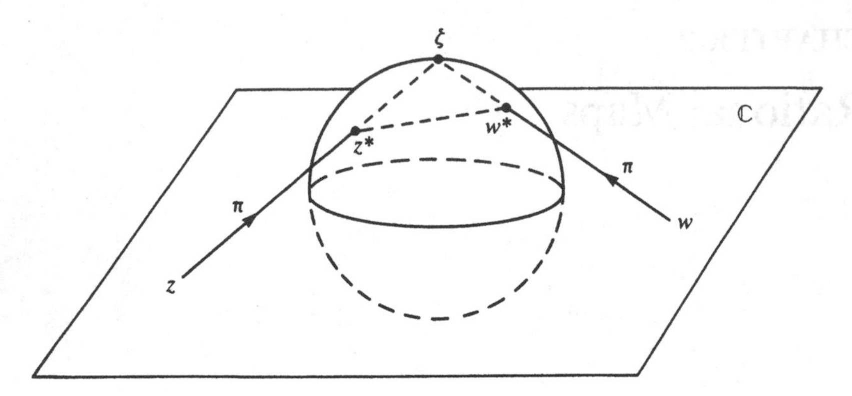


Figure 2.1.1. Stereographic projection.

We now use the bijection $\pi$ of $\mathbb{C}_\infty$ onto $S$ to transfer the Euclidean metric (in $\mathbb{R}^3$) from $S$ to a metric $\sigma$ on $\mathbb{C}_\infty$: this simply means that $\sigma$ is defined in the natural way by the formula

$$\sigma(z, w) = |\pi(z) - \pi(w)| = |z^* - w^*|.$$

A gentle exercise in vector geometry (which we omit) yields an explicit formula for $\sigma$, namely

$$\sigma(z, w) = \frac{2|z - w|}{(1 + |z|^2)^{1/2}(1 + |w|^2)^{1/2}} \tag{2.1.1}$$

when $z$ and $w$ are in $\mathbb{C}$, while for $z$ in $\mathbb{C}$,

$$\sigma(z, \infty) = \lim_{w \to \infty} \sigma(z, w) = \frac{2}{(1 + |z|^2)^{1/2}},$$

this being the limit of $\sigma(z, w)$ as $w \to \infty$. As $\sigma(z, w)$ is the Euclidean length of the chord joining $z^*$ to $w^*$, $\sigma$ is called the *chordal metric* on $\mathbb{C}_\infty$. From a more advanced point of view, $\mathbb{C}_\infty$ is both the topological one-point compactification of $\mathbb{C}$, and also a compact Riemann surface with $z \mapsto z^{-1}$ defining the chart at $\infty$, and although this motivates much of our thinking, the details need not concern us.

The Euclidean metric on $\mathbb{C}$ is simply not sufficiently versatile to cope with matters concerning $\infty$, but the chordal metric $\sigma$ handles all points of $\mathbb{C}_\infty$ with equal ease, and when we use $\sigma$, $\infty$ loses (as it should) any special significance. This is exhibited quite vividly by noting that the map $h: z \mapsto z^{-1}$ is actually a *$\sigma$-isometry* (that is, it preserves $\sigma$-distances) for, as a trivial manipulation shows, $\sigma(z, w) = \sigma(z^{-1}, w^{-1})$.

It is evident from elementary geometry that every rotation of the sphere $S$ induces a $\sigma$-isometry of $\mathbb{C}_\infty$ in the sense that for any rotation $\varphi$ of $S$, the conjugate map

$$\mathbb{C}_\infty \xrightarrow{\pi} S \xrightarrow{\varphi} S \xrightarrow{\pi^{-1}} \mathbb{C}_\infty$$

is a $\sigma$-isometry. In fact, the rotations of $S$ coincide (in this sense) with the class of the Möbius transformations of the form

$$z \mapsto \frac{az - \bar{c}}{cz + \bar{a}}, \qquad |a|^2 + |c|^2 = 1, \tag{2.1.2}$$

(for example, $a = 0$ and $c = i$ gives $z \mapsto z^{-1}$). Although we do not need to know this fact, it is certainly useful to observe that the maps in (2.1.2) are isometries, for these provide us with an adequate supply of isometries which we may use to simplify a given situation: the proof is a straightforward computation (Exercise 2.1.3).

There is an alternative metric on $\mathbb{C}_\infty$, namely the *spherical metric* $\sigma_0$, and this is equivalent to the chordal metric $\sigma$. The spherical distance $\sigma_0(z, w)$ between $z$ and $w$ in $\mathbb{C}_\infty$ is, by definition, the Euclidean length of the shortest path on $S$ (an arc of a great circle) between $z^*$ and $w^*$. If the chord joining $z^*$ and $w^*$ subtends an angle $\theta$ at the origin then, of course,

$$\sigma_0(z, w) = \theta, \qquad \sigma(z, w) = 2\sin(\theta/2)$$

so

$$\sigma(z, w) = 2\sin\left(\frac{\sigma_0(z, w)}{2}\right).$$

More useful than this, perhaps, are the inequalities

$$(2/\pi)\sigma_0(z, w) \leq \sigma(z, w) \leq \sigma_0(z, w) \tag{2.1.3}$$

which follow from the elementary inequalities

$$2\theta/\pi \leq \sin\theta \leq \theta, \qquad 0 \leq \theta \leq \pi/2.$$

In many instances, we can use $\sigma$ and $\sigma_0$ interchangeably (perhaps after changing some constants), and we shall frequently make this change without explicitly saying so.

Finally, we note that if $\gamma$ is any (say, continuously differentiable) curve in $\mathbb{C}_\infty$, then the spherical length of $\gamma$ is

$$\int_\gamma \frac{2|dz|}{1 + |z|^2},$$

because we can approximate the projection of $\gamma$ on the sphere by a polygonal curve in $\mathbb{R}^3$ and

$$\lim_{w \to z} \frac{|w^* - z^*|}{|w - z|} = \frac{2}{1 + |z|^2}.$$

EXERCISE 2.1

1. In the notation of the text, show that if $z = x + iy$, then

$$z^* = t(2x, 2y, |z|^2 - 1),$$

where $t = 1/(|z|^2 + 1)$. Find a formula for $z$ in terms of $z^*$.

2. Verify the formula (2.1.1) for $\sigma(z, w)$.
3. Verify that the Möbius transformation $g$ given by (2.1.2) is a $\sigma$-isometry; that is, show that $\sigma(gz, gw) = \sigma(z, w)$.
4. Prove that every rotation of $S$ arises from some map of the form (2.1.2).

## §2.2. Rational Maps

A *rational map* is a function of the form

$$R(z) = P(z)/Q(z),$$

where $P$ and $Q$ are polynomials, not both being the zero polynomial. If $P$ is the zero polynomial, then $R$ is the constant function zero: if $Q$ is the zero polynomial, then $R$ is the constant function $\infty$ (note that this is regarded as a rational function). If $Q(z) = 0$ and $P$ is not the zero polynomial, then $R(z)$ is defined to be $\infty$, and we define $R(\infty)$ as the limit of $R(z)$ as $z \to \infty$.

Suppose now that neither $P$ nor $Q$ is the zero polynomial. Strictly speaking, $R$ is not defined at the common zeros of $P$ and $Q$, but if these exist, we may cancel the corresponding linear factors and thereby assume that $P$ and $Q$ are *coprime* (that is, they have no common zeros). We shall always assume that this has been done, and then $P$ and $Q$ are each uniquely determined up to a scalar multiple by $R$, and the *degree* $\deg(R)$ of $R$ is defined by

$$\deg(R) = \max\{\deg(P), \deg(Q)\},$$

where $\deg(S)$ is the usual degree of a polynomial $S$. If $R$ is a constant map with value $\alpha$, where $\alpha \neq 0, \infty$, we have $\deg(R) = 0$, and it is convenient to define $\deg(R) = 0$ even when $\alpha$ is $0$ or $\infty$.

The rational maps can be characterized as the analytic maps of $\mathbb{C}_\infty$ into itself and, briefly, we recall the relevant facts and establish our terminology. Of course, the metrics $\sigma$ and $\sigma_0$ enable us to discuss functions defined at $\infty$, or taking the value $\infty$, and this, of course, is precisely the reason why they are preferred to the Euclidean metric. A function $f: D \to \mathbb{C}$ defined on a plane domain $D$ is *holomorphic* in $D$ if the derivative $f'$ exists at each point of $D$. The map $f: D \to \mathbb{C}_\infty$ is *meromorphic* in $D$ if each point of $D$ has a neighborhood on which either $f$ or $1/f$ is holomorphic. The poles of $f$ are points $w$ where $f(w) = \infty$, and near such points the map $z \mapsto 1/f(z)$ is holomorphic with value zero at $w$. Observe that $f$ is continuous at a pole $w$: indeed, $1/f$ is continuous at $w$ in the Euclidean metric, hence also in the chordal metric, and as $h: z \mapsto 1/z$ is a $\sigma$-isometry,

$$\begin{aligned} \sigma(f(z), f(w)) &= \sigma(1/f(z), 1/f(w)) \\ &= \sigma(1/f(z), 0) \\ &\to 0 \end{aligned}$$

as $z \to w$.

A function $f$ is said to be defined *near*, or *in some neighbourhood of* $\infty$ if it is defined on some set $\{|z| > r\} \cup \{\infty\}$, and in this case, $f$ is *holomorphic* (or *meromorphic*) at $\infty$ if the map $z \to f(1/z)$ is holomorphic (or meromorphic) near the origin. In keeping with the accepted terminology for Riemann surfaces, and to emphasize that $\infty$ and poles are not in any sense singularities, we say that a map $f\colon D_1 \to D_2$ between any two subdomains of $\mathbb{C}_\infty$ is *analytic* in $D_1$ if it is holomorphic, or meromorphic, at each point of $D_1$. The general rule, then, is that we should forget the Euclidean metric, use $\sigma$ instead, and, as a consequence, cease to distinguish $\infty$ as a special point of either the domain or the range. As a trivial example, let

$$P(z) = a_0 + a_1 z + \cdots + a_n z^n,$$

where $n > 0$ and $a_n \neq 0$. Then $P(\infty) = \infty$, and as the map

$$z \mapsto \frac{1}{P(1/z)} = \frac{z^n}{a_0 z^n + \cdots + a_n}$$

is holomorphic near 0 with the value 0 there, $P$ fixes $\infty$ and is analytic there. In particular, $\infty$ *is a fixed point of every non-constant polynomial.* More generally, it is well known (and easy to see) that each rational function is analytic throughout $\mathbb{C}_\infty$, and that the rational functions are the only maps with this property.

It is a crucial fact that if $R$ is a rational function of positive degree $d$, then $R$ is a *$d$-fold* map of $\mathbb{C}_\infty$ *onto* itself: that is, for any $w$ in $\mathbb{C}_\infty$, the equation $R(z) = w$ has *precisely $d$* solutions in $z$ (counting multiplicities). This result is so fundamental that we feel obliged to remind some of our readers of the proof.

We suppose that $R$ is not constant and that

$$R(z) = P(z)/Q(z), \qquad \deg(P) = n, \qquad \deg(Q) = m,$$

with $P$ and $Q$ coprime. If $n = m$, then all the zeros and poles of $R$ lie in $\mathbb{C}$; there are $n$ of each (the zeros of $P$ and $Q$ respectively) and $R$ has the same number, namely $\deg(R)$, of zeros as poles. If $n \neq m$, we need the usual convention for counting the zeros and poles at $\infty$. Suppose, for example, that $n > m$. Then $R$ has $n$ zeros and $m$ poles in $\mathbb{C}$ and $R(\infty) = \infty$. The number of poles at $\infty$ is, by definition, the number of zeros of the map $z \to 1/R(1/z)$ at zero and, as the reader can easily calculate, this is $n - m$. A similar argument copes with the case when $m > n$ and we find that in all cases, $R$ has precisely $\deg(R)$ zeros and $\deg(R)$ poles in $\mathbb{C}_\infty$.

For any $w$ in $\mathbb{C}$, the number of solutions of $R(z) = w$ is, by definition, the number of zeros of $R(z) - w$. Now

$$R(z) - w = \frac{P(z) - wQ(z)}{Q(z)},$$

and so $R(z) - w$ and $R(z)$ have the same degree (but only because of the fact that if $P$ and $Q$ are coprime, then so are $P - wQ$ and $Q$). We deduce that the equation $R(z) = w$ has precisely $\deg(R)$ solutions in $z$ regardless of the value of $w$ (in $\mathbb{C}_\infty$), and so $R$ is a $d$-fold map of $\mathbb{C}_\infty$ onto itself.

As the domain and range of $R$ coincide, we can apply $R$ repeatedly and so obtain the $n$-th iterate $R^n$ of $R$. We need to know that

$$\deg(R^n) = [\deg(R)]^n$$

and, clearly, this follows from the more general relation

$$\deg(RS) = \deg(R)\deg(S) \qquad (2.2.1)$$

between any two non-constant rational maps $R$ and $S$. To prove this, let $R$ and $S$ have degree $p$ and $q$ respectively. For all but a finite set of values for $w$, the set $R^{-1}\{w\}$ has precisely $p$ values, say $\zeta_1, \ldots, \zeta_p$. Excluding those (finitely many) $w$ for which some $S^{-1}\{\zeta_j\}$ has fewer than $q$ elements, we find that for all but a finite number of $w$, the set $(RS)^{-1}\{w\}$ has exactly $pq$ elements and this yields (2.2.1). Clearly, (2.2.1) holds if one of $R$ and $S$ are constant.

Exercise 2.2

1. For any rational function $R$, the derivative $R'$ is also rational and so has a value at $\infty$, namely $R'(\infty)$. Show that

$$R(z)/z \to R'(\infty) \quad \text{as} \quad z \to \infty.$$

Deduce that if $R(\infty) = \infty$, then $R'(\infty) \neq 0$.

2. Show that $z \mapsto \sin(1/z)$ is holomorphic at $\infty$, but that $z \mapsto \sin z$ is not.
3. Show that if $f$ has a pole at $z_0$ and if $g$ is holomorphic at $\infty$, then $gf$ is holomorphic at $z_0$.
4. Show that if $f$ is holomorphic and bounded in $\{1 < |z| < +\infty\}$, then $f$ is holomorphic at $\infty$.

## §2.3. The Lipschitz Condition

Quite generally, a continuous map from one compact metric space to another is uniformly continuous, and it follows from this that a rational map $R$ is uniformly continuous with respect to both the chordal and spherical metrics on $\mathbb{C}_\infty$. Much more is true, however, and an elementary argument shows that $R$ satisfies a Lipschitz condition with respect to these metrics. We prove

**Theorem 2.3.1.** *A rational map $R$ satisfies some Lipschitz condition*

$$\sigma_0(Rz, Rw) \leq M\sigma_0(z, w)$$

*on the complex sphere.*

**Proof.** When $R$ maps the point $z$ to $R(z)$, the change of scale (measured in the metric $\sigma_0$) at $z$ is the ratio

$$\frac{|R'(z)|(1+|z|^2)}{1+|R(z)|^2}, \tag{2.3.1}$$

so it enough to prove that this function is bounded above on the complex sphere. To do this, it is only necessary to check that it is bounded above near $\infty$ and near every pole of $R$, and this follows from standard, elementary arguments, essentially because $R$ is like some term $(z - z_0)^m$ near each of these points $z_0$. Of course, the supremum of the quantity (2.3.1) is actually the best possible value of the constant $M$ in Theorem 2.3.1. □

It will be convenient to have an explicit bound in Theorem 2.3.1 when $R$ is a Möbius map, so we extend this last result to

**Theorem 2.3.2.** *The Möbius map*

$$g(z) = \frac{az+b}{cz+d}, \qquad ad - bc = 1,$$

*satisfies the Lipschitz condition*

$$\sigma_0(gz, gw) \le \|g\|^2 \sigma_0(z, w),$$

*where*

$$\|g\|^2 = |a|^2 + |b|^2 + |c|^2 + |d|^2.$$

We remark that the best possible Lipschitz constant here is

$$[\sqrt{\|g\|^4 - 4} + \|g\|^2]/2: \tag{2.3.2}$$

for more details, see Exercise 2.3.1, or [18] (p. 42 and p. 61) where this constant is explained in terms of the action of $g$ as an isometry of hyperbolic 3-space.

**Proof of Theorem 2.3.2.** Using the proof of Theorem 2.3.1, we need only show that for all $z$,

$$\frac{1+|z|^2}{|az+b|^2 + |cz+d|^2} \le \|g\|^2.$$

The expression on the left, which we denote by $\Phi(g, z)$, measures the infinitesimal change of scale of $g$ at $z$ in the metric $\sigma_0$, so we would expect it to satisfy the Chain Rule

$$\Phi(g^{-1}, gz)\Phi(g, z) = 1:$$

in fact, it does, and the reader should now verify this by direct calculation. With this, a *lower bound* on $\Phi(g, z)$ is easy to obtain: the Cauchy–Schwarz inequality yields

$$|az+b|^2 \le (|a|^2 + |b|^2)(1+|z|^2),$$

and similarly for $|cz + d|^2$, and together these inequalities show that for any $z$,

$$\Phi(g, z) \geq \|g\|^{-2}.$$

Because of this and the Chain Rule, we find that

$$\Phi(g^{-1}, gz) \leq \|g\|^2.$$

This is true for any $g$ so, replacing $g$ by $g^{-1}$, and using the fact that $\|g\|^2 = \|g^{-1}\|^2$, we obtain

$$\Phi(g, g^{-1}z) \leq \|g\|^2$$

and hence for all complex $w$,

$$\Phi(g, w) \leq \|g\|^2.$$

This completes the proof. □

Theorem 2.3.2 implies that if the norms $\|g\|$ are bounded for all $g$ in some family $G$, then there is a uniform Lipschitz inequality for $G$. This is important, and in our next result we give a geometric criterion for this to be true.

**Theorem 2.3.3.** *Let $m$ be any positive number. Then the Möbius transformations $g$ which satisfy*

$$\sigma(g0, g1) \geq m, \qquad \sigma(g1, g\infty) \geq m, \qquad \sigma(g\infty, g0) \geq m, \tag{2.3.3}$$

*also satisfy the uniform Lipschitz condition*

$$\sigma(gz, gw) \leq (\pi/m^3)\sigma(z, w).$$

Proof. Suppose that $g$ satisfies (2.3.3): then

$$\sigma(g0, g1).\sigma(g1, g\infty).\sigma(g\infty, g0) \geq m^3. \tag{2.3.4}$$

Writing $g(z) = (az + b)/(cz + d)$ where $ad - bc = 1$, and recalling the formula (2.1.1) for $\sigma(z, w)$, we evaluate the left-hand side of (2.3.4) and obtain

$$(|a|^2 + |c|^2)(|b|^2 + |d|^2)(|a + b|^2 + |c + d|^2) \leq 1/m^3.$$

Now

$$1 = ad - bc = d(a + b) - b(c + d)$$

so, by the Cauchy–Schwarz inequality,

$$1 \leq (|b|^2 + |d|^2)(|a + b|^2 + |c + d|^2).$$

We deduce that

$$|a|^2 + |c|^2 \leq 1/m^3,$$

and an entirely similar argument shows that

$$|b|^2 + |d|^2 \leq 1/m^3.$$

We have now shown that any $g$ satisfying (2.3.3) also satisfies

$$\|g\|^2 \leq 2/m^3,$$

and Theorem 2.3.3 now follows directly from (2.1.3) and Theorem 2.3.2. □

We end this section with a result which depends on Theorem 2.3.1 (and which we shall need later) about closed curves on the sphere. Let $\gamma$ be a closed curve on the sphere: then the complement of $\gamma$ is a union of mutually disjoint domains, and we call these the *complementary components* of $\gamma$. Now suppose that $\gamma$ lies in some open hemisphere. Then exactly one of the complementary components contains a hemisphere; we denote this by $E(\gamma)$ and call it the *exterior* of $\gamma$: the other complementary components of $\gamma$ are said to be the *interior components* of $\gamma$. We can now state the result.

**Theorem 2.3.4.** *Let $R$ be a rational function. Then there is a positive number $\delta$ such that if $\gamma$ is any closed curve of $\sigma_0$-diameter less than $\delta$, then the image $R(\Omega)$ of an interior component $\Omega$ of $\gamma$ does not meet the exterior $E(R\gamma)$ of $R(\gamma)$.*

This result almost says that if the curve $\gamma$ is sufficiently small, then the interior components of $\gamma$ map to the interior components of the image curve $R(\gamma)$: however, the images of the interior components may also contain points of $R(\gamma)$, so it is necessary to express this a little more carefully in terms of the exterior of the image curve. The reader is urged to draw a diagram.

PROOF. We may assume that $R$ is not constant, and throughout the proof, the diameter of a set refers to the $\sigma_0$-diameter. First, note that any interior component $\Omega$ of $\gamma$ has diameter at most the diameter of $\gamma$ itself, and so if $M$ is as given in Theorem 2.3.1,

$$\text{diameter}(R(\Omega)) \leq M \text{ diameter}(\gamma).$$

Now choose any $\delta$ such that $M\delta < \pi$. It follows that if $\gamma$ is a closed curve of $\sigma_0$-diameter less than $\delta$, then $R(\Omega)$ has diameter less than $\pi$ and so $R(\Omega)$ certainly cannot contain $E(R\gamma)$.

We need to know that for any rational map $R$, and any domain $W$ on the sphere,

$$\partial(R(W)) \subset R(\partial W); \tag{2.3.5}$$

then, applying this to $\Omega$, we find that

$$\partial(R(\Omega)) \subset R(\partial\Omega) \subset R(\gamma).$$

This, in turn, implies that $E(R\gamma)$ is disjoint from $\partial(R(\Omega))$, and so either $E(R\gamma) \cap R(\Omega)$ is empty, or $E(R\gamma) \subset R(\Omega)$. Now we have shown that $R(\Omega)$ cannot contain $E(R\gamma)$, thus $E(R\gamma) \cap R(\Omega)$ is empty (as required) and it only remains to prove (2.3.5).

To prove (2.3.5), observe that any point $\zeta$ in $\partial R(W)$ is the limit of points

$R(z_n)$, where $z_n \in W$. Without loss of generality, $z_n \to z$, so $\zeta = R(z)$, where $z$ lies in the closure of $W$. Now $z$ cannot be in $W$, else $\zeta$ is in $R(W)$ and this, being an open set, is disjoint from $\partial R(W)$. We conclude that $z$ is in $\partial W$, and so $\zeta$ is in $R(\partial W)$. The proof is now complete. □

Exercise 2.3

1. Show that (2.3.2) gives the best Lipschitz constant for a Möbius map $g$ by regarding $g$ as a linear map (whose matrix has coefficients $a$, $b$, $c$ and $d$) from $\mathbb{C}^2$ to itself.

## §2.4. Conjugacy

The rational maps of degree one are the Möbius maps

$$z \to \frac{az + b}{cz + d}, \qquad ad - bc \neq 0,$$

and as these constitute the group of analytic homeomorphisms of $\mathbb{C}_\infty$ onto itself, they can be used to introduce conjugacy. We say that two rational maps $R$ and $S$ are *conjugate* if and only if there is some Möbius map $g$ with

$$S = gRg^{-1}.$$

It is clear that conjugacy is an equivalence relation, and the equivalence classes are the *conjugacy classes* of rational maps.

From a qualitative, or topological, point of view *it is only the conjugation invariant properties, or functions, that are of interest to us*, and perhaps the most important of these is the degree function: if $R$ and $S$ are conjugate, then

$$\deg(R) = \deg(S). \tag{2.4.1}$$

This is an immediate consequence of (2.2.1).

Another important property of conjugacy is that it respects iteration: that is, if $S = gRg^{-1}$, then

$$S^n = gR^ng^{-1}.$$

This means that we can transfer a problem concerning $R$ to a (possibly simpler) problem concerning a conjugate $S$ of $R$ and then attempt to solve this in terms of $S$. If we wish, we can rewrite the solution in terms of $R$, but it is surely better to resist the temptation to do so and to attempt, instead, to express the solution in a conjugation invariant form.

Yet another obvious property of conjugacy is that it respects fixed points: explicitly, if $S = gRg^{-1}$, then $S$ fixes $g(z)$ if and only if $R$ fixes $z$. All of these observations are elementary; nevertheless, they are important because, in general, we shall not bother to distinguish between conjugate rational functions.

As a simple illustration of the use of conjugacy, let us characterize the polynomials within the class of rational maps. Obviously, a non-constant rational map $R$ is a polynomial if and only if $R$ has a pole at $\infty$ and no poles in $\mathbb{C}$; more succinctly, if and only if $R^{-1}\{\infty\} = \{\infty\}$. More generally (and the proof is trivial), we have

**Theorem 2.4.1.** *A non-constant rational map $R$ is conjugate to a polynomial if and only if there is some $w$ in $\mathbb{C}_\infty$ with $R^{-1}\{w\} = \{w\}$.*

EXERCISE 2.4

1. (i) Prove that any Möbius map is conjugate to one of the maps $z \mapsto az, z \mapsto z + a$ for a suitable value of $a$.
   (ii) Prove that if $ab \neq 0$, the maps $z \mapsto az$ and $z \mapsto bz$ are conjugate if and only if either $b = a$ or $b = a^{-1}$.
   (iii) Prove that if $ab \neq 0$, then the maps $z \mapsto z + a$ and $z \mapsto z + b$ are conjugate.
2. Prove that any quadratic polynomial is conjugate to one and only one map of the form $z \mapsto z^2 + a$, and to one and only one map of the form $z \mapsto \lambda z(1 - z)$.
3. Show that if $d \geq 2$, then $(z^d + 1)/z^d$ is not conjugate to a polynomial.
4. Is the rational map $(2z^2 - 2z + 1)/(3z^2 - 4z + 2)$ conjugate to a polynomial?
5. Let $R(z) = z^2$. Find all Möbius map $g$ such that $gRg^{-1} = R$.

## §2.5. Valency

Consider any function $f$ that is non-constant and holomorphic near the point $z_0$ in $\mathbb{C}$. Then $f$ has a Taylor expansion at $z_0$, say

$$f(z) = a_0 + a_k(z - z_0)^k + a_{k+1}(z - z_0)^{k+1} + \cdots,$$

where $a_k \neq 0$, and the positive integer $k$ is uniquely determined by the condition that the limit

$$\lim_{z \to z_0} \frac{f(z) - f(z_0)}{(z - z_0)^k}$$

exists, is finite and is non-zero. We denote this integer $k$ by $v_f(z_0)$ and call it the *valency*, or *order*, of $f$ at $z_0$: of course, it is the number of solutions of $f(z) = f(z_0)$ at $z_0$.

The valency function $v$ satisfies the important *Chain Rule*:

$$v_{fg}(z_0) = v_f(gz_0)v_g(z_0),$$

where $z_0$, $g(z_0)$ and $fg(z_0)$ are all in $\mathbb{C}$, and the proof of this is easy. As $g$ is holomorphic and not constant near $z_0$, $g \neq g(z_0)$ on a deleted neighbourhood $N$ of $z_0$, and on $N$ we have the identity

$$\frac{fg(z) - fg(z_0)}{(z - z_0)^{kq}} = \left(\frac{fg(z) - fg(z_0)}{[g(z) - g(z_0)]^q}\right)\left(\frac{g(z) - g(z_0)}{(z - z_0)^k}\right)^q,$$

where $q = v_f(gz_0)$ and $k = v_g(z_0)$. The Chain Rule follows from this because the right-hand side, and hence the left-hand side also, has a finite non-zero limit as $z \to z_0$: thus $v_{fg}(z_0) = kq$ as required.

Observe next that $f$ is injective in some neighbourhood of $z_0$ if and only if $v_f(z_0) = 1$, and knowing this, the Chain Rule shows that the valency is preserved under a pre-application, and a post-application, of an injective function. Explicitly, if $fg$ is defined near $z$ with $g$ injective near $z$, and if $hf$ is defined near $z$ with $h$ injective near $f(z)$, then

$$v_{fg}(z) = v_f(gz), \qquad v_{hf}(z) = v_f(z). \tag{2.5.1}$$

Now (2.5.1) enables us to extend our definition of $v_f(z_0)$ to the case where $z_0 = \infty$, or $f(z_0) = \infty$ (or both). We select any Möbius maps $g$ and $h$ which map $z_0$ and $f(z_0)$ respectively to points in $\mathbb{C}$ and then define $v_f(z_0)$ by

$$v_f(z_0) = v_F(gz_0),$$

where $F = hfg^{-1}$. The essential observation, of course, is that if we repeat this procedure using different maps $g$ and $h$, then the resulting answer will be the same; thus $v_f(z_0)$ is defined independently of the choice of $g$ and $h$. In fact, a similar argument shows that the valency can be defined for any analytic map between two Riemann surfaces. Of course, if $f$ has a pole of order $k$ at $z_0$ (in $\mathbb{C}$) then $f$ has valency $k$ at $z_0$, and $R$ has $k$ zeros at $z_0$ if and only if $gRg^{-1}$ has $k$ zeros at $g(z_0)$, where $g$ is any Möbius map.

In keeping with common usage, an analytic map $f\colon D \to \mathbb{C}_\infty$ is said to be *univalent* in $D$ if it is injective there. If $f$ is univalent in $D$, then $v_f(z) = 1$ for every $z$ in $D$, but the converse *need not hold*: for example, the map $z^2$ of $\mathbb{C} - \{0\}$ onto itself has valency 1 everywhere but it is not injective in $\mathbb{C} - \{0\}$.

Finally, the fundamental fact that a non-constant rational map $R$ of degree $d$ is a $d$-fold map of $\mathbb{C}_\infty$ onto itself is easily expressed in terms of valency. For any $z_0$ in $R^{-1}\{w\}$, $v_R(z_0)$ is the number of solutions of $R(z) = w$ at $z_0$, and so for each $w$ in the complex sphere,

$$\sum_{z \in R^{-1}\{w\}} v_R(z) = \deg(R).$$

Exercise 2.5

1. Suppose that $R(z) = P(z)/Q(z)$ where

$$P(z) = a_0 + a_1 z + \cdots + a_n z^n, \qquad Q(z) = b_0 + b_1 z + \cdots + b_m z^m,$$

and where $a_n b_m \neq 0$. Show that if $m \neq n$ then $v_R(\infty) = |m - n|$. What can be said if $m = n$?

## §2.6. Fixed Points

The fixed points of a mapping play an important role in much of our later work and they will be studied and classified in Chapter 6. In this section, we simply compute the number of fixed points of a rational map. Let $R$ be a non-constant rational function, say $R = P/Q$, where $P$ and $Q$ are coprime polynomials. The location of the fixed points of $R$ is trivial. First, $R$ fixes $\infty$ if and only if $\deg(P) > \deg(Q)$. If $\zeta$ in $\mathbb{C}$ is fixed by $R$, then $Q(\zeta) \neq 0$ (else $P(\zeta) \neq 0$ and $R(\zeta) = \infty$) and so $\zeta$ satisfies

$$P(\zeta) = \zeta Q(\zeta). \tag{2.6.1}$$

Conversely, if (2.6.1) holds then $Q(\zeta) \neq 0$ (else $P$ and $Q$ vanish at $\zeta$ and are not then coprime) so $\zeta$ is fixed by $R$. Thus the fixed points of $R$ in $\mathbb{C}$ are (as expected) the solutions of

$$P(z) - zQ(z) = 0. \tag{2.6.2}$$

Note that (2.6.2) need not have any solutions in $\mathbb{C}$ (even though the left-hand side is a polynomial); for example, $z \mapsto z + z^{-1}$ has no fixed points in $\mathbb{C}$.

Having considered the location of the fixed points, our next task is to count them, and we shall show that if they are counted in the same way that we count zeros of an analytic map, then a rational map $R$ of degree $d$, $d \geq 1$, has precisely $d + 1$ fixed points in $\mathbb{C}_\infty$. It is advantageous to define the relevant quantities for general analytic maps, and we make the natural definition that if $\zeta$ is a fixed point of an analytic map $f$, and if $\zeta \neq \infty$, then $f$ has $k$ fixed points at $\zeta$ if and only if $f(z) - z$ has $k$ zeros at $\zeta$. Observe (for example), that $z + z^3$ has three fixed points at the origin but has valency 1 there, whereas $z^3$ has one fixed point at the origin and valency 3 there.

The definition above does not apply to fixed points at $\infty$, and to cope with this, we need

**Lemma 2.6.1.** *Let $\zeta$ in $\mathbb{C}$ be a fixed point of an analytic map $f$, and let $\varphi$ be any map that is analytic, injective and finite in some neighbourhood of $\zeta$. Then $\varphi f \varphi^{-1}$ has the same number of fixed points at $\varphi(\zeta)$ as $f$ has at $\zeta$.*

PROOF. Suppose that $f$ has $k$ fixed points at $\zeta$. We consider the identity

$$\frac{\varphi f \varphi^{-1}(z) - z}{[z - \varphi(\zeta)]^k}$$
$$= \left(\frac{\varphi f \varphi^{-1}(z) - \varphi\varphi^{-1}(z)}{f\varphi^{-1}(z) - \varphi^{-1}(z)}\right)\left(\frac{f\varphi^{-1}(z) - \varphi^{-1}(z)}{[\varphi^{-1}(z) - \zeta]^k}\right)\left(\frac{[\varphi^{-1}(z) - \varphi^{-1}\varphi(\zeta)]^k}{[z - \varphi(\zeta)]^k}\right)$$

and note that from this, it is sufficient to show that each term on the right tends to a finite non-zero limit as $z$ tends to $\varphi(\zeta)$. The first term on the right is of the form

$$\frac{\varphi(u) - \varphi(v)}{u - v},$$

and a simple application of Cauchy's Integral formula applied to a small circle about $\varphi(\zeta)$, shows that this tends to a finite non-zero limit, namely $\varphi'(\zeta)$, as both $u$ and $v$ tend to $\zeta$. Next, the definition of $k$ implies that as $z$ tends to $\varphi(\zeta)$, the second term on the right tends to a finite non-zero limit. Finally, the third term tends to $[\varphi'(\zeta)]^{-\kappa}$ and the proof is complete. For an alternative proof, see Exercise 2.6.5. □

The invariance expressed in Lemma 2.6.1 automatically provides a definition for the number of fixed points at $\infty$, for if $R(\infty) = \infty$, we can conjugate $R$ so that $\infty$ is transferred to a finite fixed point, say $\zeta$, and then count the number of fixed points of the conjugate function there. Clearly, Lemma 2.6.1 guarantees that the answer will be independent of the conjugation used, and so this provides the required definition. With this definition, Lemma 2.6.1 now implies

**Theorem 2.6.2.** *Let $\zeta$ be fixed point of the rational map $R$, and let $g$ be a Möbius map. Then $gRg^{-1}$ has the same number of fixed points at $g(\zeta)$ as $R$ has at $\zeta$.*

We can now prove

**Theorem 2.6.3.** *If $d \geq 1$, a rational map of degree $d$ has precisely $d + 1$ fixed points in $\mathbb{C}_\infty$.*

PROOF. Any rational map $R$ is conjugate to a rational map $S$ which does not fix $\infty$, and the number of fixed points of $S$ and of $R$ are the same, as are the degrees of $S$ and $R$. It follows that we may assume that $R$ does not fix $\infty$.

Now write $R = P/Q$ with $P$ and $Q$ coprime, and let $\zeta$ be any fixed point of $R$, so $\zeta$ is finite. As $Q(\zeta) \neq 0$, the number of zeros of $R(z) - z$ at $\zeta$ is exactly the same as the number of zeros of $P(z) - zQ(z)$ at $\zeta$: hence the number of fixed points of $R$ is exactly the number of solutions of $P(z) = zQ(z)$ in $\mathbb{C}$. As $R$ does not fix $\infty$, we have

$$\deg(P) \leq \deg(Q) = \deg(R):$$

thus the degree of $P(z) - zQ(z)$ is exactly $\deg(R) + 1$ and the proof is complete. □

To each fixed point $\zeta$ of a rational map $R$, we associate a complex number which we call the *multiplier* $m(R, \zeta)$ of $R$ at $\zeta$. If $\zeta$ is in $\mathbb{C}$, the multiplier is simply the derivative $R'(\zeta)$ and this is invariant under a conjugation (provided that the corresponding fixed point is also in $\mathbb{C}$). We must now define the multiplier $m(R, \infty)$ when $\infty$ is fixed by $R$. The procedure, of course, is to choose a Möbius map $g$ with $g(\infty)$ in $\mathbb{C}$, and then define

$$m(R, \infty) = m((gRg^{-1}, g\infty),$$

noting, by the preceding remarks, that this definition is independent of the choice of $g$. We have now defined a *conjugation invariant multiplier* $m(R, \zeta)$ of any rational map $R$ at any of its fixed points $\zeta$.

To illustrate this, suppose that

$$R(z) = \frac{a_0 + a_1 z + \cdots + a_n z^n}{b_0 + b_1 z + \cdots + b_m z^m},$$

where $a_n b_m \neq 0$ and $n > m$, so $R$ fixes $\infty$, and let us compute the multiplier $m(R, \infty)$ of $R$ at $\infty$. By definition, $m(R, \infty)$ is the derivative of

$$S: z \mapsto 1/[R(1/z)]$$

at the origin and this is found to be

$$S'(0) = \begin{cases} b_m/a_n & \text{if } n = m + 1, \\ 0 & \text{if } n > m + 1. \end{cases}$$

Another computation gives

$$R'(\infty) = \begin{cases} a_n/b_m & \text{if } n = m + 1, \\ \infty & \text{if } n > m + 1; \end{cases}$$

where, of course, by continuity,

$$R'(\infty) = \lim_{z \to \infty} R'(z)$$

(see Exercise 2.2.1). These computations show that the multiplier $m(R, \zeta)$ of a rational map $R$ at a fixed point $\zeta$ is given by

$$m(R, \zeta) = \begin{cases} R'(\zeta) & \text{if } \zeta \neq \infty; \\ 1/R'(\infty) & \text{if } \zeta = \infty. \end{cases}$$

For example, if $R(z) = 3z$, then $\infty$ is at attracting fixed point of $R$ (this is geometrically clear) and in this case, $m(R, \infty) = \frac{1}{3}$.

We end this section by considering the number of fixed points of an iterate $f^n$ at a fixed point of $f$. For this, we need

**Theorem 2.6.4.** *Suppose that*

$$f(z) = az + b_1 z^{r+1} + \cdots \tag{2.6.3}$$

*near the origin, where $a \neq 0$, $b_1 \neq 0$ and $r \geq 1$. Then*

$$f^n(z) = a^n z + b_n z^{r+1} + \cdots, \tag{2.6.4}$$

*where*

$$b_n = a^{n-1} b_1 (1 + a^r + a^{2r} + \cdots + a^{(n-1)r}).$$

Immediately, this yields

**Corollary 2.6.5.** *If $a = 1$, then $b_n = nb_1$.*

**Corollary 2.6.6.** *$b_n = 0$ if and only if $a^r \neq 1$ and $a^{nr} = 1$.*

Of course, if $a^n \neq 1$, then $a \neq 1$, so if $f^n$ has a single fixed point at 0, then so does $f$. More generally, we have

**Corollary 2.6.7.** *The iterate $f^n$ has at least as many fixed points at the origin as $f$ has, and if it has more, then $a \neq 1$ but $a^n = 1$.*

We end with the

PROOF OF THEOREM 2.6.4. It is easy to see (by induction) that if $f$ has the form (6.2.3), then $f^n$ has the form (6.2.4). Substituting the power series for $f$, $f^n$ and $f^{n+1}$ into the identity $f^{n+1}(z) = f(f^n z)$ and identifying coefficients, we obtain

$$b_{n+1} = ab_n + a^{n(r+1)}b_1,$$

and writing $\beta_n = b_n/a^n$, this gives

$$\beta_{n+1} = \beta_n + a^{nr}b_1.$$

This is easily solved for $\beta_n$ and the formula for $b_n$ follows. □

EXERCISE 2.6

1. Verify the formulae for $S'(0)$ and $R'(\infty)$ in the text.

2. By considering $z + z^{-1}$ and $z^2 + z - 1$, show that neither of the two following statements implies the other:
   (i) $R$ has all of its fixed points at $\zeta$;
   (ii) $R^{-1}\{\zeta\} = \{\zeta\}$.

3. Show that if $d \geq 2$, a polynomial of degree $d$ has $d$ fixed points in $\mathbb{C}$ and one fixed point at $\infty$.

4. Suppose that
$$R(z) = \frac{a_0 + a_1 z + \cdots + a_n z^n}{b_0 + b_1 z + \cdots + b_m z^m},$$
where $a_n b_m \neq 0$ and $n > m$. Show that $R$ has exactly one fixed point at $\infty$ if $n > m + 1$, or if $n = m + 1$ and $b_m \neq a_n$.

5. Prove Lemma 2.6.1 in the following way. Suppose that
$$f(z) = z + az^{p+1} + \cdots,$$
where $a \neq 0$, so $f$ has $p + 1$ fixed points at the origin. Now consider any power series $g(z) = \sum_{n=0}^{\infty} b_n z^n$, where $b_0 = 0$ and $b_1 \neq 0$, and let $F = gfg^{-1}$. We claim that
$$F(z) = z + Az^{p+1} + \cdots,$$
where $A \neq 0$, and this can be proved as follows.

   Consider the "commutator" $fg - gf$, and show that
$$fg(z) - gf(z) = ab_1(b_1^p - 1)z^{p+1} + O(z^{p+2}).$$
Next,
$$g^{-1}(z) = z/b_1 + O(z^2),$$
so
$$(fg - gf)(g^{-1}z) = a(1 - b_1^{-p})z^{p+1} + O(z^{p+2}),$$
whence
$$\begin{aligned} gfg^{-1}(z) &= f(z) - a(1 - b_1^{-p})z^{p+1} + O(z^{p+2}) \\ &= z + ab_1^{-p}z^{p+1} + O(z^{p+2}). \end{aligned}$$

## §2.7. Critical Points

This section is devoted to the notion of critical points and critical values even though their full significance for iteration theory will not be apparent until Chapter 9. A point $z$ is a *critical point* of a rational map $R$ if $R$ fails to be injective in any neighbourhood of $z$, and if $R$ is not constant, these points are precisely the points at which $v_R(z) > 1$. A value $w$ is a *critical value* for $R$ if it is the image of some critical point; that is, if $w = R(z)$ for some critical point $z$. These ideas are often confused and it is important to distinguish clearly between them.

If $R$ is of degree $d$ if $w$ is *not* a critical value, then $R^{-1}\{w\}$ consists of precisely $d$ distinct points, say $z_1, \ldots, z_d$. As none of the $z_j$ are critical points, there are neighbourhoods $N$ of $w$, and $N_1, \ldots, N_d$ of $z_1, \ldots, z_d$ respectively, with $R$ acting as a bijection from each $N_j$ onto $N$. It follows that for each $j$, the restriction $R_j$ of the map $R$ to $N_j$ has an inverse

$$R_j^{-1}: N \to N_j,$$

and we call these the *branches* of $R^{-1}$ at $w$.

We know that $R$ is injective in some neighbourhood of any point in $\mathbb{C}$ at which $R'$ has neither a zero nor a pole: thus for all but a finite set of $z$, $v_R(z) = 1$ and, consequently,

$$\sum_z [v_R(z) - 1] < +\infty,$$

the sum being over all $z$ in $\mathbb{C}_\infty$. This sum gives us a measure of the number of multiple roots of $R$ (and the difficulties in defining $R^{-1}$), and its actual value is given by the *Riemann–Hurwitz relation*:

**Theorem 2.7.1.** *For any non-constant rational map $R$,*

$$\sum [v_R(z) - 1] = 2\deg(R) - 2. \tag{2.7.1}$$

The general term in the sum is positive only when $z$ is a critical point of $R$ so, as the valency is an integer, this provides us with an estimate of the number of critical points of $R$. For a non-constant polynomial $P$, we have $v_P(\infty) = \deg(P)$, so, in the same way, we obtain an estimate of the number of critical points of $P$ in $\mathbb{C}$. These two estimates are given in

**Corollary 2.7.2.** *A rational map of positive degree $d$ has at most $2d - 2$ critical points in $\mathbb{C}_\infty$. A polynomial of positive degree $d$ has at most $d - 1$ critical points in $\mathbb{C}$.*

We shall see later that these estimates play a vital role in iteration theory. Usually, we allocate a multiplicity to a critical point $z$: this multiplicity is $v_R(z) - 1$ and with this convention, a rational map of degree $d$ has exactly $2d - 2$ critical points.

A topological proof of Theorem 2.7.1, using the Euler characteristic, is available and has the advantage that, suitably modified, it is valid for any analytic map between compact Riemann surfaces: this is given in Chapter 5. However, at this point some readers may prefer the following more elementary proof.

Proof of Theorem 2.7.1. First, we note that both sides of (2.7.1) are invariant under conjugation, thus it is sufficient to prove (2.7.1) for any conjugate of $R$. Now select a point $\zeta$ such that $R(\zeta) \neq \zeta$, $v_R(\zeta) = 1$, and that $R(z) = \zeta$ has $d$ distinct solutions, and then construct a Möbius transformation $g$ that maps $\zeta$ to $\infty$, and $R(\zeta)$ to 1. If we now write $S = gRg^{-1}$, then translate the properties of $R$ into properties of $S$, and then relabel $S$ as $R$, we find that we may assume:

(i) $R(\infty) = 1$;
(ii) $R$ has distinct simple poles $z_1, \ldots, z_d$ (all in $\mathbb{C}$); and
(iii) $v_R(\infty) = 1$.

These conditions imply that the valency of $R$ at $\infty$, and at each $z_j$, is one, hence the sum (2.7.1) is the same as

$$\sum [v_R(z) - 1], \tag{2.7.2}$$

summed over all $z$ in $\mathbb{C}$ except for the points $z_j$. For all such $z$, $R(z)$ is in $\mathbb{C}$ and so the value of (2.7.2) is the number of zeros of $R'(z)$.

Now write $R = P/Q$, say, in reduced form. Then

$$R'(z) = \frac{P'(z)Q(z) - P(z)Q'(z)}{Q(z)^2},$$

and this is also in reduced form for if not, the numerator and denominator have a common zero (which must be some $z_j$), and then

$$0 = P'(z_j)Q(z_j) = P(z_j)Q'(z_j):$$

but then either $P(z_j) = 0$ (which is false as $P/Q$ is in reduced form) or $Q'(z_j) = 0$ (which is false as the $z_j$ are simple zeros of $Q$). It now follows that as (2.7.2) is the number of zeros of $R'(z)$, it is also the degree of $P'Q - PQ'$, or, equivalently, the degree of the polynomial $Q(z)^2R'(z)$.

We shall compute the degree of this polynomial by finding its order of growth at $\infty$. First, it is clear that $Q(z)^2/z^{2d}$ tends to a finite non-zero limit as $z \to \infty$. Next, the fact that $v_R(\infty) = 1$ means that $R$ is injective in some neighbourhood at $\infty$ and so

$$R(1/z) = 1 + Az + \cdots$$

near the origin, where $A \neq 0$. Differentiating both sides of this, and replacing $z$ by $1/z$, we find that $z^2R'(z)$ tends to a finite non-zero limit at $\infty$ and so, finally, (2.7.2) is $2d - 2$ as required. □

We end this section with a simple result which locates the critical values of the iterate $R^n$ in terms of the critical points of $R$.

**Theorem 2.7.3.** *Let $C$ be the set of critical points of a rational map $R$. Then the set of critical values of $R^n$ is*

$$R(C) \cup \cdots \cup R^n(C). \tag{2.7.3}$$

PROOF. The proof is a simple consequence of the Chain Rule for valencies. Suppose first that $z$ lies in the set (2.7.3); then there is a sequence

$$z_0,\ z_1 = R(z_0), \ldots, z = R(z_{n-1})$$

with one of the $z_j$ in $C$. It follows from the Chain Rule for valencies that the valency of $R^n$ at $z_0$ exceeds one; hence $z$ is a critical value of $R^n$. Conversely, it $z$ is a critical value of $R^n$, there must be some such sequence with the valency of $R^n$ at $z_0$ exceeding one, and hence some $z_j$ is in $C$. □

EXERCISE 2.7

1. Show that a rational map $R$ of degree 2 has precisely two critical points and has valency 2 at each of these points.
2. Verify the details of the proof of Theorem 2.7.1 in the text that because $v_R(\infty) = 1$, $z^2R'(z)$ tends to a finite non-zero limit as $z \to \infty$.
3. Let $R$ be a non-constant rational map. Show that
$$\deg(R) - 1 \leq \deg(R') \leq 2\deg(R).$$
Show also:
(i) there is equality in the lower bound if and only if $R$ is a polynomial; and
(ii) there is equality in the upper bound if and only if all of the poles of $R$ are simple poles in $\mathbb{C}$.
4. Suppose that a rational map $R$ maps an open set $D$ into itself. Show that if $R^n$ has a critical point in $D$, then so does $R$.

# §2.8. A Topology on the Rational Functions

Let $\mathscr{C}$ be the class of continuous maps of $\mathbb{C}_\infty$ into itself and let $\mathscr{R}$ be the subclass of rational functions. The metric

$$\rho(R, S) = \sup \sigma_0(Rz, Sz)$$

on $\mathscr{C}$ is the metric of uniform convergence on $\mathbb{C}_\infty$. As the chordal and spherical metrics only differ by a bounded factor, we could equally well use the chordal metric here (the metric would change but the topology would not). Now $\mathscr{R}$ is a closed subset of $\mathscr{C}$ because if the rational functions $R_n$ converge uniformly to $R$ on the complex sphere, then $R$ is analytic on the sphere and so it too is rational. More generally (and again we could use either the chordal or the spherical metric), we have

**Theorem 2.8.1.** *Suppose that each $f_n$ is analytic in a domain $D$ on $\mathbb{C}_\infty$ and that $f_n$ converges uniformly on $D$ to $f$ with respect to $\sigma_0$. Then $f$ is also analytic in $D$.*

PROOF. The proof is elementary and we merely remind the reader of the main points. We need to check that $f$ is analytic in some neighbourhood of each point $\zeta$ of $D$. If neither $\zeta$ nor $f(\zeta)$ are $\infty$, then the uniform convergence of the $f_n$ near $\zeta$ ensures that $f$ is bounded in some neighborhood $N$ of $\zeta$. This, together with Cauchy's integral formula, shows that $f$ is holomorphic in $N$. If one, or both, of $\zeta$ and $f(\zeta)$ are $\infty$, we compose $f$ (before, or after, or both) with the $\sigma_0$-isometry $h: z \to 1/z$. As composition with a $\sigma_0$-isometry preserves the local uniform convergence, this case is reduced to the situation already discussed above. Finally, recall that if $f(\infty) = \infty$, then $f$ analytic at $\infty$ simply means that $z \mapsto 1/f(1/z)$ is holomorphic at the origin.

We return now to discuss rational functions. The degree function

$$\deg: R \mapsto \deg(R)$$

maps $\mathscr{R}$ onto $\{0, 1, 2, \ldots\}$ and we shall need to know that this is continuous. □

**Theorem 2.8.2.** *The map* $\deg: \mathscr{R} \to \{0, 1, \ldots\}$ *is continuous. In particular, if the rational functions $R_n$ converge uniformly on the complex sphere to the function $R$, then $R$ is rational and for all sufficiently large $n$,* $\deg(R_n) = \deg(R)$.

PROOF. Again, the ideas are elementary and we only sketch the proof. First, the uniform convergence of $R_n$ to $R$ guarantees that $R$ is analytic throughout the sphere and so is rational. As $\deg: \mathscr{R} \to \mathbb{Z}$ is a map between metric spaces, it is enough to work with sequences: thus we assume that $R_n$ converges uniformly on the complex sphere to $R$ and we wish to show that for sufficiently large $n$, $\deg(R_n) = \deg(R)$. This is clear if $R$ is constant, so we may assume that $\deg(R) \geq 1$.

We may assume that $R(\infty) \neq 0$ for otherwise, we can replace $R$ and $R_n$ by $1/R$ and $1/R_n$. With this assumption, $R$ has distinct zeros, say $z_1, \ldots, z_t$ and these all lie in $\mathbb{C}$. Construct small, mutually disjoint discs $D_j$ about the $z_j$, ensuring that no $D_j$ contains a pole of $R$. Let $K$ be the complement of the union of the $D_j$. For all sufficiently large $n$, $R_n$ is uniformly close to $R$, and so has no poles, in the $D_j$. As $R_n$ and $R$ are uniformly close on the circles bounding the $D_j$, Rouché's Theorem shows that $R_n$ and $R$ have the same number of zeros in each $D_j$. Finally, $R$ is bounded away from zero on the compact $K$, hence so are the $R_n$ (for large $n$): thus for all sufficiently large $n$, $R_n$ and $R$ have the same number of zeros and so they have the same degree. □

As a consequence of Theorem 2.8.2, the class $\mathscr{R}_n$ of rational maps of degree $n$ is an open subset of $\mathscr{R}$ for it is the inverse image of the open subset $\{n\}$ of $\mathbb{Z}$ under the continuous map deg. It is easy to see that each $\mathscr{R}_n$ is connected (for it $R$ and $S$ lie in $\mathscr{R}_n$, we can simply "slide" the zeros and poles of $R$ to those of $S$ while maintaining the same degree) so we obtain

**Corollary 2.8.3.** *The $\mathscr{R}_n$ are the components of $\mathscr{R}$.*

We shall not need to use this result, and we leave the interested reader to complete the details of the argument.

We end with a few remarks concerning the relation between the topological structure of $\mathscr{R}_d$ and the coefficients of rational maps (and again, these can be safely ignored). There is a natural map

$$\Phi: \mathbb{C}^{2d+2} - \{0\} \to \mathscr{R}_0 \cup \cdots \cup \mathscr{R}_d,$$

(not the opposite direction) namely the map

$$(a_0, \ldots, a_d, b_0, \ldots, b_d) \mapsto P/Q$$

where

$$P(z) = a_0 + a_1 z + \cdots + a_d z^d$$

and similarly for $Q$ (with coefficients $b_j$). As non-zero scalar multiples of the given vector lead to the same rational map $P/Q$, we can view $\Phi$ as a map from the complex projective space $\mathbb{CP}^{2d+1}$ to the space $\mathscr{R}_0 \cup \cdots \cup \mathscr{R}_d$ of rational maps of degree at most $d$ ($\mathbb{CP}^{2d+1}$ is the space obtained from $\mathbb{C}^{2d+2} - \{0\}$ by identification of vectors which are non-zero scalar multiplies of each other).

The map $\Phi$ is not as attractive as one might hope; for example:

(i) $\mathbb{CP}^{2d+1}$ is connected whereas $\mathscr{R}_0 \cup \cdots \cup \mathscr{R}_d$ is not (its components are the $\mathscr{R}_j$);
(ii) the coefficients may converge, but in the limit, the degree of the rational map may drop (for example, $(z + 1/n)/z \to 1$ in the sense of coefficients but not uniformly on the sphere); and
(iii) in passing from $\mathbb{C}^{2d+2}$ to $\mathbb{CP}^{2d+1}$, we identify some of the pairs of vectors which give the same rational map, but not all of them (for example, the coefficient vector of $z/z$ is not identified with the coefficient vector of the constant map 1).

Despite these facts, the space $\mathscr{R}_d$ of rational maps of degree $d$ is embedded homeomorphically in $\mathbb{CP}^{2d+1}$. A rational map $R$ of degree $d$ determines its $d$ zeros and $d$ poles, hence it determines its coefficients up to scalar multiples: in other words, there is a map

$$\Psi: \mathscr{R}_d \to \mathbb{CP}^{2d+1}$$

with left inverse $\Phi$. In fact, the map $\Psi$ is a homeomorphism of $\mathscr{R}_d$ onto its image $\Psi(\mathscr{R}_d)$, and this image is an open dense subset of $\mathbb{CP}^{2d+1}$: we omit the details.

**Exercise 2.8**

1. Let $R$ be a rational map other than the identity. Show that if $R^n$ converges uniformly to some function $g$ on a domain $D$, then $g$ is constant. [*Hint*: From Theorem 2.3.1, $R^{n+1}$ converges uniformly to $Rg$ on $D$.]

   Show, however, that a subsequence of $(R^n)$ can converge uniformly on a domain $D$ to a non-constant function $g$.

2. Complete the proof of Corollary 2.8.3.

CHAPTER 3

# The Fatou and Julia Sets

The informal discussion in Chapter 1 suggested that the dynamics of a rational map $R$ induces a subdivision of the complex sphere into two sets, namely the Fatou set $F$ and the Julia set $J$ of $R$. Here, we introduce the notion of equicontinuity of a family of maps, and then formally define the Fatou and Julia sets in terms of the equicontinuity of the family $\{R^n\}$. In addition, we consider conditions which imply that a family of analytic maps is equicontinuous (or a normal family) in a domain.

## §3.1. The Fatou and Julia Sets

Although we shall assume that the reader is familiar with the basic ideas of metric spaces (for example, uniform continuity and compactness), the definition of continuity will serve to motivate the crucial notion of equicontinuity. A map $f$ from a metric space $(X, d)$ to a metric space $(X_1, d_1)$ is continuous at the point $x_0$ in $X$ if, for every positive $\varepsilon$, there is some positive $\delta$ such that for every $x$,

$$d(x_0, x) < \delta \quad \text{implies} \quad d_1(f(x_0), f(x)) < \varepsilon.$$

Clearly, $\delta$ depends on $f$, $x_0$ and $\varepsilon$, but if $\delta$ can be found so that this holds for all $x$, and for all $f$ in some family $\mathcal{F}$ of maps of $X$ into $Y$, then we say that the family $\mathcal{F}$ is *equicontinuous at* $x_0$. It is important to realize that the equicontinuity of $\mathcal{F}$ at $x_0$ is the formal expression for the idea of preservation of proximity which was introduced informally in Chapter 1: indeed, it implies that every function in $\mathcal{F}$ maps the open ball with centre $x_0$ and radius $\delta$ into a ball of radius at most $\varepsilon$. For emphasis, we give a formal definition.

**Definition 3.1.1.** A family $\mathcal{F}$ of maps of $(X, d)$ into $(X_1, d_1)$ is *equicontinuous at* $x_0$ if and only if for every positive $\varepsilon$ there exists a positive $\delta$ such that for all $x$ in $X$, and for all $f$ in $\mathcal{F}$,

$$d(x_0, x) < \delta \quad \text{implies} \quad d_1(f(x_0), f(x)) < \varepsilon.$$

The family $\mathcal{F}$ is *equicontinuous on a subset* $X_0$ *of* $X$ if it is equicontinuous at each point $x_0$ of $X$.

If the family $\mathcal{F}$ is equicontinuous on each of the subsets $D_\alpha$ of $X$, then it is automatically equicontinuous on the union $\bigcup D_\alpha$. Taking the collection $\{D_\alpha\}$ to be the class of all open subsets of $X$ on which $\mathcal{F}$ is equicontinuous, this leads to the following general principle.

**Theorem 3.1.2.** *Let $\mathcal{F}$ be any family of maps, each mapping $(X, d)$ into $(X_1, d_1)$. Then there is a maximal open subset of $X$ on which $\mathcal{F}$ is equicontinuous. In particular, if $f$ maps a metric space $(X, d)$ into itself, then there is a maximal open subset of $X$ on which the family of iterates $\{f^n\}$ is equicontinuous.*

This (at last) provides us with a formal definition for the Fatou and Julia sets of a rational map $R$, and the following terminology is chosen to honour the creators of the theory.

**Definition 3.1.3.** Let $R$ be a non-constant rational function. The *Fatou set* of $R$ is the maximal open subset of $\mathbb{C}_\infty$ on which $\{R^n\}$ is equicontinuous, and the *Julia set* of $R$ is its complement in $\mathbb{C}_\infty$.

Although the use of the term "Julia set" is standard, the use of "Fatou set" was suggested as late as 1984 (in [22]). It seems appropriate, but the reader should be familiar with the common alternatives, namely *the stable set*, and *the set of normality*. This latter term is a reference to normal families of analytic functions (which we discuss in §3.3), but we have preferred to base our definition on equicontinuity because of its more immediate geometric appeal.

We denote the Fatou set of a rational map $R$ by $F$, and the Julia set by J, although sometimes, when we need to mention $R$ explicitly, we use $F(R)$ and $J(R)$ instead (in many earlier papers, the Julia set is denoted by $F$). Note that, by definition, *$F(R)$ is open, and $J(R)$ is compact.*

We end this section with two simple, but important, properties of $F$ and $J$. A rational map, and in particular, a Möbius map and its inverse, satisfy a Lipschitz condition with respect to the spherical (and chordal) metric on $\mathbb{C}_\infty$ (Theorem 2.3.1). Thus if we conjugate $R$ with respect to a Möbius map, equicontinuity is transferred in the obvious way and we have

**Theorem 3.1.4.** *Let $R$ be a non-constant rational map, let $g$ be a Möbius map, and let $S = gRg^{-1}$. Then $F(S) = g(F(R))$ and $J(S) = g(J(R))$.*

A similar argument (which we give below) leads to

**Theorem 3.1.5.** *For any non-constant rational map R, and any positive integer p, $F(R^p) = F(R)$ and $J(R^p) = J(R)$.*

PROOF. Let $S = R^p$. First, as $\{S^n: n \geq 1\}$ is a subfamily of $\{R^n: n \geq 1\}$, it is equicontinuous wherever $\{R^n: n \geq 1\}$ is: thus $F(R) \subset F(S)$. Next, as each $R^k$ satisfies a Lipschitz condition, the family

$$\mathscr{F}_k = \{R^k S^n: n \geq 0\}$$

is equicontinuous wherever $\{S^n: n \geq 1\}$ is. In particular, each $\mathscr{F}_k$ is equicontinuous on the Fatou set $F(S)$ of $S$, and hence so too is the *finite* union $\mathscr{F}_0 \cup \mathscr{F}_1 \cup \cdots \cup \mathscr{F}_{p-1}$. As this union is $\{R^n: n \geq 0\}$, the family $\{R^n: n \geq 1\}$ is equicontinuous on $F(S)$ and so $F(S) = F(R)$. [*Remark*: A similar argument makes use of uniform continuity instead of the Lipschitz condition.] □

EXERCISE 3.1

1. Show (rigorously) that the Julia set of $z^d$, where $d \geq 2$, is the unit circle.
2. Suppose that $R$ is a rational map, and that the iterates $R^n$ converge uniformly to some constant on a domain $D$. Prove that $D \subset F(R)$. Deduce that an attracting fixed point of $R$ lies in the Fatou set.
3. Prove that a repelling fixed point of $R$ lies in the Julia set.
4. Give the details of the proof of Theorem 3.1.4.

## §3.2. Completely Invariant Sets

In this section we discuss the consequences of a set being invariant under a mapping, and we begin by defining the different types of invariance. If $g$ is a map of a set $X$ into itself, a subset $E$ of $X$ is:

(i) *forward invariant* if $g(E) = E$;
(ii) *backward invariant* if $g^{-1}(E) = E$;
(iii) *completely invariant* if $g(E) = E = g^{-1}(E)$.

If $g$ is surjective (that is, if $g(X) = X$) then the concepts of backward invariance and complete invariance coincide. Indeed, by surjectivity (but not necessarily in the general case),

$$g(g^{-1}(E)) = E$$

which, with backward invariance, yields $g(E) = E$. The requirement of surjectivity here is crucial and without it, there is a difference between backward and complete invariance: for example, $\mathbb{C}$ is backward invariant, but not completely invariant, under the map $z \mapsto \exp(z)$. As our main concern is rational maps, we recall that these map $\mathbb{C}_\infty$ onto itself: thus for rational maps, backwards invariance coincides with complete invariance.

We illustrate these ideas with the following result which will be of importance later.

**Theorem 3.2.1.** *Let $R$ be a rational map of degree at least two and suppose that a finite set $E$ is completely invariant under $R$. Then $E$ has at most two elements.*

PROOF. We suppose that $E$ has $k$ elements. Because $E$ is finite, and because $R$ maps $E$ into $E$, $R$ must act as a permutation of $E$ and so for a suitable integer $q$, $R^q$ is the identity map of $E$ into itself. Now suppose that $R^q$ has degree $d$. It follows that for every $w$ in $E$, the equation $R^q(z) = w$ has $d$ solutions which are all at $w$, and so by the Riemann–Hurwitz relation (applied to $R^q$), we have

$$k(d-1) \leq 2d - 2.$$

As $d \geq 2$, we have $k \leq 2$ as required. □

It is convenient to collect together several general, but elementary, properties of completely invariants sets and throughout, *we shall assume that $g$: $X \to X$ is surjective* so that we may use complete invariance and backward invariance interchangeably. First, we mention that complete invariance behaves in the most obvious way under conjugation: if $E$ is completely invariant under $g$, and if $h$ is a bijection of $X$ onto itself, then $h(E)$ is completely invariant under $hgh^{-1}$.

Next, we apply the standard construction which is used to generate, for example, subgroups, subspaces and topologies. The operator $g^{-1}$ commutes with the intersection operator, that is, for any collection $\{E_\alpha\}$ of sets,

$$g^{-1}\left(\bigcap E_\alpha\right) = \bigcap g^{-1}(E_\alpha),$$

and, because of this, *the intersection of a family of completely invariant sets is itself completely invariant*. This means that we can take any subset $E_0$ and form the intersection, say $E$, of all those completely invariant sets which contain $E_0$: obviously, $E$ is then the *smallest completely invariant set that contains* $E_0$ and we say that $E_0$ *generates* $E$.

We now introduce an equivalence relation which greatly facilitates the discussion of completely invariant sets. For any $x$ and $y$ in $X$, we define the relation $\sim$ on $X$ by $x \sim y$ if and only if there exist non-negative integers $n$ and $m$ with

$$g^n(x) = g^m(y). \tag{3.2.1}$$

Obviously, the relation $\sim$ is symmetric and reflexive, and it is also transitive for (3.2.1) and $g^p(y) = g^q(z)$ imply that

$$g^{n+p}(x) = g^{m+q}(z):$$

thus $\sim$ is an equivalence relation on $X$. We denote the equivalence class containing $x$ by $[x]$, and we call this the *orbit* of $x$ (some authors call this the "great orbit", but see Exercise 3.2.1, [102] and [96], p. 14). It is easy to identify the orbit $[x]$, and we have

**Theorem 3.2.2.** *Let $g: X \to X$ be any map of $X$ onto itself. Then $[x]$ is the completely invariant set generated by the singleton $\{x\}$.*

PROOF. For the moment, let $\langle x \rangle$ denote the completely invariant set generated by $\{x\}$ (of course, this notation will be abandoned in favour of $[x]$ after the proof is completed). First, we take any $y$ in $[x]$. Then for some $m$ and $n$, $g^m(y) = g^n(x)$, so

$$y \in g^{-m}g^n\{x\} \subset g^{-m}g^n\langle x \rangle = \langle x \rangle,$$

equality holding here as $\langle x \rangle$ is completely invariant. We deduce that $[x] \subset \langle x \rangle$.

It only remains to show that $[x]$ is completely invariant, for once this has been done, the minimality of $\langle x \rangle$ implies that $\langle x \rangle \subset [x]$ and equality follows. As $y \sim g(y)$, we see that $x \sim y$ if and only if $x \sim g(y)$ or, in terms of set membership,

$$y \in [x] \quad \text{if and only if} \quad g(y) \in [x].$$

This shows that $y \in [x]$ if and only if $y \in g^{-1}([x])$ thereby proving the complete invariance of $[x]$, and completing the proof. □

It is a direct consequence of Theorem 3.2.2 that a set $E$ is completely invariant if and only if it is a union of equivalence classes $[x]$. If this is the case, then its complement must also be a union of equivalence classes and, therefore, also completely invariant. This suggests that we should also look at the interior, closure and boundary operators.

**Theorem 3.2.3.** *Let $g$ be any continuous, open map of a topological space $X$ onto itself and suppose that $E$ is completely invariant. Then so are the complement $X - E$, the interior $E^0$, the boundary $\partial E$, and the closure $\overline{E}$, of $E$.*

PROOF. We have already proved that $X - E$ is completely invariant. As $g$ is continuous on $X$, $g^{-1}(E^0)$ is an open subset of $g^{-1}(E)$, and hence (by invariance) of $E$. Thus $g^{-1}(E^0) \subset E^0$. Similarly, as $g$ is an open map, $g(E^0)$ is an open subset of $E$ and so $g(E^0) \subset E^0$. Thus

$$E^0 \subset g^{-1}g(E^0) \subset g^{-1}(E^0),$$

and so $E^0$ is completely invariant.

We now know that the complete invariance of $E$ implies that of $X - E$ and $E^0$, and the usual topological arguments guarantee that the closure and the boundary of $E$ are also completely invariant. This completes the proof. □

We now prove that the division of the sphere into Fatou and Julia sets is completely invariant: this is fundamental.

**Theorem 3.2.4.** *Let $R$ be any rational map. Then the Fatou set $F$ and the Julia set $J$ are completely invariant under $R$.*

PROOF. By Theorem 3.2.3, it is sufficient to prove that $F$ is completely invariant, and because $R$ is surjective, we need only prove that it is backwards invariant. First, we take any $z_0$ in $R^{-1}(F)$ and let $w_0 = R(z_0)$: thus $w_0$ is in $F$. It follows that given any positive $\varepsilon$, there is a positive $\delta$ such that if $\sigma(w, w_0) < \delta$, then for all $n$, $\sigma(R^n(w), R^n(w_0)) < \varepsilon$. By continuity, there is also a positive $\rho$ such that if $\sigma(z, z_0) < \rho$, then $\sigma(R(z), w_0) < \delta$, and hence $\sigma(R^{n+1}(z), R^{n+1}(z_0)) < \varepsilon$. This shows that $\{R^{n+1}: n \geq 1\}$ is equicontinuous at $z_0$, and clearly, the addition of one function, namely $R$, does not destroy this fact. Thus $\{R^n: n \geq 1\}$ is equicontinuous at $z_0$, and hence on $R^{-1}(F)$. As $R^{-1}(F)$ is open, we deduce that $R^{-1}(F) \subset F$.

To prove the opposite inclusion, take any $z_0$ in $F$ and again, let $w_0 = R(z_0)$. Because $z_0$ is in $F$, given any positive $\varepsilon$ there is a positive $\delta$ such that for all $n$, if $\sigma(z, z_0) < \delta$ then $\sigma(R^{n+1}(z), R^{n+1}(z_0)) < \varepsilon$. The set of $z$ satisfying $\sigma(z, z_0) < \delta$ is an open neighbourhood $N$, say, of $z_0$, and so $R(N)$ is an open neighbourhood of $w_0$. If $w$ is in $R(N)$, then $w = R(z)$ for some $z$ in $N$, and so

$$\sigma(R^n(w), R^n(w_0)) = \sigma(R^{n+1}(z), R^{n+1}(z_0)) < \varepsilon.$$

This shows that $w_0$ is in $F$, so $F \subset R^{-1}(F)$ and the proof is complete. □

We can say more than Theorem 3.2.4 when $R$ is a polynomial, and we have

**Theorem 3.2.5.** *Let $P$ be a polynomial of degree at least two. Then $\infty$ is in $F(P)$, and the component $F_\infty$ of $F$ containing $\infty$ is completely invariant under $P$.*

PROOF. It is clear that there is some neighbourhood, say $W$, of $\infty$ on which $P^n \to \infty$ uniformly. Thus given any positive $\varepsilon$, there is an integer $N$ such that if $n \geq N$, and if $z$ and $w$ are in $W$, then

$$\sigma(P^n z, P^n w) \leq \sigma(P^n z, \infty) + \sigma(\infty, P^n w) < \varepsilon,$$

so $\{P^n\}$ is equicontinuous in $W$, and $\infty$ is in $F(P)$.

To show that $F_\infty$ is completely invariant, observe first that $P(F_\infty)$ both contains $\infty$ and is a connected subset of $F$ (Theorem 3.2.4). Thus $P$ maps $F_\infty$ into itself and so $F_\infty \subset P^{-1}(F_\infty)$. Now suppose that $z$ is in $P^{-1}(F_\infty)$: then, by Theorem 3.2.4, $z$ lies in some component $F_1$ of $F$ and by the argument above, $P$ maps $F_1$ into $F_\infty$. If $P(F_1) \neq F_\infty$, then there is some point $\zeta$ in $\partial F_1$ with $P(\zeta)$ in $F_\infty$, and this cannot be so as $\zeta$ is in $J$, and $J$ is completely invariant. We deduce that $P(F_1) = F_\infty$, and hence that $F_1$ contains some $w$ with $P(w) = \infty$. But then $w = \infty$, $F_1 = F_\infty$, $z \in F_\infty$ and the proof is complete. □

We end this section with a simple observation which will be used several times later. Suppose that a topological space $X$ has only a finite number of components, say $X_1, \ldots, X_t$, and suppose also that $f$ is a continuous map of $X$ onto itself. As each $f(X_j)$ is connected, $f$ induces a map $\tau$ of $\{1, \ldots, t\}$ into itself defined by $\tau(i) = j$ where $f(X_i) \subset X_j$. As $f$ is surjective so is $\tau$, hence $\tau$ is necessarily a permutation of $\{1, \ldots, t\}$. It follows that $\tau$ has finite order, say $m$, so, for all $j$,

$$f^m(X_j) = X_j.$$

A simple argument (using the fact that the $X_j$ are pairwise disjoint) now shows that each $X_j$ is completely invariant under $f^m$, and we have proved.

**Proposition 3.2.6.** *Let $f$ be a continuous map of a topological space $X$ onto itself, and suppose that $X$ has only a finite number of components $X_j$. Then for some integer $m$, each $X_j$ is completely invariant under $f^m$.*

This, of course, was the essential idea in the proof of Theorem 3.2.1.

EXERCISE 3.2

1. Let $g$ be a bijection of a space $X$ onto itself, and let $G$ be the cyclic group $\langle g \rangle$. Show that

$$[x] = \{g^n(x): n \in \mathbb{Z}\},$$

that is, $[x]$ is the orbit of $x$ under the group $G$.

## §3.3. Normal Families and Equicontinuity

We have defined the Fatou and Julia sets in terms of the equicontinuity of the family $\{R^n\}$. However, equicontinuity is closely related to normal families of analytic functions, and once we have understood this connection (which is expressed in the Arzelà–Ascoli Theorem), we can exploit the more powerful results about normal families of analytic functions to derive further information about the Fatou and Julia sets. This is our programme for this section.

We begin with the definitions relevant to normal families. A sequence $(f_n)$ of maps from a metric space $(X_1, d_1)$ to a metric space $(X_2, d_2)$ *converges locally uniformly on* $X_1$ to some map $f$ if each point $x$ of $X_1$ has a neighbourhood on which $f_n$ converges uniformly to $f$. In these circumstances, the convergence is then uniform on each compact subset of $X_1$.

**Definition 3.3.1.** A family $\mathscr{F}$ of maps from $(X_1, d_1)$ to $(X_2, d_2)$ is said to be *normal*, or a *normal family*, in $X_1$ if every infinite sequence of functions from $\mathscr{F}$ contains a subsequence which converges locally uniformly on $X_1$.

We give the usual reminder that the limit of the convergent subsequence need not lie in $\mathscr{F}$. Although we have formulated these definitions for general metric spaces, we shall only be concerned with subdomains of the complex sphere with the chordal (or spherical) metric, and we come now to a statement of the Arzelà–Ascoli Theorem: this is proved in many texts (see, for example, [3], p. 222) and we omit the proof.

**Theorem 3.3.2.** *Let $D$ be a subdomain of the complex sphere, and let $\mathscr{F}$ be a family of continuous maps of $D$ into the sphere. Then $\mathscr{F}$ is equicontinuous in $D$ if and only if it is a normal family in $D$.*

As an illustration of the direct use of normality rather than equicontinuity, we consider Vitali's Theorem: this exploits analytic continuation and so enables us derive information about the limit $f$ of a sequence $(f_n)$ throughout a domain $D$ simply from a knowledge of $f$ on only a small part of $D$.

**Theorem 3.3.3** (Vitali's Theorem). *Suppose that the family $\{f_1, f_2, \dots\}$ of analytic maps is normal in a domain $D$, and that $f_n$ converges pointwise to some function $f$ on some non-empty open subset $W$ of $D$. Then $f$ extends to a function $F$ which is analytic on $D$, and $f_n \to F$ locally uniformly on $D$.*

PROOF. As $\{f_n\}$ is normal in $D$, there is some subsequence of $(f_n)$ which converges locally uniformly in $D$ to some function $F$. It follows that $F$ is analytic in $D$, and that $F = f$ on $W$.

Now suppose that $(f_n)$ fails to converge locally uniformly on $D$ to $F$. Then there is a compact subset $K$ of $D$, a positive $\varepsilon$, and a subsequence, say $(g_n)$, of $(f_n)$ such that for all $n$,

$$\sup_K \sigma(g_n z, Fz) \geq \varepsilon.$$

However, by normality, there is some subsequence, say $(h_n)$, of $(g_n)$ which converges locally uniformly to some function $h$ on $D$. Clearly, $h = f = F$ on $W$, and as $h$ is analytic in $D$, we must have $h = F$ throughout $D$. It follows that

$$\sup_K \sigma(h_n z, Fz) \to 0,$$

and as $(h_n)$ is a subsequence of $(g_n)$, this violates the preceding inequality (which holds for all $n$). The proof is complete. □

We turn now to obtain conditions which guarantee the normality of a family of analytic functions. It is an elementary consequence of Cauchy's Integral Formula that if $f$ is holomorphic and satisfies $|f| \leq M$ in a domain $D$ in $\mathbb{C}$, then $|f'|$ is bounded above on each compact subset $K$ of $D$, the upper bound depending on $M$, $K$ and $D$ *but not on* $f$. This implies that if $\mathscr{F}$ is a family of functions which are holomorphic and uniformly bounded in $D$, then $\mathscr{F}$ is equicontinuous, and hence normal in $D$. It is possible to obtain a variety of mild generalizations of this result without much effort, but all such results are hopelessly inadequate for our purposes. Instead, we pass these by and go straight to one of the most powerful results of all concerning normality: this result is absolutely essential for the study of iteration of rational maps, yet its substantial proof can safely be omitted without prejudicing an understanding of iteration theory. The result we seek is

**Theorem 3.3.4.** *Let $D$ be a domain on the complex sphere $\mathbb{C}_\infty$, and let $\Omega$ be the domain $\mathbb{C}_\infty - \{0, 1, \infty\}$. Then the family $\mathcal{F}$ of all analytic maps $f\colon D \to \Omega$ is normal in $D$.*

The hypothesis of Theorem 3.3.4 is that each function in $\mathcal{F}$ fails to take the values 0, 1 and $\infty$ at any point of $D$. We shall discuss the proof presently, but first, we show how Theorem 3.3.4 can be used to produce two variations on the same theme. In the first of these, we assume only that each function $f$ fails to take three values which now *may depend on $f$*. It is too much to expect that three values can be completely arbitrary, and we must add the requirement that they are uniformly separated on the sphere. The precise result is

**Theorem 3.3.5.** *Let $\mathcal{F}$ be a family of maps, each analytic in a domain $D$ on the complex sphere. Suppose also that there is a positive constant $m$ and, for each $f$ in $\mathcal{F}$, three distinct points $a_f$, $b_f$ and $c_f$ in $\mathbb{C}_\infty$ such that:*

(i) *$f$ in $\mathcal{F}$ does not take the values $a_f$, $b_f$ and $c_f$ in $D$; and*
(ii) $\min\{\sigma(a_f, b_f), \sigma(b_f, c_f), \sigma(c_f, a_f)\} \geq m.$

*Then $\mathcal{F}$ is normal in $D$.*

Of course, Theorem 3.3.5 contains Theorem 3.3.4 as a special case, for we can take $a_f = 0$, $b_f = 1$ and $c_f = \infty$ for every $f$. There is another variation of Theorem 3.3.4 that we shall need: this is

**Theorem 3.3.6.** *Let $D$ be a domain, and suppose that the functions $\varphi_1$, $\varphi_2$ and $\varphi_3$ are analytic in $D$, and are such that the closures of the domains $\varphi_j(D)$ are mutually disjoint. Now let $\mathcal{F}$ be a family of functions, each analytic in $D$, such that for every $z$ in $D$, and every $f$ in $\mathcal{F}$, $f(z) \neq \varphi_j(z)$, $j = 1, 2, 3$. Then $\mathcal{F}$ is normal in $D$.*

In view of our total dependence on these theorems, and anticipating that many readers will not wish to examine the long proof of Theorem 3.3.4 in detail, we shall now give a sketch of the proof of Theorem 3.3.4, and complete derivations of Theorems 3.3.5 and 3.3.6 from Theorem 3.3.4. The hope is that the outline of the proof of Theorem 3.3.4 will provide adequate insight and reassurance for those readers who choose not to examine the details: those who do wish to see the details should consult Appendix I of this chapter.

The Proof of Theorem 3.3.5. For each $f$ in $\mathscr{F}$, define the Möbius transformation $g_f$ by

$$g_f(0) = a_f, \qquad g_f(1) = b_f, \qquad g_f(\infty) = c_f.$$

The condition (ii) in Theorem 3.3.5 is precisely the hypothesis of Theorem 2.3.3, so the family $\{g_f: f \in \mathscr{F}\}$ satisfies a uniform Lipschitz condition on $\mathbb{C}_\infty$, say

$$\sigma(g_f(z), g_f(w)) \leq M\sigma(z, w),$$

where $M$ is independent of $f$.

Now let $\Omega = \mathbb{C}_\infty - \{0, 1, \infty\}$, let $\Omega_f = \mathbb{C}_\infty - \{a_f, b_f, c_f\}$, and define $h_f$ to be the map $g_f^{-1}f$. Thus $f: D \to \Omega_f$ is the composition $g_f h_f$ of the maps $h_f$: $D \to \Omega$ and $g_f: \Omega \to \Omega_f$. Now by Theorem 3.3.4, the family $\{h_f\}$ is normal in $D$, and hence equicontinuous there. Finally,

$$\mathscr{F} = \{g_f h_f: f \in \mathscr{F}\},$$

and it is immediate that the equicontinuity of the family $\mathscr{F}$ is inherited from the equicontinuity of the family $\{h_f\}$ and the uniform Lipschitz condition on $\{g_f\}$. This completes the derivation of Theorem 3.3.5 from Theorem 3.3.4. □

The Proof of Theorem 3.3.6. For each $w$ in $D$ we define a Möbius map $g_w$ by the conditions

$$g_w(0) = \varphi_1(w), \qquad g_w(1) = \varphi_2(w), \qquad g_w(\infty) = \varphi_3(w),$$

so by Theorem 2.3.3, the family $\{g_w: w \in D\}$ satisfies a uniform Lipschitz condition on the sphere, say with constant $M$.

Now suppose that $\mathscr{F} = \{f_\alpha: \alpha \in A\}$, and for each $\alpha$, define

$$F_\alpha(w) = g_w^{-1} f_\alpha(w) = \frac{[f_\alpha(w) - \varphi_1(w)][\varphi_2(w) - \varphi_3(w)]}{[f_\alpha(w) - \varphi_3(w)][\varphi_2(w) - \varphi_1(w)]},$$

with the usual conventions when any of these points are $\infty$. Clearly, each $F_\alpha$ is analytic in $D$, and does not take the values 0, 1 or $\infty$ there. Thus from Theorem 3.3.4, $\{F_\alpha\}$ is normal in $D$.

Now take any sequence $f_n$ from $\mathscr{F}$, and let $F_n$ be the corresponding functions defined as above. By normality, there is a subsequence of $(F_n)$ converging to some analytic function $F$, uniformly on compact subsets of $D$, and for this subsequence we have the inequalities

$$\begin{aligned}\sigma(f_n(w), g_w F(w)) &= \sigma(g_w F_n(w), g_w F(w)) \\ &\leq M\sigma(F_n(w), F(w)).\end{aligned}$$

As this last term tends to zero uniformly on compact subsets of $D$, we now see that the original family $\mathscr{F}$ is normal on $D$. □

We end this section with

A SUMMARY OF THE PROOF OF THEOREM 3.3.4. First, normality is a local property so we need only prove that for every $z$ in $D$, the family $\mathscr{F}$ is normal in some disc centred at $z$ and lying within $D$. Thus we may assume that $D$ is a disc, and a simple change of variable enables us to further assume that $D$ is the unit disc $\Delta$ in $\mathbb{C}$. We want to prove, then, that the family $\mathscr{F}_0$ of analytic maps $f: \Delta \to \Omega$ is normal in $\Delta$.

Our proof relies on hyperbolic geometry. First, we view $\Delta$ as the hyperbolic plane with the hyperbolic metric $\rho$ derived from the line element

$$ds = 2|dz|(1 - |z|^2)^{-1}.$$

Now because $\Delta$ is the universal covering space of $\Omega$, the metric $\rho$ transfers to a metric, say $\rho_\Omega$, on $\Omega$ (the hyperbolic metric on $\Omega$). It is not too difficult to analyse the metric geometry of $(\Omega, \rho_\Omega)$ and to show that the hyperbolic distance from any point in $\Omega$ to any of the boundary points 0, 1 and $\infty$ is infinite. Thus, given any positive number $d$, there are three neighbourhoods, say $\mathscr{N}_0$, $\mathscr{N}_1$ and $\mathscr{N}_\infty$ of 0, 1 and $\infty$ respectively, such that the distance between any two of these neighbourhoods exceeds $d$.

Now it is an important fact that any analytic map $f: (\Delta, \rho) \to (\Omega, \rho_\Omega)$ does not increase hyperbolic distances (this is a deeper version of the classical Schwarz Lemma). Further, any compact subset $K$ of $\Delta$ has finite $\rho$-diameter, say $d_0$, and so we can conclude that its image $f(K)$ has $\rho_\Omega$-diameter at most $d_0$: it is important to note that this bound depends on $K$ *but not on* $f$ in $\mathscr{F}$. Now take a compact subset $K$ of $\Delta$ of $\rho$-diameter at most $d$: it follows that for every $f$ in $\mathscr{F}$, the image $f(K)$ can meet at most one of the neighbourhoods $\mathscr{N}_0$, $\mathscr{N}_1$, $\mathscr{N}_\infty$.

If we are now given any sequence of functions chosen from $\mathscr{F}_0$, we can pass to a subsequence, say $f_n$, with the property that each $f_n(K)$ is disjoint from one particular $\mathscr{N}_j$, say from $\mathscr{N}_0$. Now let $g$ be a Möbius transformation which maps $\mathscr{N}_0$ onto the exterior of $\Delta$. Then, for each $f_n$, $gf_n$ maps $K$ into $\Delta$ and so the sequence $(gf_n)$ is normal in the interior of the compact set $K$. It follows easily (as $g$ is a Lipschitz map on the sphere) that the family $\mathscr{F}_0$ is normal in the interior of $K$, and as normality is a local property, $\mathscr{F}$ is normal in $\Delta$. This completes our sketch of the proof of Theorem 3.3.4: for more details, see Appendix I to this chapter. □

EXERCISE 3.3

1. Let $D$ be a subdomain of $\mathbb{C}_\infty$. Show that there exist compact subsets $K_1, K_2, \ldots$ of $D$, such that $D$ is the union of their interiors, and deduce that each compact subset of $D$ is contained in a finite union of the $K_n$.

   Now consider the family $\mathscr{F}$ of continuous maps $f: D \to \mathbb{C}_\infty$, and for each integer $n$, and each $f$ and $g$ in $\mathscr{F}$, define

$$\tau_n(f, g) = \sup_{K_n} \sigma(f(z), g(z)) \leq 2,$$

and

$$\tau(f, g) = \sum_{n=1}^{\infty} 2^{-n} \tau_n(f, g).$$

Show that $\tau(f, g)$ is the metric of local uniform convergence on $D$ in the sense that for functions $f_n$ and $f$ in $\mathscr{F}$, $f_n \to f$ locally uniformly on $D$ if and only if $\tau(f_n, f) \to 0$.

2. Let $P$ be a polynomial of degree at least two. Use Vitali's Theorem to show that $P^n \to \infty$ on the component $\mathscr{F}_\infty$ of $F(P)$ which contains $\infty$.

3. Let $P(z) = z^2 - 2$ (this was discussed in §1.4, but here is an alternative treatment).
   (i) Show that $[-2, 2]$, and hence also the complement $\Omega$ of $[-2, 2]$, is completely invariant under $P$.
   (ii) Use Theorem 3.3.4 to show that $\{P^n\}$ is normal in $\Omega$, so $F(P) \supset \Omega$, and $J(P) \subset [-2, 2]$.
   (iii) Use (ii) to show that $F(P)$ is connected, and then use Vitali's Theorem (or the previous exercise) to show that $P^n \to \infty$ on $F(P)$.
   (iv) Deduce that $J(P) = [-2, 2]$.

4. Let $D$ be a subdomain of the complex sphere and suppose that the complement of $D$ contains at least three points. Let $f$ be an analytic function which maps $D$ into itself and suppose that $f$ has an attracting fixed point $\zeta$ in $D$. Show that the iterates $f^n$ converge locally uniformly to $\zeta$ in $D$.

# Appendix I. The Hyperbolic Metric

This Appendix provides an opportunity to introduce the reader to some basic ideas relating hyperbolic geometry to complex analysis. As described in §3.3, we view the unit disc $\Delta$ as the hyperbolic plane with the metric $\rho(z_1, z_2)$ being defined as the infimum of

$$L(\gamma) = \int_\gamma ds, \qquad ds = \frac{2|dz|}{1 - |z|^2},$$

taken over all curves $\gamma$ that join $z_1$ to $z_2$ in $\Delta$. Much is known about this geometry (for example, the geodesics are the arcs of circles orthogonal to the boundary of $\Delta$) and for more details, we refer the reader to [18].

For our purposes it is best to transform this to the upper half-plane model in which the space is

$$H^+ = \{x + iy: y > 0\},$$

and where the corresponding metric $\rho_H$ is derived from $ds = |dz|/y$. We begin with the open region $\Sigma$ lying in the upper half-plane $H^+$ and illustrated in Figure 1.

Now let $\Gamma$ be the group of transformations of the form

$$g(z) = \frac{az + b}{cz + d}, \tag{1}$$

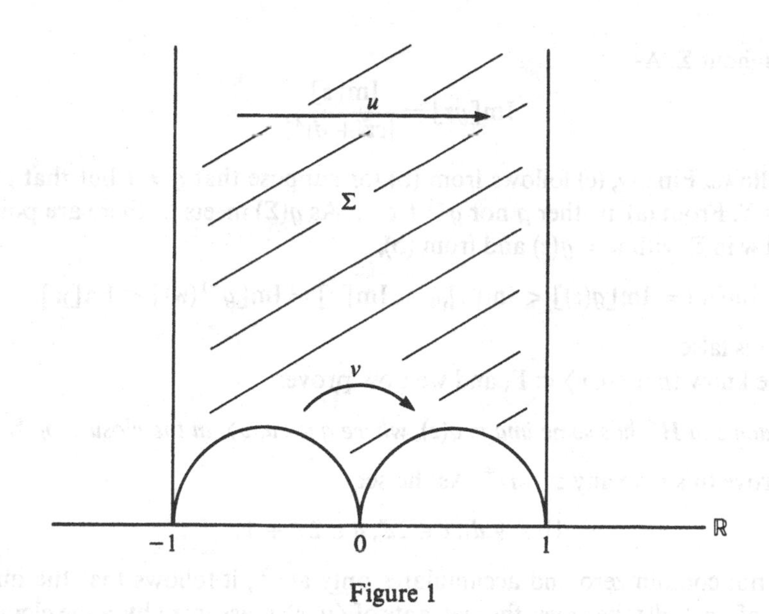


Figure 1

where

$$a, d \in 2\mathbb{Z} + 1, \qquad b, c \in 2\mathbb{Z}, \qquad ad - bc = 1 \tag{2}$$

(so $\Gamma$ consists of those Möbius maps whose matrix is congruent modulo 2 to the $2 \times 2$ identity matrix). The maps

$$u(z) = z + 2, \qquad v(z) = \frac{z}{2z + 1},$$

are in $\Gamma$ and they pair the sides of $\Sigma$ as indicated in Figure 1. We prove

**Theorem A.** $\Gamma$ *is generated by* $u$ *and* $v$, *and the* $\Gamma$-*images of* $\Sigma$ *tesselate* $H^+$.

PROOF. First, we take $g$ in $\Gamma$ and prove:

(a) *if* $g\infty = \infty$, *then* $g(z) = z + \beta$ *for some even integer* $\beta$;
(b) *if* $g\infty \neq \infty$ *and* $z \in \Sigma$, *then* $\mathrm{Im}[gz] < \mathrm{Im}[z]$;
(c) *if* $g$ *is not the identity map* $I$, *then* $g(\Sigma) \cap \Sigma = \varnothing$.

To prove these, we take $g$ as in (1) and (2). If $c = 0$ then from (2), $ad = 1$ so $a = d = \pm 1$ and (a) holds. Note that if $c = 0$, then $g(\Sigma) \cap \Sigma = \varnothing$. In (b), we have $c \neq 0$ so

$$Q_g = \{z: |cz + d| < 1\}$$

is an open disc with its centre on the real axis. Now if $z$ is any integer, then $cz + d$ is an odd integer so $z$ cannot lie in $Q_g$. It follows that 0, 1 and $-1$ lie outside $Q_g$, so $\Sigma$ does not meet the closure of $Q_g$ and $|cz + d| > 1$

throughout $\Sigma$. As

$$\operatorname{Im}[gz] = \frac{\operatorname{Im}[z]}{|cz + d|^2}, \tag{3}$$

(b) follows. Finally, (c) follows from (b) for suppose that $g \neq I$ but that $g(\Sigma)$ meets $\Sigma$. From (a), neither $g$ nor $g^{-1}$ fix $\infty$. As $g(\Sigma)$ meets $\Sigma$, there are points $z$ and $w$ in $\Sigma$ with $w = g(z)$ and from (b),

$$\operatorname{Im}[w] = \operatorname{Im}[g(z)] < \operatorname{Im}[z], \qquad \operatorname{Im}[z] = \operatorname{Im}[g^{-1}(w)] < \operatorname{Im}[w],$$

which is false.

We know that $\langle u, v\rangle \subset \Gamma$, and we now prove:

(d) *each $z$ in $H^+$ has some image $g(z)$, where $g \in \langle u, v\rangle$, in the closure of $\Sigma$.*

To prove this take any $z$ in $H^+$. As the set

$$\{|cz + d| : c \in 2\mathbb{Z}, d \in 2\mathbb{Z} + 1\}$$

does not contain zero and accumulates only at $\infty$, it follows that the minimum of $|cz + d|$ taken over the elements of $\langle u, v\rangle$ is assumed by some element $g$ of $\langle u, v\rangle$, and replacing $g$ by $u^m g$ for a suitable integer $m$ (this leaves $|cz + d|$ unaltered), we may assume that

$$|\operatorname{Re}[g(z)]| \leq 1. \tag{4}$$

Now (3) (which holds for all Möbius maps of $H^+$ onto itself) and the minimizing property of $g$ shows that for every $h$ in $\langle u, v\rangle$, $\operatorname{Im}[h(z)] \leq \operatorname{Im}[g(z)$. In particular,

$$\operatorname{Im}[g(z)] . |2g(z) + 1|^{-2} = \operatorname{Im}[vg(z)] \leq \operatorname{Im}[g(z)]$$

and

$$\operatorname{Im}[g(z)] . |2g(z) - 1|^{-2} = \operatorname{Im}[v^{-1}g(z)] \leq \operatorname{Im}[g(z)],$$

and these and (4) show that $g(z)$ lies in the closure of $\Sigma$.

It is now easy to complete the proof of Theorem 1. Take any $h$ in $\Gamma$ and any $w$ in the open set $h(\Sigma)$. By (d), there is some $g$ in $\langle u, v\rangle$ such that $g(w)$ lies in the closure of $\Sigma$, and so $gh(\Sigma)$ meets $\Sigma$. By (c), $h = g^{-1}$, whence $h \in \langle u, v\rangle$ and so $\Gamma \subset \langle u, v\rangle$. As the opposite inclusion holds, we have $\Gamma = \langle u, v\rangle$ and having proved this, (c) and (d) show that the $\Gamma$-images of $\Sigma$ tesselate $H^+$. □

It is more convenient to continue the argument from a slight variation of Theorem A, so let $\Sigma_0$ be as illustrated in Figure 2, and let $\alpha$, $\beta$ and $\gamma$ be reflections in the sides of $\Sigma_0$. The next result follows easily from Theorem A (and its proof is omitted).

**Theorem B.** *Let $\Gamma_0 = \langle \alpha, \beta, \gamma\rangle$. Then:*

(i) *the $\Gamma_0$-images of $\Sigma_0$ tesselate $H^+$; and*
(ii) *$\Gamma$ is a subgroup of index two in $\Gamma_0$, and $\Gamma_0 = \Gamma \cup \Gamma\alpha$.*

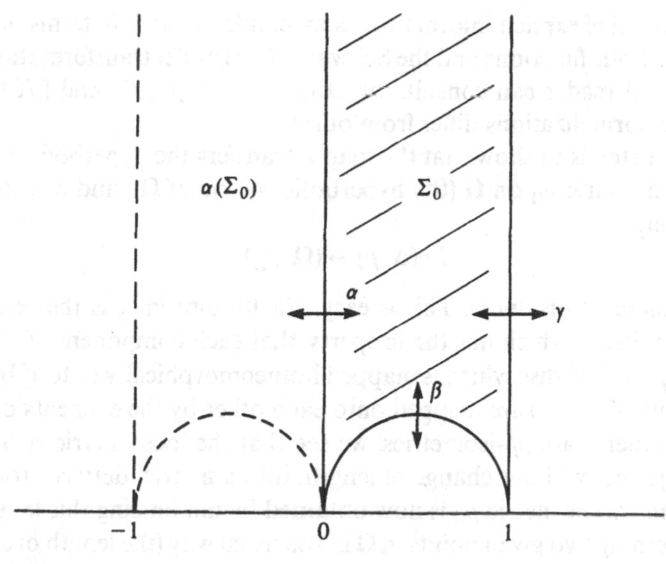


Figure 2

The next step is to construct the quotient space $H^+/\Gamma$ by means of an analytic map; more precisely, we need to construct an analytic map $\lambda$: $H^+ \to \Omega$, where $\Omega = \mathbb{C}_\infty - \{0, 1, \infty\}$, such that (up to conformal equivalence) $\lambda$ is the quotient map and $\Omega$ is the quotient space. Thus we prove

**Theorem C.** *There exists an analytic map $\lambda$ of $H^+$ onto $\Omega$ with the properties*:

(i) *for all $z$ in $H^+$, $\lambda'(z) \neq 0$; and*
(ii) *$\lambda(z) = \lambda(w)$ if and only if $w = h(z)$ for some $h$ in $\Gamma$.*

The proof is standard, and we only sketch the main ideas. The Riemann Mapping Theorem shows that there is a conformal mapping $\lambda$ of $\Sigma_0$ onto $H^+$, and the standard theory of boundary correspondence under conformal mappings shows that $\lambda$ extends to a homeomorphism between the closure of these domains. By combining $\lambda$ with a suitable Möbius map of $H^+$ onto itself, we may assume that $\lambda$ fixes 0, 1 and $\infty$. We deduce that $\lambda$ maps the positive imaginary axis onto the interval $(-\infty, 0)$ so, by the Reflection Principle, $\lambda$ is a conformal mapping of $\Sigma$ onto $\mathbb{C} - \{x: x \geq 0\}$. In a similar way we can analytically continue $\lambda$ across the other two sides of $\Sigma_0$.

It is clear from this construction that if $z$ and $w$ are any two points lying in the closure of $\Sigma$ (and also in $H^+$), then $\lambda(z) = \lambda(w)$ if and only if $w = g(z)$ for some $g$ in $\Gamma$. We now extend the definition of $\lambda$ to $H^+$ by the formula $\lambda(gz) = \lambda(z)$ for all $z$ in $H^+$ and all $g$ in $\Gamma$, and it is easy enough to verify that this defines a single-valued function on (the simply connected) $H^+$ which satisfies the requirements of Theorem C. □

A great deal of explicit information is available about $\lambda$ (in terms of infinite products, elliptic functions and the Schwarz–Christoffel transformations) and the interested reader can consult, for example, [3], p. 277, and [76], p. 318 (where the normalizations differ from ours).

The next step is to show that the map $\lambda$ transfers the hyperbolic metric $\rho_H$ on $H^+$ to a metric $\rho_\Omega$ on $\Omega$ (the hyperbolic metric of $\Omega$), and also that any analytic map

$$F: (\Delta, \rho) \to (\Omega, \rho_\Omega)$$

does not increase distances. This is easy. Each point in $\Omega$ is the centre of a small open disc $V$ which has the property that each component of $\lambda^{-1}(V)$ in $H^+$ is a topological disc which is mapped homeomorphically onto $V$ by $\lambda$. The components of $\lambda^{-1}(V)$ are mapped onto each other by the elements of $\Gamma$ and, as these elements are $\rho_H$-isometries, we see that the local metric structure of $H^+$ is projected, without change of length, into a metric (derived from a line element) on $\Omega$. The metric $\rho_\Omega$ is now obtained by minimizing this length over all paths joining two given points in $\Omega$ in the usual way (the length of any path being computed a small piece at a time by transferring each small piece back to $H^+$ and measuring it there). It is now clear that $\lambda$ is a map from $(H^+, \rho_H)$ onto $(\Omega, \rho_\Omega)$ which preserves the lengths of paths, and so does not increase distances.

Now let $F$ be any analytic map of $\Delta$ into $\Omega$. We can follow $F$ by some branch of $\lambda^{-1}$ defined near $F(0)$ and then analytically continue this composition $F_0$ over all curves in $\Delta$: the continuation exists by virtue of the properties of $\lambda$, and it is single valued because $\Delta$ is simply connected. This provides us with an analytic map $F_0: \Delta \to H^+$ such that $F(z) = \lambda(F_0(z))$ (for this holds initially). The Schwarz–Pick Lemma implies that $F_0$ is distance decreasing, and as $\lambda$ is also distance decreasing, so too is $F$.

The last step is to show that the $\rho_\Omega$-distance between any two of the boundary points 0, 1 and $\infty$ of $\Omega$ is infinite, and to do this we refer the problem back to $H^+$. Now the $\rho_H$-distance between the horizontal lines $y = y_2$ and $y = y_1$ in $H^+$ is $|\log(y_2/y_1)|$, and this tends to $+\infty$ as $y_2 \to +\infty$. In fact, for sufficiently large $y_1$, the image of $\{x + iy: y > y_1\}$ under $\lambda$ is a neighbourhood of $\infty$ in $\Omega$ (we omit the details) and the desired result follows as the situation is symmetric in 0, 1 and $\infty$.

## General References for Chapter 3

The classical references are [44], [45], [58] and [83]: more recent references are [22], [28], [35], [51], [67], [71], [87] and [96].

# CHAPTER 4

# Properties of the Julia Set

This chapter is devoted to developing the basic and most easily available properties of the Julia set of a rational map. Other properties of the Julia set occur throughout the text.

## §4.1. Exceptional Points

We begin now to exploit the fundamental Theorem 3.3.4. Given a rational map $R$, we have the induced equivalence relation $\sim$ discussed in §3.2, and the equivalence class $[z]$ is the smallest completely invariant set which contains $z$. It is easy to see that $[z]$ can be finite only in rather special circumstances, so we enshrine this idea in a formal definition.

**Definition 4.1.1.** A point $z$ is said to be *exceptional* for $R$ when $[z]$ is finite, and the set of such points is denoted by $E(R)$.

First, we justify the terminology by showing that such points are indeed exceptional.

**Theorem 4.1.2.** *A rational map $R$ of degree at least two has at most two exceptional points. If $E(R) = \{\zeta\}$, then $R$ is conjugate to a polynomial with $\zeta$ corresponding to $\infty$. If $E(R) = \{\zeta_1, \zeta_2\}$, where $\zeta_1 \neq \zeta_2$, then $R$ is conjugate to some map $z \mapsto z^d$, where $\zeta_1$ and $\zeta_2$ correspond to $0$ and $\infty$.*

From this and Theorems 3.1.5 and 3.2.5, we have immediately

**Corollary 4.1.3.** *If $\deg(R) \geq 2$, then the exceptional points of $R$ lie in $F(R)$.*

PROOF OF THEOREM 4.1.2. It is clear that $E(R)$ is completely invariant under $R$ and so, by Theorem 3.2.1, $R$ has at most two exceptional points. It follows that after a suitable conjugation, there are four possibilities to consider, namely;

(i) $E(R) = \varnothing$;
(ii) $E(R) = \{\infty\} = [\infty]$;
(iii) $E(R) = \{0, \infty\}, [0] = \{0\}, [\infty] = \{\infty\}$;
(vi) $E(R) = \{0, \infty\} = [0] = [\infty]$.

There is nothing more to say about (i). If (ii) holds, then, by Theorem 2.4.1, $R$ is a polynomial. If (iii) holds, then $R$ is again a polynomial but this time of the form $z \mapsto az^d$ for some positive integer $d$. Finally, if (iv) holds, then $R(0) = \infty$, $R(\infty) = 0$, and $R$ has all of its zeros and poles in $\{0, \infty\}$ so it is of the form $z \mapsto az^d$ for some negative integer $d$. In all cases, the exceptional points lie in $F$ and this completes the proof. Observe that this argument shows that, in an obvious sense, *most rational maps have no exceptional points.* □

We turn now to another characterization of exceptional points. For any $z$, the *backward orbit* of $z$ is the set

$$\begin{aligned} O^-(z) &= \{w: \text{for some } n \geq 0, R^n(w) = z\} \\ &= \bigcup_{n \geq 0} R^{-n}\{z\}, \end{aligned}$$

and we call the points in $O^-(z)$ the *predecessors* of $z$. As $O^-(z) \subset [z]$, any exceptional point $z$ has a finite backward orbit, and we show now that the converse is also true.

**Theorem 4.1.4.** *The backward orbit $O^-(z)$ of $z$ is finite if and only if $z$ is exceptional.*

PROOF. We need only prove that if $O^-(z)$ is finite, then $z$ is exceptional. To do this, define the non-empty sets $B_n$ by

$$B_n = \bigcup_{m \geq n} R^{-m}\{z\},$$

so $R^{-1}(B_n) = B_{n+1}$ and

$$[z] \supset O^-(z) = B_0 \supset B_1 \supset B_2 \supset \cdots.$$

We now assume that $O^-(z)$ is finite. Then each $B_n$ is finite, so there is some $m$ with $B_m = B_{m+1}$; this means that $R^{-1}(B_m) = B_m$ and so $B_m$ is completely invariant. It follows that $B_m$ contains some equivalence class $[w]$, and as it is a subset of $[z]$, it must be $[z]$: thus $[z]$ is finite as required. In fact, $[z] = O^-(z)$. □

As an application of this result, suppose that some iterate $R^k$ of $R$ is conjugate to a polynomial. Then there is some $w$ with $R^{-k}\{w\} = \{w\}$, and using

Theorem 4.1.4, this implies that $w$ is exceptional for $R$. Now such a map $R$ need not be conjugate to a polynomial (consider, for example, $z \mapsto z^{-2}$) but from Theorem 4.1.2, $R^2$ must be, so we have proved

**Theorem 4.1.5.** *If for some positive integer $k$, $R^k$ is conjugate to a polynomial, then $R^2$ is also conjugate to a polynomial.*

EXERCISE 4.1

1. Use Theorem 3.3.4 directly to show that $R$ can have at most two exceptional points.
2. Suppose for some positive $k$, $R^{-k}\{w\} = \{w\}$. Prove that each $R^{-n}\{w\}$ is a singleton, say $\{w_n\}$, and that $w_n$ is exceptional, and deduce that $R^{-2}\{w\} = \{w\}$.
3. Develop the theory of exceptional points for rational maps of degree one.

# §4.2. Properties of the Julia Set

We shall continue to exploit Theorem 3.3.4 as we explore the structure of the Julia set. We have not yet considered whether the Julia set can be non-empty, and our first task will be to settle this issue. The case when the rational map $R$ has degree one is trivial and of little interest, but in all other cases, $J$ is not empty. We prove

**Theorem 4.2.1.** *If $\deg(R) \geq 2$, then $J(R)$ is infinite.*

PROOF. We begin by showing that $J$ is not empty. If $J$ is empty, then the family $\{R^n\}$ is normal on the entire complex sphere, and so, by Theorem 2.8.2, there is some subsequence of $R^n$ in which each map has the same degree. However, $\deg(R^n) = [(\deg(R)]^n$ and this implies that $R$ has degree one, contrary to our assumption.

We know now that $J$ contains some point $\zeta$. Now $J$ is completely invariant (Theorem 3.2.4), so if $J$ is finite, then $\zeta$ must be an exceptional point. This cannot be so, however, as exceptional points lie in $F$ (Corollary 4.1.3); thus $J$ is infinite. □

It is clear from Theorem 3.3.4 that the integer three has a significant role to play and it appears yet again in a decisive way in the next result.

**Theorem 4.2.2.** *Let $R$ be a rational map with $\deg(R) \geq 2$, and suppose that $E$ is a closed, completely invariant subset of the complex sphere. Then either:*

(i) *$E$ has at most two elements and $E \subset E(R) \subset F(R)$; or*
(ii) *$E$ is infinite and $E \supset J(R)$.*

PROOF. We know that either $E$ has at most two points or it is infinite (Theorem 3.2.1). If $E$ is finite, then it contains only exceptional points and Theorem 4.1.2 and Corollary 4.1.3 then give (i). We suppose now that $E$ is infinite. As $E$ is completely invariant, so is its complement $\Omega$; hence each $R^n$ maps the open set $\Omega$ into itself. We now apply Theorem 3.3.5 to the family $\{R^n\}$ on $\Omega$, choosing the corresponding points $a_f$, $b_f$ and $c_f$ to be three given points in $E$: this shows that $\{R^n\}$ is normal in $\Omega$, and so $\Omega \subset F$. It follows that $J$ is contained in $E$ and (ii) holds. □

It is usual, and convenient, to express the conclusion of Theorem 4.2.2 in the form: *$J$ is the smallest closed, completely invariant set with at least three points*, and we shall refer to this property as the *minimality* of $J$. The proof of Theorem 4.2.2 seems surprisingly short in view of the interesting corollaries which flow from it but, of course, this is merely a reflection of the potency of Theorem 3.3.4. We now begin to develop these corollaries.

First, the extended plane $\mathbb{C}_\infty$ is the disjoint union of the interior $J^0$ of $J$, the boundary $\partial J$ of $J$, and the Fatou set $F$. We know that $F$ is completely invariant: likewise, $J$ is and hence so are $J^0$ and $\partial J$ (Theorem 3.2.3). If $F$ is not empty, then $F \cup \partial J$ is an infinite, closed, completely invariant set and so, by the minimality of $J$, it contains $J$. In these circumstances, $J \subset \partial J$ and this yields

**Theorem 4.2.3.** *Either $J = \mathbb{C}_\infty$ or $J$ has empty interior.*

We shall consider the case when $J = \mathbb{C}_\infty$ in the next section.

Next, let $J_0$ be the derived set of $J$ (that is, the set of accumulation points of $J$). We claim that $J_0$ is an infinite, closed, completely invariant subset of $J$ and given this, the minimality of $J$ implies that $J \subset J_0$, and hence that $J = J_0$. We deduce, then, that *$J$ has no isolated points.*

First, as $J$ is infinite (Theorem 4.2.1), $J_0$ is non-empty, and, as a derived set, $J_0$ is automatically closed. Next, as $R$ is continuous and of finite degree, it is clear that $R(J_0) \subset J_0$, hence $J_0 \subset R^{-1}(J_0)$. In addition, as $R$ is an open map it is easy to see that $R^{-1}(J_0) \subset J_0$, and we deduce that $J_0$ is completely invariant. It follows from Theorem 4.2.2 that $J_0$ is infinite and we have verified the claims above. □

Because $J$ has no isolated points, it is necessarily uncountable (this is a consequence of Baire's Category Theorem, [85]) and, in the usual terminology, $J$ is perfect. Thus we have proved

**Theorem 4.2.4.** *Let $R$ be a rational map with* $\deg(R) \geq 2$. *Then $J$ is a perfect set and so is uncountable.*

In the next chapter we shall prove a stronger, but similar, result for the components of $J$ (Theorem 5.7.1).

We continue to explore the consequences of Theorem 3.3.4, and the next result says that the successive iterates of $R$ increasingly expand a neighbourhood of any point of $J$ (we have seen this phenomenon before in some of the examples studied in Chapter 1). Given an open set $W$ which meets $J$, the result describes the extent to which the images $R^n(W)$ cover the sphere. We recall that $E(R)$ has at most two elements, and we prove

**Theorem 4.2.5.** *Let $R$ be a rational map of degree at least two, and let $W$ be any non-empty open set which meets $J$. Then:*

(i) $\bigcup_{n=0}^{\infty} R^n(W) \supset \mathbb{C}_\infty - E(R)$; *and*
(ii) *for all sufficiently large integers $n$, $R^n(W) \supset J$.*

PROOF. Let $W_0 = \bigcup_{n=0}^{\infty} R^n(W)$, and let $K$ be the complement of $W_0$. If $K$ contains three distinct points, say $\zeta_1$, $\zeta_2$ and $\zeta_3$, then we can apply Theorem 3.3.5 to the family $\{R^n\}$ on $W$, with the points $a_f$, $b_f$, $c_f$ being the $\zeta_j$. It follows that $\{R^n\}$ is normal in $W$, and hence $W \subset F$, a contradiction. This shows that $W_0$ contains every point of the sphere with at most two exceptions. Now consider a point $z$ that is not an exceptional point of $R$: then $z$ has an infinite backward orbit (Theorem 4.1.3) and, by the preceding remarks, this must meet $W_0$. Thus for some point $w$, and some non-negative integers $p$ and $q$, $R^p(w) = z$ and $w \in R^q(W)$. It follows that $z$ is in $R^{p+q}(W)$ and so (i) follows.

To prove (ii), take three open sets, say $W_1$, $W_2$ and $W_3$, each meeting $J$, and a positive chordal distance apart from each other. First, we claim that for each $j$ in $\{1, 2, 3\}$, some forward image of $W_j$ covers some $W_k$; that is, for each $j$ there are some integers $k$ and $n$ with

$$R^n(W_j) \supset W_k. \tag{4.2.1}$$

Indeed, suppose not. Then for some $j$, $R^n(W_j)$ fails to cover any of $W_1$, $W_2$ and $W_3$ and so, by Theorem 3.3.5, $\{R^n\}$ is normal in $W_j$. This cannot be so, however, as $W_j$ meets $J$, so we have verified (4.2.1).

We have shown that for each $j$ in $\{1, 2, 3\}$, there is some $k$ and some $n$ such that $R^n(W_j) \supset W_k$. Denote any such $k$ by $\pi(j)$: then $\pi$ maps $\{1, 2, 3\}$ into itself and so some iterate of $\pi$ has a fixed point: in other words, for some $j$ and some $n$, $R^n(W_j) \supset W_j$.

Now put $S = R^n$: then $S(W_j) \supset W_j$ and so the sequence of sets $S^m(W_j)$ is increasing with $m$. Applying (i) to $S$ and $W_j$, we find now that the sets $S^m(W_j)$ form an increasing open cover of the compact set $J$ so a finite union, and hence one of them, covers $J$. Clearly, we can choose all of the $W_k$ to lie in $W$, and if this is done, then for some $n$,

$$J \subset R^n(W_j) \subset R^n(W).$$

Of course, for all such $n$,

$$J = R(J) \subset R^{n+1}(W),$$

and (ii) follows by induction. □

The next three results show how the Julia set $J$ is realized as the limit, or the closure, of some countable set. A point $\zeta$ is a *periodic point* of $R$ if it is fixed by some iterate $R^n$ and, clearly, there are only a countable number of such points. These points will be examined in detail in Chapter 6 but here, we prove

**Theorem 4.2.6.** *Suppose that* $\deg(R) \geq 2$. *Then $J$ is contained in the closure of the set of periodic points of $R$.*

*Remark.* This implies that $R$ has infinitely many periodic points, but we shall prove much more than this in Chapter 6.

PROOF OF THEOREM 4.2.6. We consider any open set $\mathcal{N}$ which meets $J$, and we shall prove that $\mathcal{N}$ contains some periodic point of $R$. We choose a point $w$ in $J \cap \mathcal{N}$, and we may assume that $w$ is not a critical value of $R^2$, for if it is, we can replace it with another nearby point of $J$ which is not. As $\deg(R) \geq 2$, and as $w$ is not a critical value of $R^2$, there are at least four distinct points in $R^{-2}\{w\}$. Choose three of them, say $w_1$, $w_2$ and $w_3$, distinct from $w$, and construct neighbourhoods $\mathcal{N}_0, \mathcal{N}_1, \mathcal{N}_2, \mathcal{N}_3$ (with pairwise disjoint closures) about $w, w_1, w_2, w_3$ respectively, such that $\mathcal{N}_0 \subset \mathcal{N}$, and that $R^2$ is a homeomorphism from each $\mathcal{N}_j$ onto $\mathcal{N}_0$ (see Figure 4.2.1).

Now let $S_j\colon \mathcal{N}_0 \to \mathcal{N}_j$ be the inverse of $R^2\colon \mathcal{N}_j \to \mathcal{N}_0$. If for all $z$ in $\mathcal{N}_0$, all $j = 1, 2, 3$, and all $n \geq 1$, we have

$$R^n(z) \neq S_j(z),$$

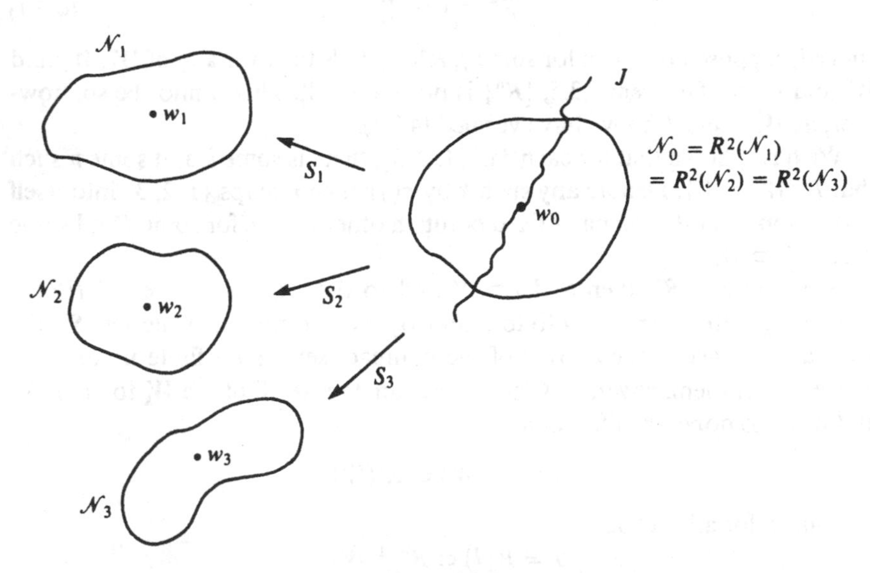


Figure 4.2.1

then, from Theorem 3.3.6, $\{R^n\}$ is normal in $\mathcal{N}_0$, and this is false as $\mathcal{N}_0$ meets $J$. We deduce that there exists some $z$ in $\mathcal{N}_0$, some $j$ in $\{1, 2, 3\}$, and some $n \geq 1$, such that $R^n(z) = S_j(z)$. This implies that

$$R^{2+n}(z) = R^2 S_j(z) = z,$$

and so $z$ is a periodic point in $\mathcal{N}$, as required. □

We now consider the relationship between $J$ and an orbit $[z]$ of a non-exceptional point $z$. As the closure $\overline{[z]}$ of $[z]$ is both closed and completely invariant, the minimality of $J$ implies that *if $z$ is not exceptional, then $J \subset \overline{[z]}$*. If $z$ is in $J$, then it is not exceptional, so $[z]$, and hence also $\overline{[z]}$, is contained in $J$: thus *if $z$ is in $J$, then $J = \overline{[z]}$*. In fact, we can replace $\overline{[z]}$ by the smaller set $O^-(z)$ and still prove

**Theorem 4.2.7.** *Let $R$ be a rational map with* $\deg(R) \geq 2$.

(i) *If $z$ is not exceptional, then $J$ is contained in the closure of* $O^-(z)$.
(ii) *If $z \in J$, then $J$ is the closure of* $O^-(z)$.

*Remark.* This shows that for any non-exceptional $z$ the backward orbit $O^-(z)$ is a countable set which accumulates at every point of $J$, and this is often used as the basis of a computer program to illustrate $J$. Roughly speaking, one selects a non-exceptional $z$ and then computes successive inverse images of $z$ to plot the set $J$ where these accumulate. There are difficulties, however; for example, if $\deg(R) = d$, then the number of inverse images at the $n$-th stage is $d^n$ and this grows too rapidly for convenient computations so some choices need to be made (see [80] and [81] for more details).

Proof of Theorem 4.2.7. Consider any non-exceptional $z$ and any non-empty open set $W$ which meets $J$. As $W$ meets $J$, Theorem 4.2.5 imples that $z$ lies in some $R^n(W)$ and so $O^-(z)$ meets $W$. This proves (i). If $z$ is in $J$, then the closed, completely invariant set $J$ contains the closure of the backward orbit $O^-(z)$, and in conjunction with (i), this yields (ii). □

If $z$ is in $F$ (and is non-exceptional), then the closure of $O^-(z)$ contains $J$ but, of course, is strictly larger that $J$. Instead of looking at the entire inverse orbit $O^-(z)$, we can also look at the $n$-th inverse image $R^{-n}(z)$ (which contains at most $d^n$ points) and ask when (in some sense) do these points converge to $J$? The same question can be asked of a set $E$ instead of the singleton $\{z\}$, and as this is a more useful result to have available, we prove

**Theorem 4.2.8.** *Let $R$ be a rational map of degree at least two, and let $E$ be a compact subset of the complex sphere with the property that for all $z$ in $F(R)$, the sequence $\{R^n(z): n \geq 1\}$ does not accumulate at any point of $E$. Then given any open set $U$ which contains $J(R)$, $R^{-n}(E) \subset U$ for all sufficiently large $n$.*

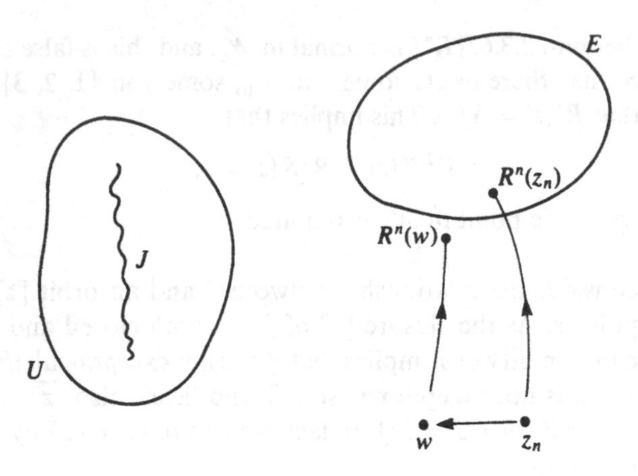


Figure 4.2.2

*Remark.* If the conclusion of the theorem holds we say that $R^{-n}(E)$ converges to $J$ and we write $R^{-n}(E) \to J$. Examples of such sets $E$ are easy to find: for example, if $F_0$ is a component of the Fatou set that contains an attracting fixed point $\zeta$, then any compact subset of $F_0 - \{\zeta\}$ has the required property. In Chapters 7 and 8, we shall analyse completely the possible forward images of $\{R^n(z): n \geq 1\}$ of a point $z$ in $F$ and after that, Theorem 4.2.8 will be even more significant.

PROOF OF THEOREM 4.2.8. We suppose that the conclusion is false: then there exists some open neighbourhood $U$ of $J$, and, for $n$ in some sequence $\{n_1, n_2, \ldots\}$, points $z_n$ in $R^{-n}(E)$, but not in $U$ (see Figure 4.2.2). Now without loss of generality, the points $z_n$ converge to $w$, say, and as $U$ is open, $w$ is not in $U$. But $J \subset U$; hence $w$ is in $F$.

Now take any positive $\varepsilon$, As $w \in F$, $\{R^n\}$ is equicontinuous in some neighbourhood of $w$, so there is a positive $\delta$ such that for all $n$, $\sigma(z, w) < \delta$ implies $\sigma(R^n(z), R^n(w)) < \varepsilon$. Thus for large enough $n$ (of the form $n_j$),

$$\sigma(R^n(z_n), R^n(w)) < \varepsilon$$

and as $R^n(z_n) \in E$, this shows that the sequence $R^n(w)$ accumulates at $E$, contrary to our assumption. The proof is complete. □

We end this section with the following generalization of Theorem 3.1.5 (but see Exercise 4.2.1).

**Theorem 4.2.9.** *Suppose that $R$ and $S$ are rational maps, each of degree at least two. If $R$ and $S$ commute, then $J(R) = J(S)$.*

PROOF. For any set $E$, we denote the diameter of $E$ computed using the metric $\sigma$ by $\text{diam}[E]$. From Theorem 2.3.1, there is some positive number $M$ such that for all $z$ and $w$,

$$\sigma(Sz, Sw) \le M\sigma(z, w).$$

Now take $w$ in $F(R)$. By the equicontinuity of $\{R^n\}$ at $w$, given any positive $\varepsilon$, there is a positive $\delta$ such that for all $n$,

$$\mathrm{diam}[R^n(D(w, \delta))] < \varepsilon/M,$$

where $D(w, \delta)$ is the $\sigma$-disc with centre $w$ and radius $\delta$. As $R$ and $S$ commute we deduce that

$$\begin{aligned} \mathrm{diam}[R^n S(D(w, \delta))] &= \mathrm{diam}[SR^n(D(w, \delta))] \\ &\le M \,\mathrm{diam}[R^n(D(w, \delta))] \\ &< \varepsilon. \end{aligned}$$

It follows that $\{R^n\}$ is a normal family on the open set $S(D(w, \delta))$ so, in particular, $S(w) \in F(R)$. This proves that $S$, and hence each $S^n$, maps $F(R)$ into itself, so $\{S^n\}$ is a normal family in $F(R)$. We conclude that $F(R) \subset F(S)$, and so, by symmetry, $F(R) = F(S)$. For more information on this topic, see [14] and [19]. □

EXERCISE 4.2

1. Show that the maps $z \mapsto 2z$ and $z \mapsto z/2$ commute, but have different Julia sets.
2. Suppose that $C$ is a circle such that $R^{-1}(C) \subset C$. Show that either $J = C$, or $J$ is a totally disconnected subset of $C$. This applies to the case when $R$ is a finite Blaschke product (a finite product of Möbius transformations, each leaving the unit disc $\Delta$ invariant).

## §4.3. Rational Maps with Empty Fatou Set

Our first objective is to show that for the rational map

$$z \mapsto \frac{(z^2 + 1)^2}{4z(z^2 - 1)}, \tag{4.3.1}$$

*the Julia set is the entire complex sphere*: this example (the first of its kind) was given by Lattès in 1918, [62]. After this, we shall make some general observations about other rational functions with this property.

We shall use the technique illustrated in §1.4, although here, we use a group of translations with two generators, and we require considerably more complex analysis. Let $\lambda$ and $\mu$ be complex numbers that are not real multiples of each other, and let $\Lambda$ be the corresponding lattice, that is,

$$\Lambda = \{m\lambda + n\mu : n, m \in \mathbb{Z}\}. \tag{4.3.2}$$

A *period parallelogram* for $\Lambda$ is any closed parallelogram of the form

$$\Omega = \{z + s\lambda + t\mu: 0 \leq s \leq 1, 0 \leq t \leq 1\}.$$

A non-constant function $f$ is an *elliptic function* for $\Lambda$ if it is meromorphic on $\mathbb{C}$, and if each $\omega$ in $\Lambda$ is a period of $f$; that is, if for all $z$ in $\mathbb{C}$ and all $\omega$ in $\Lambda$, $f(z + w) = f(z)$. Of course, any such function maps $\mathbb{C}$ into $\mathbb{C}_\infty$ and, because $f(\mathbb{C}) = f(\Omega)$, we see that $f(\mathbb{C})$ is a compact subset of $\mathbb{C}_\infty$. However, by the Open Mapping Theorem, $f(\mathbb{C})$ is also an open subset of $\mathbb{C}_\infty$: thus

$$f(\Omega) = f(\mathbb{C}) = \mathbb{C}_\infty.$$

Our argument is based on the properties of the Weierstrass elliptic function

$$\wp(z) = \frac{1}{z^2} + \sum{}^* \left[\frac{1}{(z+\omega)^2} - \frac{1}{\omega^2}\right]$$

where $\sum^*$ denotes summation over non-zero elements $\omega$ in $\Lambda$ (see, for example, [3], pp. 272–276, [40] or [61]). We shall only give the main steps in the argument but, for the convenience of some readers, we give an expanded version in the Appendix to this chapter. Now $\wp$ and its derivatives satisfy certain algebraic identities, and among these there is the *addition formula*

$$\wp(2z) = R(\wp(z)), \tag{4.3.3}$$

where $R$ is the rational function given by

$$R(z) = \frac{z^4 + g_2 z^2/2 + 2g_3 z + (g_2/4)^2}{4z^3 - g_2 z - g_3}, \tag{4.3.4}$$

and where $g_2$ and $g_3$ are known quantities defined in terms of the lattice $\Lambda$. Note that (4.3.3) is analogous to the double angle formulae in elementary trigonometry.

Now let $D$ be any disc in $\mathbb{C}$, let $U = \wp^{-1}(D)$, and define $\varphi(z) = 2z$. As $U$ is open, and as $\varphi^n(U)$ is the set $U$ expanded by a factor $2^n$, it is clear that for sufficiently large $n$, $\varphi^n(U)$ contains a period parallelogram $\Omega$ of $\wp$. Using this with (4.3.3), we deduce that for these $n$,

$$R^n(D) = R^n(\wp(U)) = \wp(2^n U) = \mathbb{C}_\infty:$$

roughly speaking, this implies that the functions $R^n$ explode any small disc $D$ onto the whole sphere. Clearly, as $D$ is arbitrary, (4.3.5) implies that the family $\{R^n\}$ is not equicontinuous on *any* open subset of the complex sphere, and so we deduce that $J(R) = \mathbb{C}_\infty$.

This argument gives a family of rational maps whose Julia set is the sphere; indeed, we always have

$$g_2^3 - 27g_3^2 \neq 0, \tag{4.3.5}$$

and any pair $(g_2, g_3)$ satisfying this is realized by some lattice (see [61], p. 39). Using this, we can construct $\Lambda$ so that $g_2 = 4$ and $g_3 = 0$; then $R$ given by

(4.3.4) is the function (4.3.1). However, there is an easier construction than this. First, we let $\tau$ be a positive number and we construct the square lattice $\Lambda$, and the corresponding function $\wp$, by taking $\lambda = \tau$ and $\mu = i\tau$. This choice leads easily to $g_3 = 0$ (see the Appendix), and hence to a simplification of (4.3.4), namely

$$R(z) = \frac{16z^4 + 8g_2z^2 + (g_2)^2}{16z(4z^2 - g_2)}.$$

Now from (4.3.6), as $g_3 = 0$, we have $g_2 \neq 0$, so by taking $h(z) = 2z/\sqrt{g_2}$, we find that $hRh^{-1}$ (which also has $\mathbb{C}_\infty$ as its Julia set) is the function given in (4.3.1). For further information on this and related examples, see [22], [53], [67] and the Appendix to this chapter.

There are other interesting dynamical aspects of this example and briefly, we shall discuss some of these. For each positive integer $k$, we denote the set of points $\omega/2^k$, $\omega \in \Lambda$, by $2^{-k}\Lambda$, and observe that if $z$ is in $2^{-k}\Lambda$, then

$$R^k(\wp(z)) = \wp(2^kz) = \infty.$$

This shows that the inverse orbit $O^-(\infty)$ of $\infty$ under $R$ is $\wp(\bigcup_k 2^{-k}\Lambda)$, the union being over all positive integers $k$. As $\bigcup_k 2^{-k}\Lambda$ is dense in $\mathbb{C}$, its $\wp$-image is dense in $\mathbb{C}_\infty$, and so by Theorem 4.2.7, we again find that $J(R) = \mathbb{C}_\infty$.

Next, the formula (4.3.3) shows that if $m = 2^k - 1$, then

$$R^k(\wp(\omega/m)) = \wp(2^k\omega/m) = \wp(\omega + \omega/m) = \wp(\omega/m),$$

and so the points $\wp(\omega/m)$ are periodic points of $R$. By considering all $\omega$ and all $k$, we now see that *the periodic points of $R$ are dense in $\mathbb{C}_\infty$*. Although we are not yet in a position to claim that this shows that $J(R) = \mathbb{C}_\infty$, it is intuitively clear that it must (and it will follow from some of our later results).

Finally, we can show that $J(R) = \mathbb{C}_\infty$ by considering the critical points of $R$. In general, a critical point of a rational map $R$ is said to be *pre-periodic* if it is not periodic, but a forward image of it is, and in Chapter 9 (Theorem 9.4.4) we shall prove

**Theorem 4.3.1.** *If every critical point of the rational map $R$ is pre-periodic, then $J = \mathbb{C}_\infty$.*

The rational map $R$ in (4.3.4) is of degree four, so it has six critical points, say $c_1, \ldots, c_6$. We shall show now that each $c_j$ is finite, and that each is mapped to $\infty$ (a fixed point of $R$) by $R^2$: then each $c_j$ is pre-periodic and we have $J = \mathbb{C}_\infty$. Again, we only sketch the main ideas and refer the reader to the Appendix for more details.

We shall now find the critical points of $R$ in (4.3.4). Let $\zeta_1, \ldots, \zeta_6$ be the six points

$$\lambda/4,\ \mu/4,\ (\lambda + \mu)/4,\ (3\lambda + \mu)/4,\ (2\lambda + \mu)/4,\ (3\lambda + 2\mu)/4,$$

and let $\eta_1, \ldots, \eta_6$ be the six points

$$3\lambda/4,\ 3\mu/4,\ (3\lambda + 3\mu)/4,\ (\lambda + 3\mu)/4,\ (2\lambda + 3\mu)/4,\ (\lambda + 2\mu)/4:$$

note that these twelve points are distinct. As $\wp$ is an even function, and as $\eta_j + \zeta_j \in \Lambda$, we find that

$$\wp(\eta_j) = \wp(-\zeta_j) = \wp(\zeta_j),$$

and from this, and the fact that modulo $\Lambda$, each point of $\mathbb{C}_\infty$ has precisely two pre-images under $\wp$, we see that the six values $\wp(\zeta_j)$ are distinct. Now $\wp'(z) = \infty$ if and only if $\wp(z) = \infty$, that is, if and only if $z$ is in $\Lambda$, and it is also known that $\wp'(z) = 0$ if and only if $z$ is in $2^{-1}\Lambda$ but not in $\Lambda$. It follows that

$$\wp'(\zeta_j) \neq 0, \infty, \qquad \wp'(2\zeta_j) = 0,$$

and as

$$R'(\wp(z))\wp'(z) = 2\wp'(2z),$$

we see that the six values $\wp(\zeta_j)$ are the six critical points of $R$. Finally, it is immediate from (4.3.3) that $R^2$ maps each of these to $\infty$.

Theorem 4.3.1 enables us to produce other examples of maps whose Julia set is the sphere: for example, the map

$$R(z) = (z - 2)^2/z^2$$

has critical points at 0 and 2, and as $2 \to 0 \to \infty \to 1 \to 1$ under an application of $R$, Theorem 4.3.1 implies that for this function too, $J(R)$ is the complex sphere. It is easy to see that this $R$ is conjugate to the map

$$z \mapsto 1 - 2/z^2$$

and in fact, many functions in the family $1 + \omega/z^2$ have this same property: see [67], §2.4, for further details.

We end this section with a characterization of those rational maps whose Julia set is the entire sphere [53].

**Theorem 4.3.2.** *Let $R$ be a rational map. Then $J(R) = \mathbb{C}_\infty$ if and only if there is some $z$ whose forward orbit $\{R^n(z)\colon n \geq 1\}$ is dense in the complex sphere.*

Proof. First, let $\{B_n\colon n \geq 1\}$ be a countable base for the topology on $\mathbb{C}_\infty$ (so every open subset of $\mathbb{C}_\infty$ is a union of some of the $B_n$). Next, let $D$ be the set of $z$ such that the forward orbit $O^+(z)$ is dense in the sphere: thus $z$ is in $D$ if and only if for all $k$, there exists some $n$ with $R^n(z) \in B_k$, and this implies that

$$D = \bigcap_{k \geq 1} \bigcup_{n \geq 1} R^{-n}(B_k).$$

Suppose now that $D = \varnothing$. We write $A_k = \mathbb{C}_\infty - B_k$ and

$$E_k = \bigcap_{n \geq 1} R^{-n}(A_k);$$

then, as $D = \emptyset$, we have

$$\mathbb{C}_\infty = \bigcup_{k\geq 1} E_k.$$

Now by Baire's Theorem (for example, [85]), $\mathbb{C}_\infty$ is not the countable union of nowhere dense sets: thus for some $k$, the closure of $E_k$ has a non-empty interior, say $W$. However, $E_k$ is closed, so there is some non-empty open subset $W$ of $E_k$. This means that for all $n$, $R^n(W) \subset A_k$: thus in $W$, the functions $R^n$ do not take values in $B_k$, and so $W \subset F(R)$. This shows that if $J(R) = \mathbb{C}_\infty$, then $D \neq \emptyset$, so there is some $z$ whose forward orbit is dense in the sphere.

Now suppose that $J$ is not the entire sphere, so $F(R) \neq \emptyset$, and also that there is some $z$ whose forward orbit $O^+(z)$ is dense in the sphere. First, $z$ is not in $J$ for if it were, then $O^+(z) \subset J$ and so $O^+(z)$ could not be dense in the sphere. It follows that $z$ lies in some component $\Omega$ of the Fatou set $F(R)$, so consider now the components

$$\Omega, R(\Omega), R^2(\Omega), \ldots$$

of $F(R)$. As $O^+(z)$ is dense in the sphere there must be some $N$ such that $R^N(\Omega)$ meets, and hence is, $\Omega$, and we may assume that $N$ is the minimal such integer. Further, if $\Omega_1$ is any component of $F(R)$, then for some $n$, $R^n(z) \in \Omega_1$, and so $R^n(\Omega) = \Omega_1$. We deduce that

$$F(R) = \Omega \cup R(\Omega) \cup \cdots \cup R^{N-1}(\Omega),$$

and that the sets on the right are mutually disjoint. It follows that $\Omega$ is completely invariant under $R^N$, and that the set $\{R^{kN}(z): k \geq 1\}$ is dense in $\Omega$. Again, we are not yet in a position to explain why this cannot be so, but it cannot, and we shall see why after we have analysed the structure of forward invariant components of the Fatou set in Chapter 7. □

# Appendix II. Elliptic Functions

Details concerning the function $\wp$ defined in §4.3 can be found, for example, in [3], [40] or [61] as well as in many other texts: however, for the reader's convenience, we sketch some of the details here. Our main concern is to justify the formula (4.3.3), but first, we shall discuss the problem from a slightly broader perspective.

Given the lattice $\Lambda$ in (4.3.2), let

$$S_n(\Lambda) = \sum\nolimits^* \omega^{-n}, \tag{1}$$

where, as before, this sum is over the non-zero elements of $\Lambda$. This series converges for $n \geq 3$, and as the general term in the series for $\wp$ is $O(\omega^{-3})$, we see that $\wp$ is meromorphic on $\mathbb{C}$.

The quotient space $\mathbb{C}/\Lambda$ (which is also a quotient group) is topologically a torus, and as every element of $\Lambda$ is a period of $\wp$, we see that $\wp$ induces a map

$\wp_0$ of $\mathbb{C}/\Lambda$ onto $\mathbb{C}_\infty$. As these are compact Riemann surfaces, $\wp_0$ is an $N$-fold covering map for some $N$, and by considering the poles of $\wp$, we see that $N = 2$. It follows that for each $w$ in $\mathbb{C}_\infty$, there are exactly two solutions (modulo $\Lambda$) of $\wp(z) = w$ in $\mathbb{C}$. Note also that $\wp$ is an even function, that is, $\wp(-z) = \wp(z)$.

There is a good reason why one might expect a formula of the type $\wp(2z) = R(\wp(z))$ to hold. Given any $w$ in $\mathbb{C}_\infty$, there are only two solutions $z$ (modulo $\Lambda$) of the equation $\wp(z) = w$, and these can be taken to be, say, $u$ and $(\lambda + \mu) - u$. As

$$\wp(2[\lambda + \mu - u]) = \wp(2u),$$

we can define a map $w \mapsto \wp(2u)$ of $\mathbb{C}_\infty$ onto itself which is independent of the choice of $u$. It is easy to see that this map is analytic, thus it must be a rational map $R$, and so we can now see why there is a formula of the type

$$R(\wp(u)) = \wp(2u).$$

Of course, we can replace 2 here by any other integer (and indeed by certain other numbers as well; see [53] and [61]).

We shall now sketch the details of the proof of (4.3.3). It is clear that the derivative $\wp'$ has triple poles at, and only at, the points in $\Lambda$, and because of this, it is not hard to see that one can construct a cubic polynomial $P$ such that the elliptic function $\wp'(z)^2 - P(\wp(z))$ has no poles at the origin, and hence no poles in $\mathbb{C}$. Such an elliptic function must be constant (by Liouville's Theorem, for it is bounded on any period parallelogram, and hence also on $\mathbb{C}$), and a computation of $P$ leads to the relation

$$\wp'(z)^2 = 4\wp(z)^3 - g_2\wp(z) - g_3, \tag{2}$$

where the $g_j$ are given by

$$g_2 = 60S_4(\Lambda), \qquad g_3 = 140S_6(\Lambda). \tag{3}$$

Now select distinct points $u$ and $v$ in $\mathbb{C}$ at which $\wp$ has different values and determine $A$ and $B$ so that

$$\wp'(u) = A\wp(u) + B, \qquad \wp'(v) = A\wp(v) + B.$$

It is clear that the elliptic function

$$f(z) = \wp'(z) - A\wp(z) - B$$

has three poles at (and only at) each point in $\Lambda$, and as a consequence of this and the ideas sketched above, $f$ must have three zeros. By construction, two of these zeros are at $u$ and $v$. Now (in general) if an elliptic function has poles $p_i$ and zeros $z_j$ in a period parallelogram, then $\sum p_i$ differs from $\sum z_j$ by an element of $\Lambda$. In our case, all of the poles of $f$ occur at the origin, thus the zeros of $f$ must sum to zero (modulo $\Lambda$) and so must be $u$, $v$ and $-(u + v)$ and their translates by $\Lambda$. However, as

$$[f(z) + A\wp(z) + B]^2 = \wp'(z)^2$$
$$= 4\wp(z)^3 - g_2\wp(z) - g_3,$$

we find that $\wp(u)$, $\wp(v)$ and $\wp(-[u+v])$ are (distinct) solutions of the equation

$$(Az + B)^2 = 4z^3 - g_2 z - g_3,$$

and so

$$\wp(u) + \wp(v) + \wp(-[u+v]) = \frac{A^2}{4} = \frac{1}{4}\left(\frac{\wp'(u) - \wp'(v)}{\wp(u) - \wp(v)}\right)^2.$$

Letting $u \to v$, and using the fact that $\wp$ is an even function, we now obtain

$$2\wp(v) + \wp(2v) = \tfrac{1}{4}[\wp''(v)/\wp'(v)]^2. \tag{4}$$

Finally, differentiating both sides of (2) gives an expression for $\wp''(z)$, and using this expression together with (2) and (4), we obtain the addition formula given by (4.3.3) and (4.3.4).

With $\lambda = \tau$ and $\mu = i\tau$, where $\tau > 0$, we have $i\Lambda = \Lambda$ and so, from (1),

$$S_n(\Lambda) = S_n(i\Lambda) = i^{-n}S_n(\Lambda).$$

Putting $n = 6$, we find from (3) that $g_3 = 0$, and this completes the sketch of the details of the first part of §4.3.

Returning to the general lattice $\Lambda$, it remains only to show that the derivative $\wp'$ is zero at each of the points $\lambda/2$, $\mu/2$ and $(\lambda + \mu)/2$. This is not difficult: indeed for each $z$, $\wp(\lambda - z) = \wp(z)$ and equating the derivative of each side at $\lambda/2$ we find that $\wp'(\lambda/2) = 0$. The same argument holds for the points $\mu/2$ and $(\lambda + \mu)/2$. It is usual to put

$$e_1 = \wp(\lambda/2), \qquad e_2 = \wp(\mu/2), \qquad e_3 = \wp([\lambda + \mu]/2),$$

then from (2) and (4.3.5) we see that $e_1$, $e_2$ and $e_3$ are the distinct zeros of the cubic equation

$$4z^3 - g_2 z - g_3 = 0.$$

In particular,

$$e_1 + e_2 + e_3 = 0, \qquad e_2e_3 + e_3e_1 + e_1e_2 = -g_2/4, \qquad e_1e_2e_3 = g_3/4,$$

which simplifies considerably when $g_3 = 0$. Indeed, in this case some $e_j$ is zero and (4.3.4) reduces to $(z^2 + \alpha^2)^2/4z(z^2 - \alpha^2)$.

CHAPTER 5

# The Structure of the Fatou Set

We begin with the simple properties of the Fatou set of a rational map $R$. As $R$ is a branched covering map of the sphere onto itself with the critical points as branch points, the position of the critical points influences the structure of the Fatou set $F$ through the Euler characteristic. We discuss these ideas and then apply them to study the structure of $F$.

## §5.1. The Topology of the Sphere

Some results on the structure of the Fatou and Julia sets are simple consequences of facts concerning the topology of the complex sphere and so require almost nothing of the theory of iteration. It seems best to present these results in this light and so our first task is to collect together the relevant topological results. Throughout, the underlying space is the complex sphere $\mathbb{C}_\infty$, and the complement of a set $E$ is denoted by $\mathbb{C}_\infty - E$. The *connectivity* $c(D)$ of a domain $D$ is the number of components of $\partial D$, and $D$ is *simply connected* if every closed curve in $D$ can be deformed (in $D$) to a point of $D$.

The first three results are standard and we omit their proofs (for more details see, for example, [3] or [77].

**Proposition 5.1.1.** *The closure of a connected set is connected.*

**Proposition 5.1.2.** *A compact set $K$ on the sphere is disconnected if and only if there exists a Jordan curve $\gamma$ which separates $K$ (that is, $K$ is disjoint from $\gamma$ and meets both components of the complement of $\gamma$).*

**Proposition 5.1.3.** *A domain $D$ is simply connected if and only if its complement is connected.*

We will also find the following two variations of Proposition 5.1.3 useful.

**Proposition 5.1.4.** *A domain $D$ is simply connected if and only if its boundary $\partial D$ is connected.*

**Proposition 5.1.5.** *Let $D$ be an open subset of the complex sphere. Then $\mathbb{C}_\infty - D$ is connected if and only if each component of $D$ is simply connected.*

To illustrate our earlier remarks, we record an immediate (though elementary) consequence of Proposition 5.1.5.

**Theorem 5.1.6.** *Let $R$ be a rational map. Then $J(R)$ is connected if and only if each component of $F(R)$ is simply connected.*

We now give the straightforward proofs of the last two propositions.

Proof of Proposition 5.1.4. First, if $D$ is not simply connected there is a simple closed curve $\gamma$ in $D$ which separates the complement of $D$, and it follows from this that $\partial D$ is disconnected. Suppose now that $\partial D$ is disconnected: then, by Proposition 5.1.2, there is a simple closed curve $\gamma$ which separates $\partial D$ into two disjoint subsets $A$ and $B$. There are points of $D$ arbitrarily close to $A$ and to $B$ so, as $D$ is arcwise connected, $D$ meets $\gamma$. By construction, $\gamma$ does not meet $\partial D$: thus $\gamma$ lies in $D$, $A$ and $B$ lie in different components of the complement of $D$ and hence $D$ cannot be simply connected. □

Proof of Proposition 5.1.5. Suppose first that the complement $K$ of $D$ is disconnected. By Proposition 5.1.2, we can separate $K$ by a Jordan curve $\gamma$ lying in $D$ and, as $\gamma$ is connected, it must lie in some component of $D$. Clearly, this component is not simply connected: thus if each component of $D$ is simply connected, then $K$ is connected.

Now suppose that $K$ is connected and consider any component $V$ of $D$. Let $Q$ be the union of $K$ and all other components, say $D_n$, of $D$ so $Q$ (the complement of $V$) is compact. Because $\partial D_n \subset K$, each set $K \cup D_n$ is connected, hence so is their union, namely $Q$. As $Q$ is connected, we deduce from Proposition 5.1.3 that $V$ is simply connected and the proof is complete. □

Finally (because we use this idea often), we include

**Proposition 5.1.7.** *Let $\{D_\alpha\}$ be a collection of simply connected domains that is linearly ordered by inclusion. Then $\bigcup_\alpha D_\alpha$ is a simply connected domain.*

PROOF. Let $D = \bigcup_\alpha D_\alpha$, so $D$ is open. The assumption that $\{D_\alpha\}$ is linearly ordered by inclusion means that for every pair of suffices $\alpha$ and $\beta$, either $D_\alpha \subset D_\beta$ or $D_\beta \subset D_\alpha$. It is clear that $D$ is connected for if not, it would be the union of two disjoint non-empty open sets $A$ and $B$, and there would then be some $D_\alpha$ contained in $A$ and some $D_\beta$ contained in $B$. Thus $D$ is a domain.

To show that $D$ is simply connected, take any closed curve $\gamma$ in $D$. The family $\{D_\alpha\}$ is an open cover of $\gamma$, so $\gamma$ lies in the union of a finite number of the $D_\alpha$, and hence (by the linear ordering) in one of them, say in $D_\beta$. It follows that $\gamma$ can be deformed in $D_\beta$ to a point of $D_\beta$, and as $D_\beta \subset D$, $D$ is simply connected. □

EXERCISE 5.1

1. Find a simply connected domain in $\mathbb{C}$ whose complement in $\mathbb{C}$ is not connected. (Compare this with Proposition 5.1.3.)
2. Construct a simply connected domain whose boundary in $\mathbb{C}_\infty$ is *not* arcwise connected.
3. Let $\Gamma$ be a closed curve in $\mathbb{C}_\infty$. Show that every component of the complement of $\Gamma$ is simply connected.
4. Show how to construct *three* pairwise disjoint, simply connected domains $D_1$, $D_2$ and $D_3$ in $\mathbb{C}$ with the property that
$$\partial D_1 = \partial D_2 = \partial D_3.$$
(See [56], p. 143.)
5. Show that Proposition 5.1.5 includes Proposition 5.1.3.

## §5.2. Completely Invariant Components of the Fatou Set

We begin our discussion of the structure of the Fatou set $F$ by studying the completely invariant components of $F$ (and in Chapter 7, we study the forward invariant components of $F$). The basic result in this section is

**Theorem 5.2.1.** *Suppose that* $\deg(R) \geq 2$ *and that* $F_0$ *is a completely invariant component of* $F$. *Then*:

(i) $\partial F_0 = J$;
(ii) $F_0$ *is either simply connected or infinitely connected*;
(iii) *all other components of* $F$ *are simply connected; and*
(iv) $F_0$ *is simply connected if and only if* $J$ *is connected.*

Applying this directly to the case when the Fatou set itself is connected, we obtain

**Corollary 5.2.2.** *If* $\deg(R) \geq 2$ *and if* $F$ *is connected, then either*:

(i) $F$ *is simply connected and* $J$ *is connected; or*
(ii) $F$ *is infinitely connected and* $J$ *has an infinite number of components.*

In the case of a polynomial $P$, $\{\infty\}$ is completely invariant and hence so too is the unbounded component of the Fatou set $F(P)$. With this, another direct application of Theorem 5.2.1 yields

**Theorem 5.2.3.** *Let* $F$ *be the Fatou set of a non-linear polynomial. Then*:

(i) *the unbounded component of* $F$ *is either simply connected or infinitely connected; and*
(ii) *each bounded component of* $F$ *is simply connected.*

Examples of the two possibilities described in Theorem 5.2.3(i) occur in Chapters 1 and 9, and Theorem 5.2.3 verifies a visually obvious feature of the illustrations of Julia sets given earlier. We end this section with the

PROOF OF THEOREM 5.2.1. As $F_0$ is completely invariant, so is its closure $\overline{F}_0$, and so, by the minimality of $J$, $J \subset \overline{F}_0$. As $J$ is disjoint from $F_0$, we conclude that $J = \partial F_0$ and this is (i).

To prove (ii), we assume that $F_0$ is a completely invariant component with finite connectivity $c$ and we denote the components of the complement of $F_0$ by $E_1, \ldots, E_c$. By Proposition 3.2.6, there is an integer $m$ such that each $E_j$ is completely invariant under $R^m$ and as $J$ is infinite, one of the $E_j$, say $E_1$, is infinite. The minimality of $J(R^m)$ implies that it lies in $E_1$, and so

$$J(R) = J(R^m) \subset E_1.$$

However, by (i) each $E_j$ meets $J(R)$; thus $c = 1$ and (ii) follows.

To prove (iii), observe that from (i), $J \cup F_0$ is the closure of $F_0$ and so it is connected (Proposition 5.1.1). By Proposition 5.1.5, the components of its complement are simply connected and as these components are just the components of $F$ other than $F_0$, (iii) follows. Finally, (iv) is a direct consequence of (i) and Proposition 5.1.4. The proof is now complete. □

## §5.3. The Euler Characteristic

We devote this section to a brief and informal introduction to the Euler characteristic (a detailed treatment can be found in any one of a large number of standard texts on the subject). Let $S$ be a compact, or a bordered, surface: for our applications, it is enough to consider $S$ to be either the sphere, or a plane domain together with its boundary, provided that this boundary consists only of a finite number of simple closed curves. A *triangulation* $T$ of $S$ is a partition of $S$ into a finite number of mutually disjoint subsets called *ver-*

*tices*, *edges* and *faces* with the following properties:

(i) each vertex is a point of $S$;
(ii) for each edge $e$, there is a homeomorphism $\varphi$ of a closed interval $[a, b]$ in $\mathbb{R}$ into $S$ which maps the open interval $(a, b)$ onto $e$, and the end-points $a$ and $b$ to vertices of $T$;
(iii) for each face $f$, there is a homeomorphism $\varphi$ of a closed triangle $Q$ in $\mathbb{C}$ into $S$ which maps the edges and vertices of $Q$ (in the usual sense) to edges and vertices of $T$, and such that $f$ is the $\varphi$-image of the interior of $Q$.

We stress that $T$ partitions $S$ into mutually disjoint subsets of $S$. Each such subset is either a vertex, an edge or a face and we call each of these a *simplex* of $T$ (of dimension 0, 1 and 2 respectively). For any simplex $s$ of dimension $m$, we define the *Euler characteristic* $\chi(s)$ as $(-1)^m$: more generally, if $S_0$ is any subset of $S$ comprising a union of simplices, say $s_1, \ldots, s_r$ where $s_j$ has dimension $n_j$, we define

$$\chi(S_0) = \sum_j \chi(s_j) = \sum_j (-1)^{n_j}.$$

In particular, if the triangulation $T$ contains $F$ faces, $E$ edges and $V$ vertices respectively, then the Euler characteristic $\chi(S)$ of $S$ is, by definition,

$$\chi(S) = F - E + V.$$

The crucial, and well-known, fact (which we accept here without proof) is that $\chi(S)$ is a topological invariant which is independent of the particular triangulation $T$ used: thus we can compute $\chi(S)$ by using any convenient triangulation we choose. When we refer to a triangulation and Euler characteristic of a domain, we are implicitly assuming that the closure of the domain can be triangulated so that the domain itself is a union of simplices.

Calculations of $\chi$ can often be simplified by using the following simple idea. Suppose that $S_1$ and $S_2$ are subsets of $S$, each comprising a union of some of the simplices in $T$: then, simply by counting the contribution of each simplex to each side of the equation, we obtain

$$\chi(S_1 \cup S_2) + \chi(S_1 \cap S_2) = \chi(S_1) + \chi(S_2). \tag{5.3.1}$$

Let us illustrate this with some simple examples. The boundary $\partial S$ of $S$ must be a union of edges and vertices (for clearly, no face can meet $\partial S$) so if $S_0$ denotes the interior of $S$, we have

$$\chi(S) = \chi(S_0) + \chi(\partial S).$$

It is easy to see that a simple closed curve $C$ is a union of the same number of edges and vertices, so $\chi(C) = 0$. From this, we see that if the boundary of a surface $S$ consists of a finite number of mutually disjoint simple closed curves, then $S$ and its interior $S_0$ have the same Euler characteristic. Next, by constructing explicit triangulations it is easy to see that $\chi(\mathbb{C}_\infty) = 2$, and that $\chi(D) = 1$ for any open or closed disc $D$: these computations are simple but important. Now suppose that $D$ is the complement (in $\mathbb{C}_\infty$) of $k$ mutually disjoint topological closed discs $Q_1, \ldots, Q_k$, each being bounded by a Jordan curve, so $D$ is of connectivity $k$. We can triangulate the sphere such that each of $D, Q_1, \ldots, Q_k$ is a union of simplices, and then (5.3.1) yields

$$2 = \chi(\mathbb{C}_\infty) = \chi(D) + \sum \chi(Q_j)$$

so $\chi(D) = 2 - k$. Note that for any such subdomain $D$ of the sphere:

(i) $\chi(D) = 2$ if and only if $D$ is the sphere $\mathbb{C}_\infty$;
(ii) $\chi(D) = 1$ if and only if $D$ is simply connected, but not $\mathbb{C}_\infty$;
(iii) $\chi(D) = 0$ if and only if $D$ is doubly connected;

while in all other cases, $\chi(D)$ is negative. Later, when we apply these ideas to a component of the Fatou set, we shall see that there is an important distinction to be made between the cases $\chi(D) \geq 0$ and $\chi(D) < 0$.

EXERCISE 5.3

1. Let $S$ be the surface $\{z: |z| = 1\} \times [0, 1]$ (so $S$ is homeomorphic to a closed annulus). Show that $\chi(S) = 0$.
2. Let $D$ be a domain obtained by removing $2g$ discs from the sphere and let $C_1, \ldots, C_{2g}$ be the boundary curves of $D$. Let $S_1, \ldots, S_g$ be tubes (each topologically equivalent to $S$ in Exercise 5.3.1) and adjoin the boundary circles of $S_j$ to $C_j$ and $C_{j+g}$ to construct the surface $\Sigma$ which is a sphere with $g$ handles attached: we say that $S$ is a surface of genus $g$. Show that $\chi(\Sigma)$ is $2 - 2g$.
3. Let $S$ be a surface of genus $g$ from which $k$ discs have been removed. Show that $\chi(S) = 2 - 2g - k$ and describe the topologically distinct surfaces of this type with $\chi(S) \geq 0$.

## §5.4. The Riemann–Hurwitz Formula for Covering Maps

We are going to obtain the Riemann–Hurwitz formula for the situation described briefly by

$$R: U \to V,$$

where $R$ is a rational map, and $U$ and $V$ are domains in $\mathbb{C}_\infty$ (and so generalize the Riemann–Hurwitz formula described in §2.7 which is the case $U = V = \mathbb{C}_\infty$). First, for any domain $U$, we have

$$\partial R(U) \subset R(\partial U),$$

so, if $R$ maps $U$ onto $V$, then

$$\partial V \subset R(\partial U); \tag{5.4.1}$$

of course, there may be strict inclusion here.

Suppose now that $U$ is a component of $R^{-1}(V)$. Then each point $\zeta$ of $\partial U$ must map to $\partial V$ (for certainly, $R(\zeta)$ must be in the closure of $V$, and it cannot lie in $V$ unless $\zeta$ lies in $U$) so

$$R(\partial U) \subset \partial V. \tag{5.4.2}$$

In addition to this, $R$ maps $U$ onto $V$. Indeed, $R(U) \subset V$ and if there is not equality here, we could join a point of $R(U)$ to a point of $V - R(U)$ by a curve in $V$ and such a curve would have to meet $\partial R(U)$ in $V$, contradicting the fact that $\partial R(U) \subset R(\partial U) \subset \partial V$. It follows that if $U$ is a component of $R^{-1}(V)$, then both (5.4.1) and (5.4.2) hold and so in this case,

$$\partial V = R(\partial U).$$

These facts suggest that instead of considering a map $R$ of $U$ into $V$, we should (as is so often the case) emphasize $R^{-1}$ and study the map $R$ of a component $U$ of $R^{-1}(V)$ onto $V$. This change of emphasis is important.

In this section, our objective is to use $R$ to relate the Euler characteristics $\chi(U)$ and $\chi(V)$ of domains $U$ and $V$ as above. Any such relationship will involve a contribution from the critical points of $R$ and in order to quantify this, we introduce the *deficiency* of $R$ at $z$ as

$$\delta_R(z) = v_R(z) - 1, \tag{5.4.3}$$

where $v_R(z)$ is the valency as defined in §2.5. For any set $A$ we define the *total deficiency of $R$ over $A$* as

$$\delta_R(A) = \sum_{z \in A} \delta_R(z),$$

and this is additive, that is, for disjoint sets $A$ and $B$ we have

$$\delta_R(A \cup B) = \delta_R(A) + \delta_R(B).$$

Frequently, but only when $R$ is understood, we omit the suffix $R$ and use $\delta$ instead of $\delta_R$.

The reader will recall that (5.4.3) is the expression arising in the Riemann–Hurwitz relation (Theorem 2.7.1) and that $\delta_R(z) = 0$ for all but a finite set of $z$. In our new notation, the Riemann–Hurwitz relation is

$$\delta(\mathbb{C}_\infty) = 2\deg(R) - 2,$$

and, of course, for a polynomial $P$, $\delta(\mathbb{C}) = \deg(P) - 1$.

We are now ready to relate the quantities $\chi(U)$, $\chi(V)$, $\deg(R)$ and $\delta_R(U)$ as suggested above. We shall assume:

(1) $V$ is a domain bounded by a finite number of mutually disjoint Jordan curves;
(2) $U$ is a component of $R^{-1}(V)$; and
(3) there are no critical values of $R$ on $\partial V$;

and with these assumptions, we prove

**Theorem 5.4.1.** *With the assumptions above, there is an integer $m$ such that $R$ is an $m$-fold map of $U$ onto $V$ and*

$$\chi(U) + \delta_R(U) = m\chi(V). \tag{5.4.4}$$

In order to obtain a better appreciation of this result before giving its proof, we briefly consider some consequences of it. First, taking $U$ and $V$ to be $\mathbb{C}_\infty$, we have $\chi(U) = \chi(V) = 2$ and $m = \deg(R)$ and this recaptures the Riemann–Hurwitz relation given in §2.7. Second, if we take $V$ to be a simply connected domain other than $\mathbb{C}_\infty$, then $\chi(V) = 1$ and (5.4.4) yields

$$\delta_R(U) = (m-1) + (1 - \chi(U)).$$

Written this way, both terms on the right are non-negative and we can conclude that either:

(i) $R$ has critical points in $U$ (that is, $\delta_R(U) > 0$); or
(ii) $U$ is simply connected and $R$ is a homeomorphism of $U$ onto $V$.

If, in addition to this, $R^{-1}(V)$ is connected, then we must have $U = R^{-1}(V)$ and $m = \deg(R)$, so in this case,

$$\delta_R(R^{-1}(V)) \geq d - 1$$

with equality only when $R^{-1}(V)$ is simply connected (for example, the map $z \mapsto z^d$ of the unit disc onto itself). For future reference, we state this as

**Corollary 5.4.2.** *Let $R$ be a rational map of degree $d$, suppose that $V$ is simply connected, and that $R^{-1}(V)$ is connected. Then*

$$\delta_R(R^{-1}(V)) \geq d - 1,$$

*with equality if and only if $R^{-1}(V)$ is simply connected.*

We return now to the main result and give the

Proof of Theorem 5.4.1. We have already seen that $R$ maps $U$ onto $V$, and also $\partial U$ onto $\partial V$, and it follows from this that $U$ is a component of the complement of $R^{-1}(\partial V)$.

We show now that for some integer $m$, $R$ is an $m$-fold map of $U$ onto $V$ (that is, for each $w$ in $V$, there are exactly $m$ solutions $z$ of $R(z) = w$ in $U$). For $w$ in $V$, let $N(w)$ be the number of solutions (counting multiplicities) of $R(z) = w$ in $U$: we need only show that the map $w \mapsto N(w)$ is continuous for then, being integer valued, it is constant, say $m$, on the connected set $V$.

The proof that $w \mapsto N(w)$ is continuous is standard but, nevertheless, we remind our readers of it. It is well known that for each point $\zeta_j$ in $U$ which maps to $w$, there is an open neighbourhood $N_j$ such that $R$ is a $k$-fold map of $N_j$ onto $R(N_j)$, where $k$ is the valency of $R$ at $\zeta_j$. We remove the open sets $N_j$

from the closure of $U$ to form a compact set $E$, no points of which map to $w$: thus the compact set $R(E)$ is at a positive distance from $w$. It follows that there is an open neighbourhood $\mathcal{N}$ of $w$ that is disjoint from $R(E)$. Now for each point $w_1$ in $\mathcal{N}$, the solutions of $R(z) = w_1$ in $U$ must lie in $\bigcup_j N_j$ and so we find that $N(w_1) = N(w)$. We deduce that the function $w \mapsto N(w)$ is constant on $\mathcal{N}$ and so is continuous at $w$. As $V$ is connected, it follows that for some integer $m$, $R$ is an $m$-fold map of $U$ onto $V$.

The next step is to extend this result to the map $R$ of $\partial U$ onto $\partial V$; that is, we need to show that $R$ maps $\partial U$ in an $m$-fold manner onto $\partial V$. To show this, we take $\zeta$ on $\partial V$ and let the $d$ pre-images of $\zeta$ be $z_1, \ldots, z_d$: note that there are exactly $d$ such points, where $d = \deg(R)$, because of the assumption (3) concerning critical values. We know that none of the $z_j$ are in $U$, so we can relabel and hence assume that $z_1, \ldots, z_t$ lie in $\partial U$, while none of $z_{t+1}, \ldots, z_d$ lie in the closure of $U$.

We can now select a neighbourhood $N$ of $\zeta$ and, for each $j$, a neighbourhood $N_j$ of $z_j$, with the properties:

(4) the $N_j$ are mutually disjoint;
(5) for $j$ in $\{t + 1, \ldots, d\}$, $N_j$ is disjoint from the closure of $U$;
(6) $R$ is a homeomorphism of $N_j$ onto $N$ (we denote the inverse of this by $R_j^{-1}$); and
(7) $N \cap V$ is connected.

This last property is a local property of Jordan curves which we shall accept here without proof. Now let $j$ be in $\{1, \ldots, t\}$. Then $N_j$ meets $U$, and $R$ maps this intersection into $N \cap V$. We deduce that $R_j^{-1}(N \cap V)$ is a connected subset of $R^{-1}(V)$ which meets $U$, and hence it lies in $U$.

It is now easy to see that $m = t$. We take any point $\alpha$ in $N \cap V$. The argument in the previous paragraph shows that for each $j$ in $\{1, \ldots, t\}$, the point $R_j^{-1}(\alpha)$ lies in $U$: thus $t \leq m$. On the other hand, it is clear that for $j$ in $\{t + 1, \ldots, d\}$, $R_j^{-1}(\alpha)$ is not in $U$; thus $m \leq t$ and so $m = t$ as asserted.

Finally, we need to know that $\partial U$ is a finite union of mutually disjoint Jordan curves. Now by assumption, $\partial V$ comprises a finite number of Jordan curves, and there are no critical values of $R$ on $\partial V$. We take any $\zeta$ on one of these curves, say on $\gamma$, and find the $m$ distinct branches, say $R_j^{-1}, \ldots, R_m^{-1}$, of $R^{-1}$ defined near $\zeta$ and mapping $\zeta$ to the $m$ pre-images $z_j$ of $\zeta$ on $\partial U$. Each branch $R_j^{-1}$ can be analytically continued around $\gamma$ (there are no critical values of $R$ on $\gamma$) and this continuation induces a permutation on $\{1, \ldots, m\}$. It is clear that the assumption (3) implies that these continuations lead to simple closed curves on $\partial U$, and that each such curve is mapped in a $k$-fold manner (for some $k$) onto a curve in $\partial V$.

In remains to prove (5.4.4), and to do this, we triangulate the closure of $V$, ensuring (as we may) that all critical values of $R$ in $V$ are vertices of the triangulation (it is known that if $\zeta$ lies on a Jordan curve $\gamma$, and if $w$ does not, then there is a simple arc $\sigma$ which joins $w$ to $\zeta$ and which meets $\gamma$ only at $\zeta$). Suppose, then, that this triangulation, say $T$, has $f$ faces, $e$ edges and $v$ vertices so, by definition,

$$\chi(V) = f - e + v.$$

We claim now that the $m$-fold map $R$ of $U$ onto $V$ induces a triangulation $T_0$ of the closure of $U$, and that this is obtained as the inverse images of the simplices in $T$. The vertices of $T_0$ are the inverse images (in the closure of $U$) of the vertices in $T$ and so (by counting the number of solutions of $R(z) = w$, where $w$ is a vertex in $T$, and taking account of multiple solutions) we find that $T_0$ has precisely $mv - \delta_R(U)$ vertices. Each edge of $T$ lifts under each branch of $R^{-1}$ to an edge of $T_0$, and as there are no critical values of $R$ on any edge of $T$ (and so no multiple solutions of $R(z) = w$ for $w$ on an edge of $T$), we see that $T_0$ has precisely $me$ edges. A similar argument holds for the faces of $T$, noting that as each face, $F$ say, of $T$ is simply connected, the Monodromy Theorem ensures that each branch of $R^{-1}$ is single valued on $F$ and moreover, maps $F$ onto a simply connected subdomain of $U$. We deduce, then, that $T_0$ has precisely $mf$ faces and (5.4.4) now follows easily as

$$\begin{aligned}\chi(U) &= mf - me + (mv - \delta_R(U)) \\ &= m(f - e + v) - \delta_R(U) \\ &= m\chi(V) - \delta_R(U).\end{aligned}$$

□

We end this section with another simple consequence of Theorem 5.4.1, and we recall that for a domain $D$ bounded by $c(D)$ Jordan curves,

$$\chi(D) = 2 - c(D). \tag{5.4.5}$$

We now have

**Proposition 5.4.3.** *With the same assumptions as in Theorem* 5.4.1, $c(U) \geq c(V)$.

Proof. From (5.4.5), this inequality is equivalent to

$$\chi(U) \leq \chi(V). \tag{5.4.6}$$

As $\chi(U) \leq 2$, (5.4.6) holds when $\chi(V) = 2$. It also holds when $\chi(V) = 1$ for otherwise, $\chi(U) = 2$, whence $U$, and therefore $V$, must be $\mathbb{C}_\infty$, and then $\chi(V) = 2$. In all other cases, $\chi(V) \leq 0$ and then from Theorem 5.4.1 and the fact that $m \geq 1$, we have

$$\chi(U) \leq \chi(U) + \delta(U) = m\chi(V) \leq \chi(V).$$

□

Exercise 5.4

1. Find examples of domains $U$ and $V$, and a rational map $R$ of $U$ onto $V$, such that $\partial V$ is a proper subset of $R(\partial U)$.
2. Suppose that $R(z) = z^d$ and that $V$ is a Euclidean disc with the origin 0 not on its boundary. Show that $R^{-1}(V)$ is either simply connected (and $R$ is a $d$-fold map of it onto $V$) or it consists of $d$ simply connected domains (each homeomorphic under $R$ to $V$). What is $R^{-1}(V)$ when $V = \{z: |z - 1| < 1\}$?

3. In the circumstances described by Theorem 5.4.1, suppose that $\chi(U) = \chi(V)$. Examine the possible cases corresponding to this common value being 2, 1, 0 or negative.

## §5.5. Maps Between Components of the Fatou Set

The Euler characteristic of a domain $D$ on the complex sphere has been defined whenever its boundary $\partial D$ consists of a finite number of Jordan curves. In general, however, the boundary of a domain $D$ is much more complicated than this; we may not be able triangulate the closure of $D$ and in these circumstances, $\chi(D)$ is as yet undefined. As this is likely to be so in the case of major interest to us (when $D$ is a component of the Fatou set), this presents us with a problem which we must now tackle. We propose to show that given any domain $D$, we can define $\chi(D)$ as the limiting value of the Euler characteristic of smooth subdomains which exhaust $D$ and, once this has been done, we can then use the Euler characteristic as a tool to study the way in which $R$ maps one component of the Fatou set onto another. We shall restrict our discussion to subdomains of $\mathbb{C}_\infty$; nevertheless, the following development is closely related to the construction of the ideal boundary components of a Riemann surface (see [5], pp. 25–26 and 81–87).

Let $D$ be any domain on the sphere. A subdomain $\Omega$ of $D$ is said to be a *regular subdomain* of $D$ if:

(1) $\Omega$ is bounded by a finite union of mutually disjoint Jordan curves, say $\gamma_1, \ldots, \gamma_n$, all of which lie in $D$; and
(2) the complement of $\Omega$ consists of $n$ topological discs, say $W_1, \ldots, W_n$ (bounded by $\gamma_1, \ldots, \gamma_n$ respectively) and each $W_j$ meets the complement of $D$. For example, $\{|z| < 1\}$ is a regular subdomain of $\mathbb{C}$, whereas $\{1 < |z| < 2\}$ is not. Observe that $\chi(\Omega)$ is defined for each regular subdomain $\Omega$ of $D$.

Of course, if a subdomain $\Omega$ of $D$ satisfies (1) but not (2), we can adjoint to $\Omega$ those sets $W_j \cup \gamma_j$ which do not meet the complement of $D$ to form a regular subdomain $\Omega_0$ of $D$. Obviously, $\Omega$ has a higher connectivity that $\Omega_0$, and $\chi(\Omega_0) \geq \chi(\Omega)$.

We want to consider $D$ as the limit of regular subdomains, and as no canonical sequence of subdomains of $D$ presents itself, it is best to reject the idea of a sequential limit and to consider instead convergence with respect to the *directed set* (or *net*) of regular subdomains. There is no need for great generality here and the details are quite simple and explicit. First, we prove

**Lemma 5.5.1.** *Let D be a proper subdomain of the complex sphere. Then:*

(i) *any compact subset of D lies in some regular subdomain of D; and*
(ii) *if $\Omega_1$ and $\Omega_2$ are regular subdomains of D, then there is a regular subdomain $\Omega$ of D which contains both $\Omega_1$ and $\Omega_2$.*

PROOF. It is clearly sufficient to prove this in the case when $\infty$ is in $D$, so we shall assume that this is so. Let $n$ be a positive integer, and cover the plane with a square grid (including the axes), each square having diameter $1/2^n$. Now let $K_n$ be the union of those closed squares in the grid that contain some point of the complement of $D$, and let $D_n$ be that component of the complement of $K_n$ that contains $\infty$. Then it is easy to see that $D_n$ is an increasing sequence of regular subdomains of $D$ whose union is $D$.

With this, the proof is trivial. Given any compact subset $K$ of $D$, the family $\{D_n\}$ is an open cover of $K$ and so is covered by a finite collection of the $D_n$. As the $D_n$ are increasing with $n$, this finite collection contains a largest domain $D_m$ and this contains $K$. This proves (i). Finally, (ii) follows from (i) for if $\Omega_1$ and $\Omega_2$ are regular subdomains, then the union of their closures is a compact subset of $D$ and so by (i), it lies in some regular subdomain of $D$. □

Our next task is to show that $\chi(\Omega)$ is a monotonic function on the class $\mathscr{R}(D)$ of all regular subdomains of $D$, and so tends to a limit (which may be $-\infty$) as $\Omega$ increases to $D$ through $\mathscr{R}(D)$. The monotonicity property is expressed in

**Lemma 5.5.2.** *The Euler characteristic $\chi$ is a decreasing function on $\mathscr{R}(D)$: explicitly, if $\Omega_1$ and $\Omega_2$ regular subdomains of $D$ such that $\Omega_1 \subset \Omega_2$ then $\chi(\Omega_2) \leq \chi(\Omega_1)$.*

PROOF. Let $W_1, \ldots, W_n$ be the components of the complement of $\Omega_1$, and let $V_1, \ldots, V_m$ be the components of the complement of $\Omega_2$: because $\Omega_1 \subset \Omega_2$, we have

$$V_1 \cup \cdots \cup V_m \subset W_1 \cup \cdots \cup W_n.$$

For each $j$ in $\{1, \ldots, n\}$ choose a point $z_j$ in $W_j$, but not in $D$. As $z_j$ is not in $D$, it lies in some $V_k$, and so $V_k$ (being connected) lies in $W_j$. It follows that each $W_j$ contains some $V_k$, and hence $m \geq n$. The given inequality now follows as

$$\chi(\Omega_2) = 2 - m \leq 2 - n = \chi(\Omega_1). \qquad \square$$

Lemma 5.5.1 says that the class $\mathscr{R}(R)$ of regular subdomains of $D$ is a *net*, and Lemma 5.5.2 is the monotonicity of the function $\chi: \mathscr{R}(D) \to \{2, 1, 0, \ldots, -\infty\}$: thus these lead to

**Definition 5.5.3.** For any subdomain $D$ of $\mathbb{C}_\infty$,

$$\chi(D) = \inf\{\chi(\Omega): \Omega \text{ a regular subdomain of } D\}.$$

Quite explicitly, either:

(1) $\chi(D) = -\infty$, and there are regular subdomains $\Omega_n$ with $\chi(\Omega_n) \to -\infty$ or, equivalently, with $c(\Omega_n) \to +\infty$; or
(2) there is some regular subdomain $\Omega_0$ of $D$ such that

$$\chi(\Omega_0) = \chi(D) > -\infty,$$

and then (from Lemma 5.5.2) $\chi(\Omega) = \chi(D)$ whenever $\Omega$ is a regular subdomain which contains $\Omega_0$. Note that when $\chi(D)$ is finite, it is necessarily attained by some $\Omega_0$ as $\chi$ is integer valued.

If $D$ is a simply connected domain, then $\partial D$ is connected and each regular subdomain $\Omega$ can only have one complementary component: thus $\chi(D) = 1$ for a simply connected domain $D$, regardless of the nature of $\partial D$. More generally, if $D$ has connectivity $k$, then $\chi(\Omega) = 2 - k$ for all sufficiently large regular subdomains $\Omega$, and so $\chi(D) = 2 - k$, again irrespective of the complexity of $\partial D$.

As a culmination of these ideas, we have

**Theorem 5.5.4.** *Let $F_0$ and $F_1$ be components of the Fatou set $F$ of a rational map $R$ and suppose that $R$ maps $F_0$ into $F_1$. Then, for some integer $m$, $R$ is an $m$-fold map of $F_0$ onto $F_1$ and*

$$\chi(F_0) + \delta_R(F_0) = m\chi(F_1). \tag{5.5.1}$$

*Remark.* We call (5.5.1) the Riemann–Hurwitz relation.

Proof. First, it is easy to see that as $R(F_0) \subset F_1$, equality holds and $F_0$ is a component of $R^{-1}(F_1)$. As $R$ is locally an open $k$-fold mapping, the number $N(w)$ of solutions of $R(z) = w$ in $F_0$ is a continuous, and hence constant, function of $w$ in $F_1$; thus for some $m$, $R$ is an $m$-fold map of $F_0$ onto $F_1$.

The proof of (5.5.1) is based on the construction of regular subregions $\Omega_0$ and $\Omega_1$ of $F_0$ and $F_1$ respectively. We select a point $w$ in $F_1$ and construct a regular subregion $\Omega_0$ *of* $F_0$ which contains:

(1) all the critical points of $R$ that lie in $F_0$; and
(2) all the pre-images $z_1, \ldots, z_m$ of $w$ that lie in $F_0$.

Next, we select a regular subdomain $\Omega_1$ of $F_1$ which contains the compact set $R(\overline{\Omega}_0)$. Now each of the components of $R^{-1}(\Omega_1)$ lies in $F(R)$ and is mapped by $R$ onto $\Omega_1$ (Theorem 5.4.1). Further, one such component, say $\Omega_2$, contains the connected set $\Omega_0$ which satisfies (2), and from this we deduce that $\Omega_2$ is the only component of $R^{-1}(\Omega_1)$ that meets $F_0$; in other words,

$$\Omega_0 \subset \Omega_2 = R^{-1}(\Omega_1) \cap F_0 \subset F_0.$$

We now claim that $\Omega_2$ is a regular subregion of $F_0$, and to verify this it is sufficient to show that each component $W$ of the complement of $\overline{\Omega}_2$ meets $J(R)$. Now $W$ is bounded by some Jordan curve $\gamma$ (which separates $W$ from $\Omega_2$) so there are points in $W$ (and close to $\gamma$) which map to some component $B$ of the complement of $\overline{\Omega}_1$. As $R^{-1}(B)$ is disjoint from $\Omega_2$, $W$ contains some component, say $V$, of $R^{-1}(B)$. However, as $\Omega_1$ is a regular subregion of $F_1$, $B$ contains some point not in $F_1$, and hence some point of $J(R)$. It follows that $V$, and hence $W$ also, meets $J(R)$ so $\Omega_2$ is indeed a regular subregion of $F_0$.

Now $\Omega_0 \subset \Omega_2 \subset F_0$ so from Lemma 5.5.2,

$$\chi(\Omega_0) \geq \chi(\Omega_2) \geq \chi(F_0),$$

and from (1),

$$\delta_R(\Omega_0) = \delta_R(\Omega_2) = \delta_R(F_0).$$

Using this with Theorem 5.4.1, we now have

$$\chi(\Omega_2) + \delta_R(F_0) = m\chi(\Omega_1). \tag{5.5.2}$$

Given any $\alpha$ with $\alpha > \chi(F_0)$, we choose $\Omega_0$ such that $\chi(\Omega_0) < \alpha$: then the same holds for $\chi(\Omega_2)$ so

$$\alpha + \delta_R(F_0) \geq m\chi(\Omega_1) \geq m\chi(F_1),$$

and letting $\alpha \to \chi(F_0)$ we obtain

$$\chi(F_0) + \delta_R(F_0) \geq m\chi(F_1).$$

To obtain the opposite inequality, take any $\Omega_0$. Then, from (5.5.2), and regardless of the choice of $\Omega_1$, we have

$$\chi(F_0) + \delta_R(F_0) \leq m\chi(\Omega_1).$$

Now if $\beta > \chi(F_1)$, we can always choose $\Omega_1$ in the argument above so that $\chi(\Omega_1) < \beta$, and a similar argument shows that

$$\chi(F_0) + \delta_R(F_0) \leq m\chi(F_1).$$

The proof is now complete. □

## §5.6. The Number of Components of the Fatou Set

We suppose that $R$ has degree $d$, where $d \geq 2$, and we shall now obtain information on the number of components of $F$. Suppose first that each of $F_1, \ldots, F_k$ is a completely invariant component of $F$, where $k \geq 2$. Applying Theorem 5.2.1(iii) to each of the components $F_j$ in turn, we find (as $k \geq 2$) that every $F_j$ is simply connected. Next, we apply Theorem 5.5.4 to each $F_j$ and, noting that as each $F_j$ is simply connected, $\chi(F_j) = 1$, we obtain

$$\delta_R(F_j) = (d-1)\chi(F_j) = d-1.$$

The Riemann–Hurwitz relation now yields

$$\begin{aligned} k(d-1) &= \sum_{j=1}^{k} \delta_R(F_j) \\ &\leq \delta_R(\mathbb{C}_\infty) \\ &= 2d-2, \end{aligned}$$

and so $k \leq 2$. This argument constitutes the proof of:

**Theorem 5.6.1.** *The Fatou set $F$ of $R$ contains at most two completely invariant components, and if there are two, then each is simply connected.*

Both of the possibilities in Theorem 5.6.1 can arise: if $R(z) = z^2 - 2$, then $F$ has a single completely invariant component, and if $R(z) = z^2$, then $F$ has two completely invariant components. Theorem 5.6.1 does not by itself exclude the possibility that $F$ may have two completely invariant components *and* other additional components: however, it is a consequence of Sullivan's No Wandering Domains Theorem (see §8.1 and §9.4) that *if $F$ has two completely invariant components, then these are the only components of $F$.*

With reference to Theorem 5.6.1, it is of interest to note that for any positive integer $m$, there exist simply connected domains $D_1, \ldots, D_m$ with

$$\partial D_1 = \partial D_2 = \cdots = \partial D_m;$$

see [56], p. 143 (for this shows that we cannot prove Theorem 5.6.1 from the single topological fact that all of the completely invariant components of $F$ have a common boundary, namely $J$).

An argument similar to that used above enables us to obtain information on the total number of components of $F$. First, Chapter 1 contains a discussion of the examples

$$z \mapsto z^2 - 2, \qquad z \mapsto z^2, \qquad z \mapsto z^2 - 1,$$

and in these, $F$ has one, two, and infinitely many components respectively (see Figure 1.5.1 and also Exercise 5.6.1). In addition, we have seen (in §4.3) that it is possible for $F$ to be empty. It is possible, then, for $F$ to have exactly 0, 1, 2 or infinitely many components, and our last result in this section says that these are the only posibilities.

**Theorem 5.6.2.** *The Fatou set $F$ of a rational map $R$ has either* 0, 1, 2 *or infinitely many components.*

PROOF. This is trivial if $\deg(R) = 1$. If $\deg(R) \geq 2$, this a direct consequence of Theorem 5.6.1, for suppose that $F$ has only finitely many components, say $F_1, \ldots, F_k$. By Proposition 3.2.6, each $F_j$ is completely invariant under some iterate $R^m$ (which has the same Fatou set as $R$) and so, according to Theorem 5.6.1 (applied to $R^m$), $k \leq 2$. □

*Remark.* The proof of Theorem 5.6.1 is based on the idea that certain components of $F(R)$ contain critical points and this, in conjunction with the Riemann–Hurwitz relation, bounds the number of components. This idea is important and will be used again later to good effect.

EXERCISE 5.6

1. Let $P(z) = z^2 - 1$. Show that $0$, $-1$ and $\infty$ are attracting fixed points of $P^2$. Deduce that $F(P)$ has at least three, and hence infinitely many, components.

Let $F_{-1}$ and $F_0$ be the components of $F$ which contain $-1$ and $0$ respectively (as in Figure 1.5.1). Prove:

(1) $P$ maps $F_0$ onto $F_{-1}$ with multiplicity two;
(2) $P$ maps $F_{-1}$ onto $F_0$ with multiplicity one.

2. A component $F_0$ of $F(R)$ is *pre-periodic* if it is not forward invariant under any iterate $R^n$, but if some image $R^m(F_0)$ is. Prove that if $F(R)$ has a pre-periodic component, then it has infinitely many components.
3. Let $P$ be a polynomial of degree at least two, and suppose that the component $F_\infty$ of $F(P)$ which contains $\infty$ is infinitely connected. Show that $F_\infty$ is the only completely invariant component. [*Note*: $F$ may have infinitely many components.]

# §5.7. Components of the Julia Set

As usual, $R$ is a rational map of degree $d$, where $d \geq 2$. We have seen that no point of $J$ is isolated (Theorem 4.2.4), and it is easy to show that if $J$ is disconnected then it has infinitely many components. Indeed, if $J$ has only a finite number of components, say $J_1, \ldots, J_n$, then some $J_k$, say $J_1$, is infinite (as $J$ is). Moreover, by Proposition 3.2.6, there is some integer $m$ such that each $J_k$ is completely invariant under $R^m$ so, by the minimality of the Julia set,

$$J(R) = J(R^m) = J_1,$$

and $J(R)$ is connected. We now prove a stronger result than this, namely

**Theorem 5.7.1.** *If $J$ is disconnected, then it has uncountably many components and each point of $J$ is an accumulation point of infinitely many distinct components of $J$.*

In Chapter 11, we give an example of a rational map $R$ with the properties (i) $J$ is disconnected, and (ii) each component $F_j$ of $F$ has finite connectivity. It follows that only a countable number of components of $J$ can meet the boundary of some $F_j$; thus in this example, *there are components of $J$ which do not meet the closure of any component of $F$.*

In order to prove Theorem 5.7.1, we need

**Lemma 5.7.2.** *Let $K$ be a compact connected subset of the complex plane. Then $R^{-1}(K)$ has at most $d$ components and each is mapped by $R$ onto $K$.*

We prove Theorem 5.7.1 next (assuming Lemma 5.7.2) and then give the proof of the lemma.

Proof of Theorem 5.7.1. Let $K$ be the set of points in $J$ at which infinitely many components of $J$ accumulate: our first objective is to use the minimality of $J$ to show that $J \subset K$, and so deduce that $K = J$. We have just seen that $J$ has infinitely many components, so $K$ is not empty and obviously, it is closed.

Now take any $w$ *not* in $K$, so there is an open set $U$ containing $w$ such that $U$ meets only a finite number of components, say $J_1, \ldots, J_n$, of $J$. Now let $\zeta$ be any point of $J \cap R(U)$: then $\zeta$ belongs to one of the connected subsets $R(J_1), \ldots, R(J_n)$ of $J$, and we deduce that the open neighbourhood $R(U)$ of $R(w)$ meets only a finite number of components of $J$. This shows that $R(w)$ is not in $K$, and hence $R^{-1}(K) \subset K$.

Note that this imples that $K$ is infinite, for each point of the non-empty set $K$ is non-exceptional and so has an infinite backward orbit (Theorem 4.1.4).

Next, take $\zeta$ in $K$, so there is a sequence $J_1, J_2, \ldots$ of distinct components of $J$ which accumulate at $\zeta$. Lemma 5.7.2 implies that at most $d$ of the $J_n$ can map to any given component of $J$ so, using the continuity of $R$ at $\zeta$, it is clear that $R(\zeta)$ is also in $K$; thus $R(K) \subset K$. We have now shown that $J = K$ and so, in particular, no component of $J$ is isolated. Further, each point of $J$ is an accumulation point of distinct components of $J$.

It only remains to prove that $J$ has uncountably many components. We argue by contradiction, so suppose that $J$ has only countably many components, say $J_1, J_2, \ldots$: thus $J$ is a compact metric space which is the countable union of the $J_n$. Now by Baire's Category Theorem, $J$ is not the countable union of nowhere dense sets, so we may suppose that the closure of $J_1$ has a non-empty interior (in the relative topology on $J$). However, $J_1$, as a component of $J$, is closed in $J$, so we can conclude that $J_1$ has a non-empty interior in $J$. This violates the statement at the end of the previous paragraph, so $J$ must have uncountably many components and the proof is complete. □

We end with the

Proof of Lemma 5.7.2. Let $D$ be the complement of $K$. By Proposition 5.1.5, each component $D_j$ of $D$ is simply connected and (by the methods described in §5.4 and §5.5) we find that each component $U$ of $R^{-1}(D_j)$ is either a simply connected domain, or it is a domain of finite connectivity which contains a critical point of $R$ (for if $U$ does not contain a critical point, the Monodromy Theorem shows that $R: U \to D_j$ is a homeomorphism). It follows that $R^{-1}(D)$ is the union of a *finite* number of multiply connected domains, say $M_1, \ldots, M_t$, and a number (possibly infinite) of simply connected domains $S_j$.

When there are no multiply connected domains $M_j$ present, all of the components of $R^{-1}(D)$ are simply connected and then Proposition 5.1.5 implies that the complement of $R^{-1}(D)$, namely $R^{-1}(K)$, is connected: thus the conclusion of Lemma 5.7.2 holds in this case.

We now assume that at least one domain $M_j$ exists, and we consider the minimal, and necessarily finite, set of components $E_1, \ldots, E_q$ of $R^{-1}(K)$ such that

$$\bigcup_j \partial M_j \subset E_1 \cup \cdots \cup E_q: \tag{5.7.1}$$

we claim that $E_1, \ldots, E_q$ are all of the components of $R^{-1}(K)$. To see this, suppose that $Q$ is another component of $R^{-1}(K)$ and write $E = E_1 \cup \cdots \cup E_q$:

then $E$ and $Q$ are disjoint compact subsets of $R^{-1}(K)$, so from [77] (Theorem 5.6, p. 82), there are compact subsets $A$ and $B$ of $R^{-1}(K)$ such that

$$A \cup B = R^{-1}(K), \qquad A \cap B = \varnothing, \qquad Q \subset A, \qquad E \subset B. \tag{5.7.2}$$

We may assume that $\infty \in R^{-1}(D)$; then $A$ and $B$ are disjoint, compact subsets of $\mathbb{C}$, so we can find a bounded open set $\Omega$ which has a finite number of components $\Omega_j$, each being bounded by a finite number of Jordan curves, and which is such that

$$A \subset \Omega, \qquad B \cap \Omega = \varnothing, \qquad \partial\Omega \cap (A \cup B) = \varnothing: \tag{5.7.3}$$

see Exercise 5.7.1. Because of (5.7.2), we also have

$$\partial\Omega \subset R^{-1}(D). \tag{5.7.4}$$

Now let $\Omega_Q$ be the component of $\Omega$ that contains the connected set $Q$. Using (5.7.1) and (5.7.2), we find that for each $r$,

$$\partial M_r \subset E \subset B,$$

and so we see from (5.7.3) that $\Omega_Q$ and $\partial M_r$ are disjoint. Clearly, $\Omega_Q$ is arcwise connected, and this means that either $\Omega_Q \subset M_r$ or $\Omega_Q \cap M_r = \varnothing$. Now the first possibility cannot occur because if it does, then

$$Q \subset \Omega_Q \subset M_r \subset R^{-1}(D)$$

which violates the fact that $Q \subset R^{-1}(K)$; thus $\Omega_Q$ is disjoint from each $M_r$. As each $M_r$ is open, we deduce that the closure of $\Omega_Q$ is disjoint from $\bigcup M_r$.

As a consequence of this, each boundary component $\gamma_j$ (a Jordan curve) of $\Omega_Q$ lies in some simply connected domain $S_m$ for, by (5.7.4), it lies in $R^{-1}(D)$; thus one side of $\delta_j$ lies in $S_m$, while the other side contains $R^{-1}(K)$ and each $M_r$. It follows that for any $z_1$ in $M_r$, and any $z_2$ in $Q$,

$$n(\gamma_j, z_1) = n(\gamma_j, z_2),$$

and hence that

$$n(\partial\Omega_Q, z_1) = n(\partial\Omega_Q, z_2), \tag{5.7.5}$$

where, in general, $n(\gamma, z)$ denotes the winding number of $\gamma$ about $z$. Now

$$\Omega_Q = \{z \in \mathbb{C}: z \notin \partial\Omega_Q, n(\partial\Omega_Q, z) \neq 0\},$$

so from (5.7.5), $z_1 \in \Omega_Q$. This, however, contradicts the fact that $\Omega_Q$ and $M_r$ are disjoint: thus no such component $Q$ exists, and we have proved that

$$R^{-1}(K) = E_1 \cup \cdots \cup E_q.$$

As $R^{-1}(K)$ is compact, so is each $E_j$, and hence $R(E_j)$ also: thus $R(E_j)$ is a closed subset of $K$. We shall show that each $R(E_j)$ is relatively open in $K$: then, as $K$ is connected, we find that $R(E_j) = K$. Clearly, this implies that that $q \leq d$ and the proof of the proposition will then be complete.

To show that $R(E_j)$ is relatively open in $K$, we take any $\zeta$ in $R(E_j)$, say

$\zeta = R(w)$, where $w \in E_j$. We find a neighbourhood $N$ of $w$ not meeting any other $E_i$ (this is possible because $R^{-1}(K)$ has only finitely many components) and observe that

$$K \cap R(N) = R(E_j \cap N) \subset R(E_j).$$

This shows that $R(E_j)$ is relatively open in $K$, and the proof of the proposition is complete. □

EXERCISE 5.7

1. Let $A$ and $B$ be disjoint, non-empty, compact subsets of $\mathbb{C}$, and put a rectangular grid on $\mathbb{C}$ which is fine enough so that no square in the grid meets both $A$ and $B$. Let $\{Q_j\}$ be the set of those closed squares which meet $A$, and let $\Omega$ be the interior of $\bigcup Q_j$. Show that $\Omega$ is a bounded open set with a finite number of components $\Omega_j$, each being bounded by a finite number of Jordan curves. Show also that (5.7.3) holds.

CHAPTER 6

# Periodic Points

This chapter is devoted to an extensive discussion of the fixed and periodic points of a rational map and their role in iteration theory.

## §6.1. The Classification of Periodic Points

We have already used elementary facts about fixed points, and we turn now to a systematic and detailed discussion of the periodic points of a rational map $R$. A fixed point $\zeta$ of $R$ is classified according to the multiplier $m(R, \zeta)$ of $R$ at $\zeta$ (see §2.6) and as this is conjugation invariant, we may assume that $\zeta$ is in $\mathbb{C}$ and so $m(R, \zeta) = R'(\zeta)$. It is important to realize that as the classification of fixed points is a purely local matter, it applies to any analytic function and, in particular, to the local inverse (when it exists) of a rational map.

**Definition 6.1.1.** Suppose that $\zeta$ in $\mathbb{C}$ is a fixed point of an analytic function $f$. Then $\zeta$ is:

(a) *super-attracting* if $f'(\zeta) = 0$;
(b) *attracting* if $0 < |f'(\zeta)| < 1$;
(c) *repelling* if $|f'(\zeta)| > 1$;
(d) *rationally indifferent* if $f'(\zeta)$ is a root of unity;
(e) *irrationally indifferent* if $|f'(\zeta)| = 1$, but $f'(\zeta)$ is not a root of unity.

Some explanatory remarks may be helpful. First, the distinction between (a) and (b) is that a super-attracting fixed point is a critical point of $f$, whereas an attracting fixed point is not; thus $f$ has a local inverse near $\zeta$ if $\zeta$ is attracting, but not if it is super-attracting. The reader may recall that in §1.1, no such distinction was made and, of course, there are still occasions when we wish to embrace both cases; we do this by saying that $\zeta$ is *(super)attracting*. In a similar way, it is sometimes convenient to combine cases (d) and (e), and such a fixed point is said to be *indifferent*. For an indifferent fixed point $\zeta$, the best linear approximation to $f$ near $\zeta$ is a rotation about $\zeta$ and the distinction between (d) and (e) is simply whether this rotation has finite or infinite order.

The notion of periodic points, and the extension of the classification to these, is straightforward. A point $\zeta$ is a *periodic point* of a rational function $R$ if it is a fixed point of some iterate $R^m$. For such a point $\zeta$ there is a positive integer $n$ for which

$$\zeta, R(\zeta), R^2(\zeta), \ldots, R^{n-1}(\zeta) \tag{6.1.1}$$

are distinct but $R^n(\zeta) = \zeta$: the finite set of points in (6.1.1) is the *cycle* of $\zeta$ and the integer $n$ is the *period* of $\zeta$. Of course, the fixed points of $R$ are points of period one: more generally, $\zeta$ has period $n$ if and only if it is a fixed point of $R^n$ but not of any lower-order iterate.

A periodic point $\zeta$ of period $n$ is classified as a fixed point of $R^n$ but we can say a little more than this. By conjugation, we may assume that the cycle does not contain $\infty$ and we write

$$\zeta_m = R^m(\zeta), \qquad m = 0, 1, 2, \ldots,$$

so $\zeta_{m+n} = \zeta_m$. By $n$ applications of the Chain Rule, we now have

$$\begin{aligned}(R^n)'(\zeta_m) &= \prod_{k=0}^{n-1} R'(R^k(\zeta_m)) \\ &= \prod_{k=0}^{n-1} R'(\zeta_k),\end{aligned}$$

the second product being a re-arrangement of the first. This argument shows that the derivative $(R^n)'$ has the same value at each point $\zeta_j$ of the cycle, and so each point $\zeta_j$ is classified in exactly the same way as any other $\zeta_k$ in the cycle. As a consequence of this, we can extend our classification to cycles and speak naturally of the *multiplier of the cycle*, *attracting cycles* and so on.

A point $\zeta$ is *pre-periodic* under $R$ if it is not periodic but if some image, say $R^m(\zeta)$, is: in this case there exist positive integers $m$ and $n$ with

$$\zeta, R(\zeta), R^2(\zeta), \ldots, R^m(\zeta), \ldots, R^{m+n-1}(\zeta)$$

distinct but with $R^{m+n}(\zeta) = R^m(\zeta)$. Most periodic points have associated pre-periodic points (see Exercise 6.1.1); for example, the origin is a pre-periodic point of the polynomial $z^2 - 2$.

In the remainder of this chapter, we shall discuss how the attracting, repelling and indifferent cycles relate to the Fatou and Julia sets. Briefly, the (super)attracting cycles are in the Fatou set, while the repelling cycles are in the Julia set. The corresponding results for indifferent cycles are more delicate and require much more work.

Exercise 6.1

1. Let $C$ be the cycle $\{\zeta_1, \ldots, \zeta_m\}$ for a rational map $R$. Show that either $R^{-1}(C)$ contains pre-periodic points, or $C$ is completely invariant, has at most two elements and lies in $F$.

2. Suppose that $\zeta$ is a periodic point of $R$ of period $k$. Prove that $R^n$ fixes $\zeta$ if and only if $k$ divides $n$.

3. Show that if any point $\zeta$ is fixed by both $R^n$ and $R^m$, then $\zeta$ is fixed by $R^q$ where $q$ is the greatest common divisor of $m$ and $n$.

4. Show that $z \mapsto z^2$ has no indifferent fixed points, that $z \mapsto z^{-2}$ has no attracting fixed points, and that $z \mapsto z + z^2$ has no repelling fixed points.

5. This exercise shows that given any integer $N$, there is a polynomial with an attracting cycle of period at least $N$. We may assume that $N$ is prime. For each $c$, let $P_c(z) = z^2 + c$, so
$$P_c^{n+1}(z) = [P_c(z)]^2 + c.$$
As the coefficients of $P_c^n(z)$ are polynomials in $c$,
$$Q_n \colon c \mapsto P_c^n(c)$$
is a polynomial in $c$ which satisfies
$$Q_{n+1}(c) = [Q_n(c)]^2 + c.$$
Show that for $n \geq 1$,
$$Q_n(0) = 0, \qquad Q_n'(0) = 1, \qquad Q_n(-2) = 2,$$
and deduce that for each $n$, there is some number $\gamma$ (which depends on $n$) in $(-2, 0)$ such that $Q_n(\gamma) = 0$.
Show that $P_c^{n-1}(0) = Q_n(c)$: thus there is some $\gamma$ in $(-2, 0)$ with $P_\gamma^N(0) = 0$. This shows that the origin has period $k$ for $P_\gamma$, where $k$ divides $N$. As $N$ is prime, $k$ is 1 or $N$, and $k \neq 1$ as 0 is not fixed by $P_\gamma$: thus 0 belongs to a super-attracting cycle of length $N$.

6. Prove that the fixed points of a polynomial cannot all be attracting (see [35], p. 279).

# §6.2. The Existence of Periodic Points

In an attempt to find the periodic points of period $n$ of a rational map $R$, we naturally consider solutions of
$$R^n(z) = z. \tag{6.2.1}$$
Any solution of this equation will be periodic; however, it will not necessarily be of period $n$ for it may already be fixed by some earlier iterate $R^m$, where $m < n$. This occurs, for example, when
$$R(z) = z^2 - z: \tag{6.2.2}$$

in this case every solution of $R^2(z) = z$ is also a solution of $R(z) = z$ and so for this $R$, *there are no periodic points of period two*. The Julia set of $z^2 - z$ is shown in Figure (3.14), [22], p. 101.

This example raises a question of existence, namely: How often are periodic points absent? Could there exist, for example, a rational function with only finitely many periodic points? The answer is no. In fact, in the vast majority of cases, periodic points of a given period do exist and for polynomials, the situation is particularly striking. We have

**Theorem 6.2.1.** *Let $P$ be a polynomial of degree at least two and suppose that $P$ has no periodic points of period $n$. Then $n = 2$ and $P$ is conjugate to $z \mapsto z^2 - z$.*

In short, the example given in (6.2.2) was the only example of this type (up to conjugacy) that we could possibly have given. There is also a corresponding (though slightly weaker) result for rational functions, namely

**Theorem 6.2.2.** *Let $R$ be a rational function of degree $d$, where $d \geq 2$, and suppose that $R$ has no periodic points of period $n$. Then $(d, n)$ is one of the pairs*

$$(2, 2), (2, 3), (3, 2), (4, 2);$$

*moreover, each such pair does arise from some $R$ in this way.*

This shows, for example, that all rational functions have periodic points of period $4, 5, 6, \ldots$, and that every rational function of degree five or more has periodic points of all orders. These results are due to I.N. Baker [9], and we shall prove them in §6.8 (but see Exercise 6.2.2). Roughly speaking, if a rational map fails to have any periodic points of period $n$, then every solution of $R^n(z) = z$ must also be a solution of some equation $R^m(z) = z$ for some $m$ which divides, and is less than, $n$. This places a strong algebraic constraint on the coefficients of $R$ which, according to the two results above, is rarely satisfied; nevertheless, the known proofs of these results are function-theoretic.

The reader should note that not only is Theorem 6.2.2 weaker than Theorem 6.2.1 in that several exceptional pairs $(d, n)$ arise, but it is also weaker in that an exceptional pair $(d, n)$ does not necessarily determine a unique conjugacy class of exceptional maps $R$: see Exercise 6.2.3 for more details.

Although we shall prove Theorems 6.2.1 and 6.2.2 later, we shall give a simple argument now which guarantees the existence of at least an infinite set of periodic points for each rational map (a fact that we shall need before we prove the two theorems). This also follows from Theorem 4.2.6 but the following argument is simple and direct. We suppose that $R$ is a rational function of degree $d$, where $d \geq 2$, and that $R$ has no points of period $p$, where $p$ is a prime. Then every solution $\zeta$, say, of $R^p(z) = z$ is of some period $k$, where $k$ is less than, and divides $p$: thus $k = 1$, and $\zeta$ is a fixed point of $R$ itself. As $R$ and $R^p$ have $d + 1$ and $d^p + 1$ fixed points respectively, we deduce that there must be some fixed point $\zeta$ of $R$ such that $R^p$ has more fixed points at $\zeta$ than $R$ does, and the only way that this can arise is that $R'(\zeta) \neq 1$ but $R'(\zeta)^p = 1$ (see Corollary 2.6.7). Finally, there are at most $d + 1$ points $\zeta$ fixed by $R$ and for each of these, $R'(\zeta)^q = 1$ for at most one prime $q$. Thus *each rational $R$ has periodic points with prime period $p$ for all but at most $d + 1$ prime numbers $p$.*

If we consider this argument when $R$ is a quadratic polynomial with no periodic points of period two, we find that up to conjugacy, $R(0) = 0$ and $R'(0) = -1$. It follows that $R(z) = az^2 - z$, and just by changing the scale, this is easily seen to be conjugate to the polynomial $z^2 - z$ in (6.2.2).

For periodic points of entire functions, see (for example) [7] and [10].

## EXERCISE 6.2

1. Show that every quadratic polynomial is conjugate to one and only one polynomial of the form $z \mapsto z^2 + c$. Show also that $z^2 - z$ is conjugate to $z^2 - \frac{3}{4}$.

   Let $P(z) = z^2 + c$. Explain why $P(z) - z$ divides $P^2(z) - z$, and using this, show that if $P$ has no periodic points of period 2, then $P(z) = z^2 - \frac{3}{4}$.

2. The example $z^2 - z$ shows that the pair (2, 2) in Theorem 6.2.2 is exceptional. To show that the other pairs are exceptional (see [9] and Theorem 2.6.4), show that:
   (a) if $R(z) = z + (\omega - 1)(z^2 - 1)/2z$, where $\omega = \exp(2\pi i/3)$, then $R$ has no points of period 3;
   (b) if $R(z) = (z^3 + 6)/3z^2$, then $R$ has no points of period 2;
   (c) if $R(z) = -z(1 + 2z^3)/(1 - 3z^3)$ then $R$ has no points of period 2.

3. For $a \neq 1$, let
$$R_a(z) = \frac{z^2 - z}{1 + az}.$$
   Show that:
   (a) $R_a$ has fixed points 0, $\infty$ and $\alpha = 2/(1 - a)$;
   (b) $R_a^2$ has three fixed points at the origin;
   (c) the five fixed points of $R_a^2$ are 0, 0, 0, $\infty$ and $\alpha$;
   (d) $R_a$ has no two-cycles.

   Prove that $R_a$ is conjugate to $R_b$ if and only if $b = a$ or $b = (3 - a)/(1 + a)$. [*Hint*: If $gR_a g^{-1} = R_b$, then $R_b$ must fix $g(0)$, $g(\infty)$ and $g(\alpha)$, so these three points must coincide with 0, $\infty$, $\beta$, where $\beta = 2/(1 - b)$. Further, the multipliers at corresponding points must coincide.]

4. Suppose that $f$ is analytic on a neighbourhood of the origin and that $f(0) = 0$, $f'(0) = a$. Show that there exists an integer $N(f)$ such that for all integers $n$, the number of fixed points of $f^n$ at 0 is either 1 (if $a^n \neq 1$) or $N(f)$ (if $a^n = 1$).

5. Let $R$ be a rational map with $\deg(R) \geq 2$, and let
$$G = \{g: g \text{ Möbius}, gRg^{-1} = R\}.$$

Use the fact that $R$ has periodic points of period $p$ for some prime $p$ to show that $G$ is a *finite* group. Select a rational map of your choice and attempt to find $G$ explicitly. (For a list of all finite Möbius groups, see [18] or [46].)

6. Prove that every quadratic polynomial has points of period 3.

## §6.3. (Super)Attracting Cycles

Suppose that $\zeta$ is a (super)attracting fixed point of a rational map $R$. Then $|R'(\zeta)| < 1$ and so there is a number $\alpha$, $|R'(\zeta)| < \alpha < 1$, and a disc $D$ centred at $\zeta$ such that in $D$,

$$|R(z) - \zeta| = |R(z) - R(\zeta)| < \alpha|z - \zeta|.$$

This shows that each $R^n$ maps $D$ into itself, so $D$ lies in the Fatou set $F(R)$ and $R^n \to \zeta$ uniformly on $D$. By Vitali's Theorem (Theorem 3.3.3), $R^n \to \zeta$ locally uniformly on the component of $F(R)$ which contains $\zeta$, and as $F(R^q) = F(R)$ for any iterate $R^q$, we easily obtain

**Theorem 6.3.1.** *Let $\{\zeta_1, \ldots, \zeta_q\}$ be a (super)attracting cycle of $R$. Then each $\zeta_j$ lies in a component, say $F_j$, of the Fatou set $F(R)$, and as $n \to \infty$, $R^{nq} \to \zeta_j$ locally uniformly on $F_j$.*

In view of this, we introduce some terminology. Given a (super)attracting fixed point $\zeta$ of $R$, the component of $F(R)$ which contains $\zeta$ is called the *local*, or *immediate*, *basin* of $\zeta$. It follows that $R^n \to \zeta$ precisely when $z$ lies in some inverse image $R^{-m}(B)$, $m \geq 0$, of the local basin $B$, and we call the set of such $z$ the *basin* (or *stable set*) for $\zeta$. More generally, the *local basin of a (super)attracting cycle* $\{\zeta_1, \ldots, \zeta_q\}$ is the union of the (necessarily distinct) components $F_1, \ldots, F_q$ of $F(R)$ (as in Theorem 6.3.1) and the *basin* for the cycle is the union of the local basin and all its inverse images.

Although a rational map $R$ of degree $d$, $d \geq 2$, has infinitely many cycles, we shall see later that it can have *only finitely many (super)attracting cycles* (in fact, at most $2d - 2$). Thus, for example, a quadratic polynomial can have at most one (super)attracting cycle in $\mathbb{C}$ (the reader may recall that in §1.6, we showed that a quadratic polynomial can have at most one attracting fixed point in $\mathbb{C}$).

Our next task is to analyse the behaviour of an analytic function $f$ in a neighbourhood of an attracting fixed point $\zeta$, and this analysis will enable us to understand completely the dynamics of $f$ near $\zeta$. We suppose, then, that $f$ is analytic in some neighbourhood $\mathcal{N}$ of an attracting fixed point $\zeta$, and we may assume that $f(\mathcal{N}) \subset \mathcal{N}$. Now let $g$ be any function which is analytic and univalent on $\mathcal{N}$ with $g(\zeta) = 0$, and let $F = gfg^{-1}$. Then $F$ is defined and analytic on the neighbourhood $g(\mathcal{N})$ of the origin (see Figure 6.3.1) and, as $F'(0) = f'(\zeta)$, we see that 0 is an attracting fixed point of $F$. We call $F$ a *local*

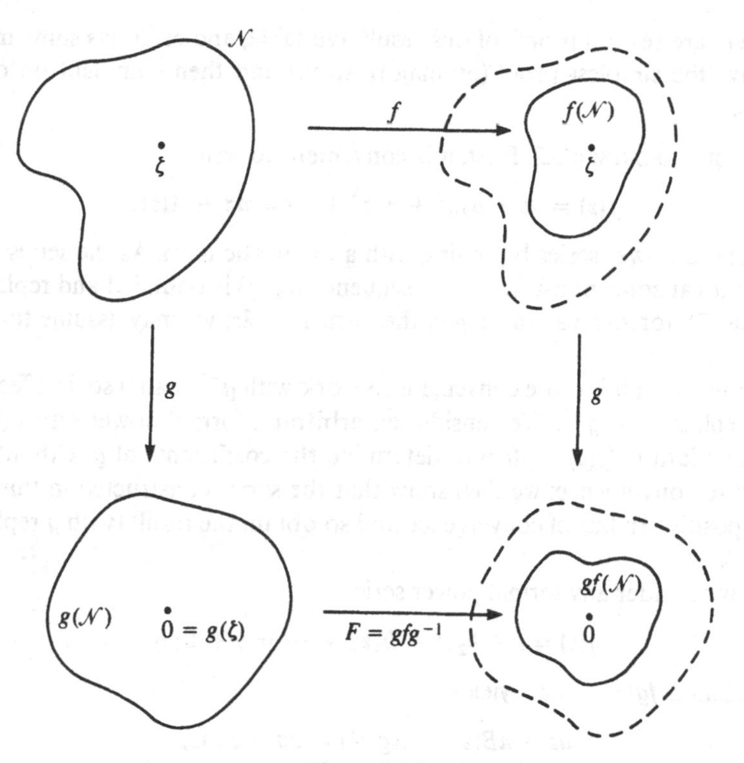


Figure 6.3.1

(*analytic*) *conjugate* of $f$ and we ask whether, by choosing $g$ suitably, we can arrange for $F$ to be considerably simpler (say, in terms of its Taylor coefficients) than $f$ is. The benefits that arise depend on how simple we can choose $F$ to be but, in any case, the iterates of $f$ (near $\zeta$) are reflected in the behaviour of the iterates of $F$ for $F^n = gf^ng^{-1}$.

We shall assume that $\zeta = 0$: this simplifies the formulae, and the corresponding results for a general fixed point $\zeta$ can easily be obtained afterwards by conjugation. The function $f$ has a Taylor expansion at 0, say

$$f(z) = az + bz^q + \cdots, \qquad 0 < |a| < 1,$$

and because $F'(0) = f'(\zeta)$, we must have

$$F(z) = az + b_2z^2 + b_3z^3 + \cdots.$$

We shall now show that by choosing a suitable $g$, we can achieve the greatest possible simplification, namely that each $b_j$ is zero and $F$ is the map $z \mapsto az$.

**Theorem 6.3.2.** *Suppose that $f$ is analytic in a neighbourhood of the origin, that $f(0) = 0$, and that $f'(0) = a$ where $0 < |a| < 1$. Then there exists a unique function $g$ which is analytic in some disc $\{|z| < r\}$ with $g(0) = 0$, $g'(0) = 1$, and which satisfies $gfg^{-1}(z) = az$ for all $z$ sufficiently near the origin.*

There are several proofs of this result available, and each has some merit. We give the simplest proof (by majorization) and then comment on other proofs.

PROOF OF THEOREM 6.3.2. First, it is convenient to write

$$f(z) = az + a_2z^2 + a_3z^3 + \cdots = az + A(z),$$

so $A(z)$ is a power series beginning with a quadratic term. As the series for $f$ converges at some non-zero $z_0$, the sequence $(a_nz_0^n)$ is bounded and replacing $f$ by $\varphi f \varphi^{-1}$, for some suitable $\varphi$ of the form $z \mapsto \lambda z$, we may assume that for $n \geq 2$, $|a_n| \leq 1$.

In this proof it is more convenient to work with $g^{-1}$ than $g$ so, in effect, we shall replace $g$ by $g^{-1}$. We consider an arbitrary, formal power series $g$ and use the relation $fg(z) = g(az)$ to determine the coefficients of $g$ without any regard to convergence: we then show that the series constructed in this way has a positive radius of convergence and so obtain the result (with $g$ replaced by $g^{-1}$).

Now consider any formal power series

$$g(z) = z + b_2z^2 + b_3z^3 + \cdots = z + B(z).$$

The relation $fg(z) = g(az)$ yields

$$az + aB(z) + A(g(z)) = az + B(az),$$

and hence

$$\sum_{n\geq 2} b_n(a^n - a)z^n = \sum_{n\geq 2} a_n(z + b_2z^2 + \cdots)^n.$$

As $|a| < 1$, we have $a^n \neq a$ for $n \geq 2$, and this determines the coefficients $b_n$ uniquely (by induction) and in the form

$$b_n(a^n - a) = P_n(a_2, \ldots, a_n, b_2, \ldots, b_{n-1}),$$

where the polynomial $P_n$ in the given variables $a_j$, $b_j$, has positive coefficients. Further, as $|a_n| \leq 1$ for $n \geq 2$, this implies that

$$|a|(1 - |a|)|b_n| \leq P_n(1, \ldots, 1, |b_2|, \ldots, |b_{n-1}|),$$

and this gives an upper bound on the rate of growth of the $b_n$.

For our comparison function, we consider the function

$$H(z) = z - \alpha(z^2 + z^3 + \cdots),$$

where $\alpha = 1/(|a| - |a|^2)$. Obviously, $H$ is analytic in the unit disc and so has a local inverse, say

$$h(z) = z + \beta_2z^2 + \beta_3z^3 + \cdots,$$

that is, analytic in some neighbourhood of the origin. Thus for all $z$ sufficiently near to the origin,

$$z = H(h(z)) = h(z) - \alpha \sum_{m\geq 2} h(z)^m.$$

This identity reduces to

$$\sum_{m\geq 2}(z+\beta_2 z^2+\cdots)^m = |a|(1-|a|)\sum_{n\geq 2}\beta_n z^n,$$

and this shows that

$$|a|(1-|a|)\beta_n = P_n(1,\ldots,1,\beta_2,\ldots,\beta_{n-1}).$$

Using induction, it is now clear that first, $\beta_n \geq 0$ for all $n\geq 2$, and second, that $|b_n|\leq \beta_n$ for all $n\geq 2$. It follows that the formal power series $g$ does indeed converge in some neighbourhood of the origin and so $g$ is analytic near the origin. Given this, the identity $fg(z) = g(az)$ is valid near the origin, and as $g$ has a local inverse near the origin, the proof is complete. □

In fact, this result implies more than we have stated so far. We consider the case when $f$ is a rational function, which we now denote by $R$, and we let $B$ be the local basin of an attracting fixed point $\zeta$ of $R$. According to Theorem 6.3.2, there is a disc $W$ centred at $\zeta$ such that $R(W)\subset W$, and a power series $g$ analytic in $W$ such that $gR(z) = ag(z)$ there. Now let $W_n$ be the component of $R^{-n}(W)$ that contains $\zeta$: we claim that

$$B = W\cup W_1\cup W_2\cup\cdots. \tag{6.3.1}$$

First, as $W_n$ is a connected subset of $F(R)$, it must lie within $B$. Next, given any $z$ in $B$, join $z$ to $\zeta$ by a curve $\gamma$ in $B$. Then by Theorem 6.3.1, there is some $n$ such that $R^n(\gamma)\subset W$, and hence $\gamma$, being connected, lies in $W_n$. This shows that $z$ is in $W_n$ and thus proves (6.3.1).

Now define $g_n$ on $W_n$ by the formula

$$g_n(z) = a^{-n}g(R^n z).$$

Clearly, $g_n$ is analytic on $W_n$ and for all $z$ sufficiently close to the origin, $g_n(z) = g(z)$. We deduce that the sequence $g_n$ provides an analytic continuation of $g$ to the entire local basin $B$, and by analytic continuation, $gR(z) = ag(z)$ throughout $B$. Finally, we observe that

$$g(W_n) = a^{-n}g(R^n(W_n)),$$

and as $R^n$ maps each component of $R^{-n}(W)$ onto $W$, we have

$$g(W_n) = a^{-n}g(W).$$

Now $g(W)$ is an open neighbourhood of the origin, and as $|a|<1$, we easily deduce that $g(B) = \mathbb{C}$, and we have proved

**Theorem 6.3.3.** *Let $\zeta$ be an attracting fixed point of the rational map $R$ with $R'(\zeta) = a$, and let $B$ be the local basin of $\zeta$. Then there exists a unique function $g$ that is analytic in $B$ with $g(\zeta) = 0$, $g'(\zeta) = 1$, and $g(B) = \mathbb{C}$, and such that in $B$, $gR(z) = ag(z)$.*

*Remark*. Unlike the local conjugacy relation $gRg^{-1}(z) = az$, the functional equation $gR(z) = ag(z)$ does not require the existence of $g^{-1}$ (indeed, by Liouville's Theorem, $g^{-1}$ cannot exist throughout $B$). The analysis of these equations and their solutions is attributed to Poincaré, Schröder and Koenigs.

The method used in the proof of Theorem 6.3.2 will be used later (in §6.7) to analyse the behaviour of $R$ near certain indifferent fixed points. Other proofs are available; see, for example, [67] where a proof is sketched using quasiconformal mappings. We end with a brief sketch of another proof of Theorem 6.3.3: this proof gives the existence of the function $g$ throughout $B$ directly, and also motivates the proof of the corresponding result for super-attracting fixed points which is given in §6.10.

A SECOND PROOF OF THEOREM 6.3.3. We may assume that $\zeta = 0$. Ignoring the details for the moment, if $g$ satisfies $g(Rz) = ag(z)$ near the origin, then for each $n$,

$$gR^n(z) = a^n g(z),$$

and hence

$$\frac{R^n(z)}{a^n} = g(z)\left(\frac{g(R^n z)}{R^n(z)}\right)^{-1} \to g(z) \tag{6.3.2}$$

because $R^n(z) \to 0$ as $n \to \infty$, and $g'(0) = 1$. This representation indicates how we can establish the existence of $g$, namely by proving that $R^n(z)/a^n$ converges to some limit as $n \to \infty$. Observe that if this limit exists on some neighbourhood $W$ of the origin, then it exists throughout $B$, for given $z$ in $B$, then say, $R^m(z)$ is in $W$ and

$$\frac{R^{n+m}(z)}{a^{n+m}} = a^{-m}\left[\frac{R^n(R^m z)}{a^n}\right].$$

Note also that if such a function $g$ exists, then it is unique, for it is determined by the limit in (6.3.2).

We now sketch the formal proof. We start with a disc $\mathcal{N}$ centred at the origin, and a number $\alpha$ such that $|a| < \alpha < 1$ and

$$|R(z)| \leq \alpha|z|$$

on $\mathcal{N}$. Next, we define the function $\Phi$ on $\mathcal{N}$ by

$$R(z) = az[1 + \Phi(z)]. \tag{6.3.3}$$

Note that $\Phi(0) = 0$ and, by decreasing $\mathcal{N}$ if necessary, we may assume that $\Phi$ is analytic on the closure of $\mathcal{N}$, and so for some $M$,

$$|\Phi(z)| \leq M|z|$$

there. Observe that on $\mathcal{N}$, we then have

$$|\Phi(R^m z)| \leq M|R^m(z)| \leq M\alpha^m|z|. \tag{6.3.4}$$

Now for $z$ in $\mathcal{N}$,

$$R^{n+1}(z) = R(R^n z) = aR^n(z)[1 + \Phi(R^n z)],$$

and so by induction, the first step being (6.3.3),

$$\frac{R^{n+1}(z)}{a^{n+1}} = z \prod_{m=0}^{n} [1 + \Phi(R^m z)]. \tag{6.3.5}$$

This, together with (6.3.4) and standard results on infinite products (see Theorem 2 in Appendix III to this chapter), shows that the infinite product

$$g(z) = z \prod_{m=0}^{\infty} [1 + \Phi(R^m)]$$

exists and is analytic on $\mathcal{N}$. Moreover, by (6.3.5), $g$ satisfies (6.3.2) on $\mathcal{N}$, and it is clear that $g(0) = 0$ and $g'(0) = 1$. Finally, from (6.3.5) we have

$$g(Rz) = \lim \frac{R^{n+1}(Rz)}{a^{n+1}} = \lim \frac{R^{n+2}(z)}{a^{n+1}} = ag(z),$$

this being valid on $\mathcal{N}$ and so throughout $B$.

The question of whether or not an analytic map $f$ is locally conjugate to its derivative (as a linear map) near one of its fixed points $\zeta$ is important, so we say that $f$ is *linearizable* at $\zeta$ if there is some function $g$ that is analytic near $\zeta$ with $g(\zeta) = 0, g'(\zeta) = 1$, and such that for all $z$ sufficiently close to the origin,

$$gfg^{-1}(z) = f'(\zeta)z. \qquad \square$$

Exercise 6.3

1. Suppose that $g(z) = z + \cdots$ is analytic near the origin. Use Leibnitz's formula (for the derivative of a product) to show that if $G(z) = g(z)^m$, then $G^{(k)}(0)$ is a polynomial in the variables $g^{(1)}(0), \ldots, g^{(k-1)}(0)$ with positive coefficients.
2. Suppose that $R$ is a rational map, that $\zeta$ is an attracting fixed point of $R$, and that the basin $B$ of $\zeta$ is a simply connected proper subdomain of $\mathbb{C}$. Prove that $R$ is conjugate on $B$ to a Blaschke product of the form $zg_1(z) \cdots g_m(z)$, where the $g_j$ are Möbius transformations mapping the unit disc onto itself with $g_j(0) \neq 0$.

# §6.4. Repelling Cycles

We begin with

**Theorem 6.4.1.** *For any rational map $R$, every repelling cycle of $R$ lies in $J(R)$.*

Proof. First, suppose that the origin is a repelling fixed point of $R$. Then, near the origin,

$$R(z) = az + \cdots,$$

where $|a| > 1$ and, consequently, as $n \to \infty$,

$$(R^n)'(0) = a^n \to \infty.$$

Suppose now that 0 is in $F(R)$. Then $\{R^n\}$ is normal on some neighbourhood $\mathcal{N}$ of 0, and so some sequence of the iterates $R^n$ converge uniformly on $\mathcal{N}$ to some analytic function $g$. Now $g(0) = 0$, so $g'(0)$ is finite. On the other hand, the uniform convergence implies that for the given sequence,

$$g'(0) = \lim(R^n)'(0) = \infty,$$

which is a contradiction. We deduce that the repelling fixed 0 is in $J$.

Of course, this implies (by conjugation) that any repelling fixed point of $R$ is in $J(R)$. If $\{\zeta_1, \ldots, \zeta_q\}$ is any repelling cycle for $R$, then each $\zeta_j$ is in $J(R^q)$, and as $J(R^q) = J(R)$, the proof is complete. □

We can interpret Theorem 6.3.3 in the context of repelling fixed points, for if $\zeta$ is a repelling fixed point of $R$, then there is a branch of $R^{-1}$ for which $\zeta$ is an attracting fixed point. Applying Theorem 6.3.3 to this branch, we find that $R$ is locally conjugate to $z \to R'(\zeta)z$ in some neighbourhood of $\zeta$.

**Exercise 6.4**

1. Use the definition of equicontinuity (instead of the derivative) to show that a repelling fixed point of $R$ is in $J(R)$.

2. Let $\zeta$ be a repelling point of $R$. Show that if $R^n(z) \to \zeta$ as $n \to \infty$, then for some $n$, $R^n(z) = \zeta$, and so $z$ is in $J(R)$.

   Show, however, that it is possible to have $z$ in $F(R)$ and $R^n(z)$ converging to a fixed point $\zeta$ of $R$ in $J$. This happens, for example, when $R(z) = z/(1 + z^2)$ and $\zeta = 0$ (see §1.8).

3. The origin is a repelling fixed point of $f(z) = 2z/(1 + z^2)$, and $f'(0) = 2$. Show that $gfg^{-1}(z) = 2z$, where $g$ is defined near the origin by $g^{-1}(z) = \tanh z$.

   Show that a similar statement is true when $f(z) = 2z(1 + z)$ and $g^{-1}(z) = (e^{2z} - 1)/2$.

## §6.5. Rationally Indifferent Cycles

Although it is not easy to decide whether a given indifferent cycle lies in the Fatou or the Julia set, the question is easily settled for rationally indifferent cycles.

**Theorem 6.5.1.** *If* $\deg(R) \geq 2$ *then every rationally indifferent cycle of* $R$ *lies in* $J(R)$.

PROOF. We recall that $\zeta$ lies in a rationally indifferent cycle of $R$ of length $m$ if $R^m$ fixes $\zeta$ and $(R^m)'(\zeta)$ is some root of unity. Suppose first that $\zeta$ is a rationally indifferent fixed point of $R$. By conjugation, we may assume that $\zeta = 0$, so

$$R(z) = az + bz^r + \cdots,$$

say, where $b \neq 0$, $r \geq 2$, and $a^k = 1$ for some positive integer $k$. Writing $S = R^k$, we have

$$S(z) = z + cz^p + \cdots$$

for some $p \geq 2$, where $c \neq 0$ (for $S$ cannot be the identity map unless $\deg(R) = 1$). By induction (or from Corollary 2.6.5),

$$S^n(z) = z + ncz^p + \cdots,$$

and so $(S^n)^{(p)}(0) \to \infty$, as $n \to \infty$. This means that $\{S^n\}$ cannot be normal in any neighbourhood of the origin (for otherwise, some sequence of $S^n$ would converge to some analytic $\varphi$ with $\varphi(0) = 0$ and $\varphi^{(p)}(0) = \infty$); consequently, $\{R^n\}$ is not normal near 0, so $0\ (= \zeta)$ is in $J(R)$.

Finally, if $\zeta$ is any point of a rationally indifferent cycle of length $m$, the preceding argument shows that $\zeta$ is in $J(R^m)$, and hence in $J(R)$. Note that if $\deg(R) = 1$, then every indifferent cycle is in $F(R)$. □

The remainder of this section is devoted to a detailed description of the dynamics of the iterates $R^n$ near rationally indifferent cycles. This is a long and difficult task, and in order to obtain a clear understanding of the global structure of the arguments, we have provided a succession of results, each (generally speaking) being an improvement on the preceding results. One of the reasons why this description is much harder to obtain than, say, the description of the dynamics near an attracting fixed point, is that a rational map $R$ is *not* locally conjugate to its (linear) derivative near a rationally indifferent fixed point $\zeta$ (for the derivative is a rotation of finite order about $\zeta$, and $R$ can have finite order only if $\deg(R) = 1$). This observation means that we have to find another way to describe the dynamics of the $R^n$ near such a cycle, and the answer lies in showing that $R$ is conjugate to a translation not, of course, on the whole neighbourhood of $\zeta$, but on certain open sets which have $\zeta$ on their boundary. We shall include examples to illustrate the ideas as we proceed, and we begin with a simple lemma which isolates the most basic idea in this section.

**Lemma 6.5.2.** *Suppose that $f$ is analytic and satisfies*

$$f(z) = z - z^{p+1} + O(z^{p+2})$$

*in some neighbourhood $N$ of the origin. Let $\omega_1, \ldots, \omega_p$ be the p-th roots of unity and let $\eta_1, \ldots, \eta_p$ be the p-th roots of $-1$. Then for sufficiently small positive numbers $r_0$ and $\theta_0$:*

(i) $|f(z)| < |z|$ *on each sector*

$$S_j = \{z: 0 < |z/\omega_j| < r_0, |\arg(z/\omega_j)| < \theta_0\};$$

*and*

(ii) $|f(z)| > |z|$ *on each sector*

$$\Sigma_j = \{z: 0 < |z/\eta_j| < r_0, |\arg(z/\eta_j)| < \theta_0\}.$$

PROOF. Write

$$f(z)/z = 1 - z^p[1 + v(z)],$$

where $v$ is analytic in $N$ with $v(0) = 0$, and also

$$S = \{\zeta: |\zeta| < \tfrac{1}{2}, |\arg \zeta| < \pi/4\}.$$

It is easy to see that for a suitable choice of $r_0$ and $\theta_0$, $z \in S_j$ implies that $z^p[1 + v(z)] \in S$, and $z \in \Sigma_j$ implies that $-z^p[1 + v(z)] \in S$. The desired inequalities now follow by elementary geometry. □

The numbers $\omega_j$ and $\eta_j$ in Lemma 6.5.2 are the $2p$-th roots of unity and they are alternately and equally spaced around the unit circle. If

$$f(z) = z + az^{p+1} + O(z^{p+2}), \qquad a \neq 0,$$

a similar result holds with the directions $\omega_j$ and $\eta_j$ replaced by the directions of the solutions of $az^p = -1$ and $az^p = 1$ respectively. Also with this $f$, $f^{-1}$ exists near the origin and

$$f^{-1}(z) = z - az^{p+1} + O(z^{p+2})$$

so (with $r_0$ and $\theta_0$ small enough) $|f^{-1}(z)| > |z|$ on each $S_j$, and $|f^{-1}(z)| < |z|$ on each $\Sigma_j$.

Next, we consider a simple example and here our discussion contains the ideas that we shall use to develop the general theory.

**Example 6.5.3.** The polynomial

$$P(z) = z - z^2$$

has a rationally indifferent fixed point at the origin. If $0 < x < 1$, then $0 < P(x) < x$, and so $P^n(x) \to 0$ (for $P^n(x)$ can only converge to a fixed point of $P$): if $x < 0$, then $P(x) < x < 0$, and so $P^n(x) \to \infty$. These observations are consistent with Lemma 6.5.2 and we shall now investigates the behaviour of $P^n(z)$ for all $z$ near the origin.

Let $\sigma(z) = 1/z$ and replace $P$ by its conjugate $Q = \sigma P \sigma^{-1}$: thus

$$Q(z) = z + 1 + 1/(z - 1).$$

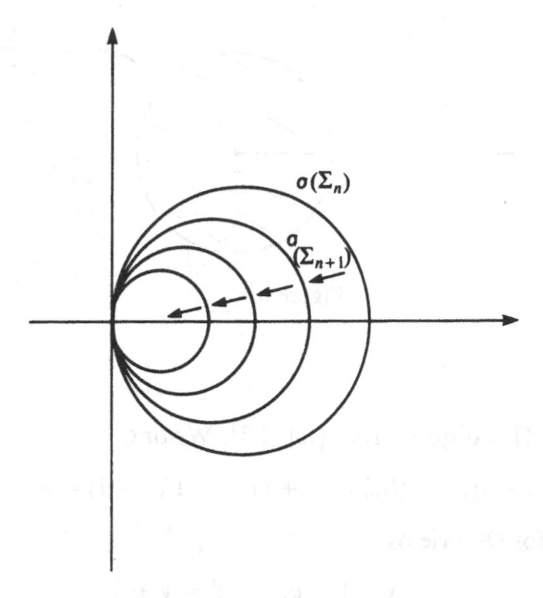


Figure 6.5.1

Now note that if $\text{Re}[z] > 1$ then $\text{Re}[(z-1)^{-1}] > 0$ so

$$\text{Re}[Q(z)] \geq \text{Re}[z] + 1.$$

In particular, if $t > 1$, then $Q$ maps $\{x + iy: x > t\}$ into $\{x + iy: x > t + 1\}$, and rewriting this in terms of $P$, we see that $P^n \to 0$ uniformly in each sufficiently small disc $\{z: |z - r| < r\}$. In fact, $P$ acts as illustrated in Figure 6.5.1 (which the reader should compare with Figure 1.7.1).

Because of the particularly simple form of $Q$, we can say more than this. Suppose that $\Pi$ is any region that is disjoint from $\{|z| \leq 3\}$, and which is mapped into itself by $Q$. Then on $\Pi$,

$$\text{Re}[Q(z)] \geq \text{Re}[z] + 1 - 1/|z - 1| \geq \text{Re}[z] + \tfrac{1}{2},$$

so

$$\text{Re}[Q^n(z)] \geq \text{Re}[z] + n/2$$

and $Q^n \to \infty$ on $\Pi$ ("drifting" to the right). One possibility for $\Pi$ is the half-plane $\{\text{Re}[z] > 3\}$, but it is easy to find a much larger region than this. Now $Q(z)$ lies in the shaded disc illustrated in Figure 6.5.2, and if $z \in \Pi$, the radius $r(z)$ of this disc is at most $\frac{1}{2}$. Further, $r(z) \to 0$ as $z \to \infty$ and this suggests that we might be able to take $\Pi$ to be bounded by some parabola. We shall now show that we can.

Let

$$\Pi = \{x + iy: y^2 > 12(3 - x)\}.$$

The boundary of $\Pi$ is a parabola which cuts the $x$-axis when $x = 3$, and it is

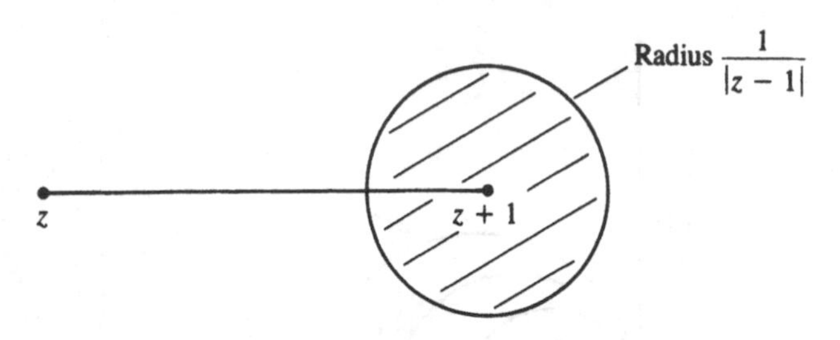


Figure 6.5.2

easy to see that $\Pi$ is disjoint from $\{|z| \leq 3\}$. Writing

$$z = x + iy, \qquad Q(z) = X + iY, \qquad 1/(z-1) = a + ib,$$

the expression for $Q(z)$ yields

$$X = x + 1 + a, \qquad Y = y + b,$$

and so for $z$ in $\Pi$,

$$\begin{aligned} Y^2 - 12(3 - X) &= (y + b)^2 - 12(3 - [x + 1 + a]) \\ &= [y^2 - 12(3 - x)] + b^2 + 2by + 12(1 + a) \\ &\geq 12 - 2|by| - 12|a|. \end{aligned}$$

However, as $|z| \geq 3$, we have

$$|a| \leq |a + ib| \leq 1/(|z| - 1) < \tfrac{1}{2}$$

and

$$|yb| \leq |z|/(|z| - 1) < \tfrac{3}{2}.$$

It follows that if $z$ is in $\Pi$, then $Y^2 > 12(3 - X)$ and hence so is $Q(z)$: we deduce that $Q$ maps $\Pi$ into itself and so $P^n \to 0$ on $\sigma(\Pi)$. The Julia set of $P$ is illustrated in Figure 6.5.3 where the black region is $\sigma(\Pi)$.

In order to analyse the situation for a general rationally indifferent fixed point, we need the notion of petals (which, roughly speaking, play the role of the region $\sigma(\Pi)$ in the example above). For each positive $t$, each positive integer $p$, and each $k$ in $\{0, 1, \ldots, p - 1\}$, we define the sets

$$\Pi_k(t) = \{re^{i\theta} : r^p < t(1 + \cos(p\theta)); |2k\pi/p - \theta| < \pi/p\}.$$

These sets are called *petals* (at the origin) and they are illustrated in Figure 6.5.4 when $p = 6$. Note that the petals are pairwise disjoint, and that each subtends an angle $2\pi/p$ at the origin (so the total angle subtended at the origin by all the petals is $2\pi$). We call the line of symmetry of $\Pi_k(t)$ (the ray $\theta = 2k\pi/p$) the *axis* of $\Pi_k$.

We come now to the first major result of this section and in this, we

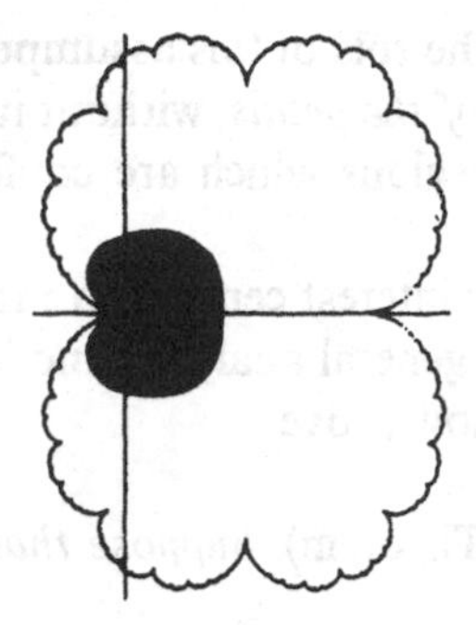

Figure 6.5.3. $z \mapsto z - z^2$. Reprinted with permission of the American Mathematical Society.

describe the dynamics of the iterates $R^n$ in the petals at a fixed point $\zeta$ where $R'(\zeta) = 1$. By conjugation, we may assume that $\zeta = 0$ and, in addition, we shall assume that $R$ has a Taylor expansion there of the form

$$R(z) = z(1 - z^p + bz^{2p} + cz^{2p+1} + \cdots).$$

Of course, we can always ensure that $z^{p+1}$ has coefficient $-1$ simply by conjugating $R$ with a map of the form $z \mapsto \alpha z$: thus *the significant part of this assumption is that there are no terms between* $z^{p+1}$ *and* $z^{2p+1}$ *in the Taylor series*

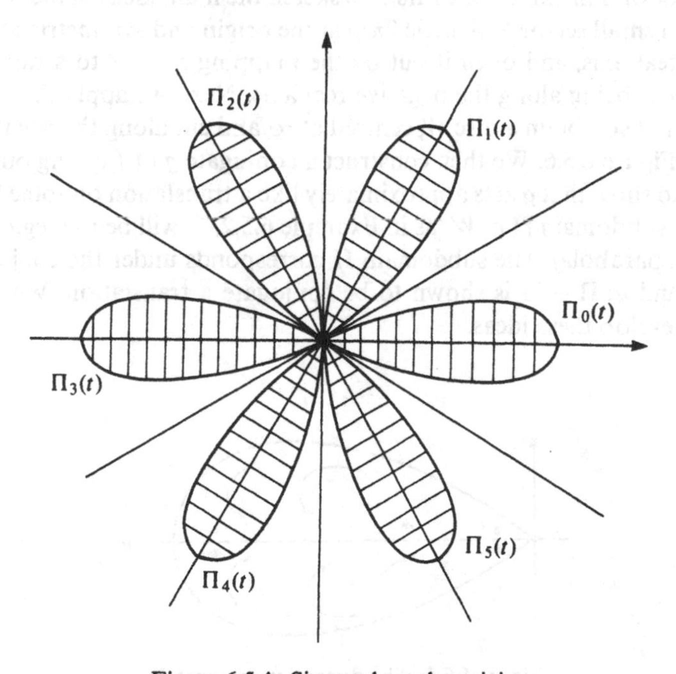


Figure 6.5.4. Six petals at the origin.

*for R*. We emphasize that the role of this assumption is that is enables us to give an *explicit description of the petals*; without it, we can only establish the existence of petal shaped regions which are conformal images of the $\Pi_k(t)$ described above.

Despite the fact that our interest centres on rational functions, it is desirable to prove the result for general analytic functions, for we can then apply it to branches of $R^{-1}$. We now prove

**Theorem 6.5.4** (The Petal Theorem). *Suppose that the analytic map $f$ has a Taylor expansion*

$$f(z) = z - z^{p+1} + O(z^{2p+1}). \tag{6.5.1}$$

*at the origin. Then for all sufficiently small $t$:*

(a) *$f$ maps each petal $\Pi_k(t)$ into itself*;
(b) *$f^n(z) \to 0$ uniformly on each petal as $n \to \infty$*;
(c) *$\arg f^n(z) \to 2k\pi/p$ locally uniformly on $\Pi_k$ as $n \to \infty$*;
(d) *$|f(z)| < |z|$ on a neighborhood of the axis of each petal*;
(e) *$f: \Pi_k(t) \to \Pi_k(t)$ is conjugate to a translation.*

By (a), the iterates $f^n$ are defined on each $\Pi_k(t)$, and (c) implies that for any $z$ in $\Pi_k(t)$, the sequence $f^n(z)$ converges to 0 along a path which is asymptotic to the axis of the petal $\Pi_k(t)$: see Figure 6.5.5.

THE PROOF OF THEOREM 6.5.4. First, we sketch the main ideas of the proof. We consider a small sector $S$ of angle $2\pi/p$ at the origin and symmetric about the positive real axis, and open it out by the mapping $z \mapsto z^p$ to a radially cut disc, the cut being along the negative real axis. Next, we apply the mapping $z \mapsto 1/z$ and so obtain a disc $W$, centred at $\infty$ and cut along the negative real axis: see Figure 6.5.6. We then construct a conjugate $g$ of $f$ acting on $W$, and proceed to show that $g$ acts approximately like a translation on some forward invariant subdomain $\Pi$ of $W$ (as in Example 6.5.2, $\Pi$ will be the region to the right of a parabola). The subdomain $\Pi$ corresponds under the conjugacy to a petal, and $g: \Pi \to \Pi$ is shown to be conjugate a translation. We proceed now to develop these ideas. □

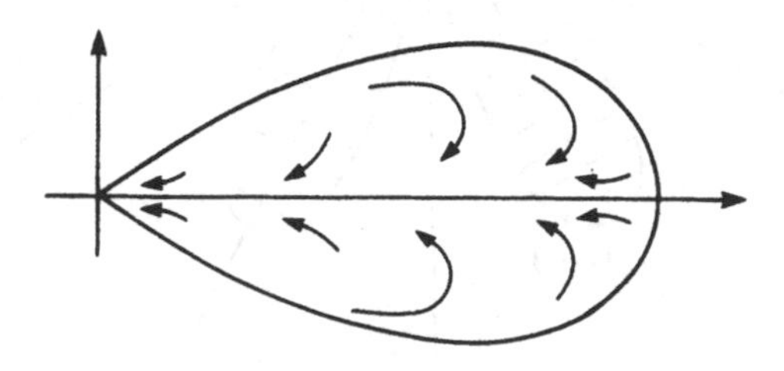

Figure 6.5.5. The dynamics on a petal.

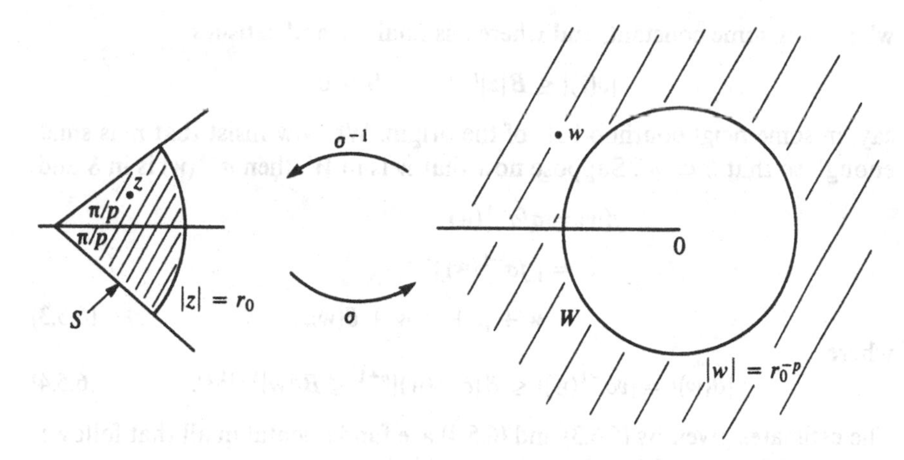


Figure 6.5.6

For each positive $r_0$, let

$$S = \{re^{i\theta}: 0 < r < r_0, |\theta| < \pi/p\}$$

and

$$W = \{re^{i\theta}: r > 1/r_0^p, |\theta| < \pi\}.$$

We assume that $r_0 < 1$, but later we shall decrease $r_0$ so as to satisfy various other conditions.

The map $\sigma: z \mapsto 1/z^p$ is an analytic map of the complex sphere onto itself, and it is also an analytic homeomorphism of $S$ onto $W$. It follows that there is an inverse map

$$\sigma^{-1}: W \to S,$$

so we can define a function $g$ on $W$ by

$$g(z) = \sigma f \sigma^{-1}(z) = [f(z^{-1/p})]^{-p}.$$

This procedure merely replaces the action of $f$ on $S$ by that of $g$ on $W$ and these maps act according to the mapping diagram

$$\begin{array}{ccc} S & \xleftarrow{\sigma^{-1}} & W \\ \downarrow f & & \downarrow g \\ \mathbb{C}_\infty & \xrightarrow[\sigma]{} & \mathbb{C}_\infty \end{array}$$

Our first task is to translate the coefficient structure of $f$ given by (6.5.1) into corresponding information about $g$. Now using (6.5.1), a computation gives the Laurent expansion of $[f(z)]^{-p}$ as

$$1/[f(z)]^p = z^{-p} + p + Az^p + v(z), \tag{6.5.2}$$

where $A$ is some constant, and where $v$ is analytic and satisfies

$$|v(z)| \leq B|z|^{p+1}, \qquad B > 0,$$

say, in some neighbourhood $\mathcal{N}$ of the origin. We now insist that $r_0$ is small enough so that $S \subset \mathcal{N}$. Suppose now that $w$ is in $W$: then $\sigma^{-1}(w)$ is in $S$ and

$$\begin{aligned} g(w) &= \sigma f \sigma^{-1}(w) \\ &= [f\sigma^{-1}(w)]^{-p} \\ &= w + p + A/w + \theta(w), \end{aligned} \tag{6.5.3}$$

where

$$|\theta(w)| = |v\sigma^{-1}(w)| \leq B|\sigma^{-1}(w)|^{p+1} \leq B/|w|^{1+(1/p)}. \tag{6.5.4}$$

The estimates given by (6.5.3) and (6.5.4) are fundamental in all that follows.

Now (6.5.3) and (6.5.4) show that if $|w|$ is large, then $g$ acts approximately like the translation $w \mapsto w + p$, and it is easy to see that if $\alpha$ is sufficiently large, $g$ maps the half-plane $\{w: \text{Re}[w] \geq \alpha\}$ into itself. However, Example 6.5.2 suggests that we can do better than this and we now show that we can replace the half-plane by a region $\Pi$ which is bounded on the left by a parabola $\gamma$.

Choose any $K$ satisfying

$$K > \max\{1/r_0^p, 3(|A| + B)\} > 1 \tag{6.5.5}$$

(as $r_0 < 1$) and let

$$\Pi = \{x + iy: y^2 > 4K(K - x)\}:$$

then $\Pi$ is bounded by a parabola, and $\Pi \subset W$: see Figure 6.5.7.

**Lemma 6.5.5.**

(a) $\Pi$ *is forward invariant under* $g$;
(b) $\text{Re}[g^n(w)] \to +\infty$ *uniformly on* $\Pi$;
(c) $g: \Pi \to \Pi$ *is conjugate to a translation.*

Lemma 6.5.5 is so closely linked to Theorem 6.5.4 that it is convenient to merge the proofs of these two results as we proceed.

Proof of Lemma 6.5.5. We write

$$w = x + iy, \qquad g(w) = X + iY, \qquad A/w + \theta(w) = a + ib, \tag{6.5.6}$$

so, from (6.5.3), we have

$$X = x + p + a, \qquad Y = y + b,$$

and hence if $w \in \Pi$,

$$\begin{aligned} Y^2 - 4K(K - X) &= (y + b)^2 - 4K(K - x - p - a) \\ &= [y^2 - 4K(K - x)] + b^2 + 2yb + 4K(a + p) \\ &\geq 4Kp - (2|yb| + 4K|a|). \end{aligned} \tag{6.5.7}$$

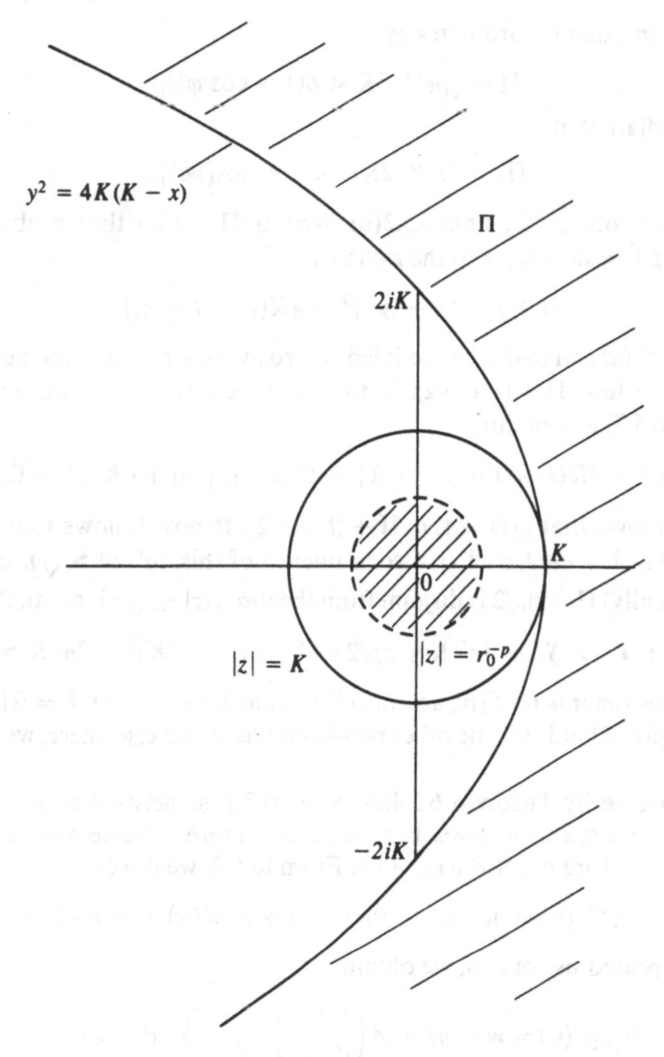


Figure 6.5.7

Assuming now that $w$ lies in $\Pi$, we have $|w| > K > 1$, and from this, (6.5.4), (6.5.5) and (6.5.6), we obtain

$$2|yb| + 4K|a| \leq 6|w||a + ib| \leq 6(|A| + B) < 2K < 4Kp.$$

Using this in (6.5.7), we find that $Y^2 - 4K(K - X) > 0$; hence $g(w) \in \Pi$, and this verifies Lemma 6.5.3(a).

To establish Theorem 6.5.2(a), we put $\Pi_0 = \sigma^{-1}(\Pi)$. It is immediate that $f$ maps $\Pi_0$ into itself, and the fact that $\Pi_0$ is a petal for a suitable $t$ is simply a matter of checking that $\sigma^{-1}$ maps the parabola to the boundary of a petal. This is simplest if polar co-ordinates are used, so we write $z = re^{i\theta}$ (in $S$) and $w = \rho e^{i\varphi}$ (in $W$), where $w = \sigma(z)$. Then $\varphi = -p\theta$ and $\rho r^p = 1$, and as $\Pi$ is expressed in polar co-ordinates by

$$\Pi = \{\rho e^{i\varphi}: 2K < \rho(1 + \cos\varphi)\},$$

it is immediate that

$$\Pi_0 = \{re^{i\theta}: 2Kr^p < 1 + \cos(p\theta)\}.$$

Next, we consider Lemma 6.5.3(b). Writing $\Pi + t$ for the set obtained by translating $\Pi$ a distance $t$ to the right, so

$$\Pi + t = \{x + iy: y^2 > 4K(K + t - x)\},$$

the proof of (a) can easily be modified to prove the stronger statement that $g$ maps $\Pi + t$ into $\Pi + (t + p/2)$. Indeed, when $z \in \Pi + t$, we can mimic the proof of (6.5.7) and obtain

$$Y^2 - 4K(K + t + p/2 - X) > 2Kp - (2|yb| + 4K|a|) > 0,$$

and this shows that $g(\Pi + t) \subset \Pi + (t + p/2)$. It now follows that if $z \in \Pi$, then $g^n(z) \in \Pi + np/2$, and as a consequence of this, $|g^n(z)| > \sqrt{n}$ on $\Pi$, for (geometrically) $\Pi + np/2$ is disjoint from the disc $\{|z| \leq \sqrt{n}\}$, or (analytically)

$$X^2 + Y^2 > X^2 + 4K(K + np/2 - X) = (X - 2K)^2 + 2npK > n.$$

This proves Lemma 6.5.5(b), and also Theorem 6.5.4(b) when $k = 0$ (virtually the same proof holds for the other petals, and here and elsewhere, we omit the details).

Next, we verify Theorem 6.5.4(c). Now (6.5.3) suggests that as a first approximation, $g^n(w) = w + np$, but we need a more delicate estimate of the asymptotic nature of $g^n(w)$ as $n \to \infty$. From (6.5.3) we have

$$g^{k+1}(w) = g(g^k w) = g^k(w) + p + A/g^k(w) + \theta(g^k w), \tag{6.5.8}$$

and by repeated use of this, we obtain

$$g^n(w) = w + np + A\left(\sum_{k=0}^{n-1} \frac{1}{g^k(w)}\right) + \sum_{k=0}^{n-1} \theta(g^k w). \tag{6.5.9}$$

We now take any compact subset $Q$ of $\Pi$ and we shall use $C_1, C_2, \ldots,$ to denote positive numbers which depend only on $Q$: for brevity, we shall implicitly assume that in any expression involving the $C_j$, the point $w$ is necessarily in $Q$. For example, if $w \in \Pi$, then

$$|A/w + \theta(w)| \leq (|A| + B)/|w| \leq (|A| + B)/K < \tfrac{1}{3},$$

so

$$\mathrm{Re}[g(w)] = \mathrm{Re}[w] + p - |A/w + \theta(w)| > \mathrm{Re}[w] + p/2.$$

From this, we obtain

$$\mathrm{Re}[g^n(w)] > \mathrm{Re}[w] + np/2$$

and so for all sufficiently large $n$,

$$|g^n(w)| \geq C_1 n. \tag{6.5.10}$$

Next, this with (6.5.4) shows that for all $n$,

$$|\theta(g^n w)| \leq B/|g^n(w)|^{1+(1/p)} \leq C_2/n^{1+(1/p)} \tag{6.5.11}$$

and, finally, from (6.5.9), (6.5.10) and (6.5.11), we obtain

$$|g^n(w) - np| \leq C_4 \log n. \tag{6.5.12}$$

This inequality controls the "vertical drift" of $g^n(z)$ as $n \to \infty$, and it is evident from this that $\arg(g^n w) \to 0$ locally uniformly on $\Pi$ as $n \to \infty$. As $g\sigma = \sigma f$ we have $g^n \sigma = \sigma f^n$, and so if $w = \sigma(z)$, then

$$g^n(w)[f^n(z)]^p = 1$$

and this yields

$$\arg(f^n z) = -(1/p)\arg(g^n w) \to 0$$

on $\Pi_0(t)$. This completes the proof of Theorem 6.5.4(c).

Next, Theorem 6.5.4(d) is a direct consequence of Lemma 6.5.2.

It remains to prove Lemma 6.5.5(c) (for Theorem 6.5.4(e) follows immediately from this) and to do this we continue to examine the asymptotic nature of $g^n(z)$. As a rough guide to how we should proceed, we replace $g^k(w)$ by $kp$ in (6.5.9) (for $k \geq 1$) and obtain

$$g^n(w) = w + np + (A/p)\log n + O(1),$$

and in view of this, we define functions $u_n$ by

$$g^n(w) = np + (A/p)\log n + u_n(w), \tag{6.5.13}$$

and proceed to prove

**Lemma 6.5.6.** *$u_n$ converges locally uniformly on $\Pi$ to some function $u$ that is holomorphic and univalent on $\Pi$.*

Assuming this for the moment, we have

$$\begin{aligned}(n+1)p + (A/p)\log(n+1) + u_{n+1}(w) &= g^{n+1}(w) \\ &= g^n(gw) \\ &= np + (A/p)\log n + u_n(gw),\end{aligned}$$

and after simplifying this and letting $n$ tend to $\infty$, we obtain

$$u(w) + p = u(gw). \tag{6.5.14}$$

As $f$ is injective near the origin, $g$ is injective on $\Pi$ (if $K$ is chosen large enough); hence so is $g^n$, and $u_n$ also. Hurwitz's Theorem now implies that $u$ is either injective or constant, and it is clearly not constant as (6.5.14) holds with $p \neq 0$. This shows that the restriction of $g$ to $\Pi$ is conjugate to the mapping $z \mapsto z + p$ of $u(\Pi)$ into itself so, subject to proving Lemma 6.5.6, we have completed the proofs of Lemma 6.5.5 and Theorem 6.5.4. □

THE PROOF OF LEMMA 6.5.6. From (6.5.13), we have

$$u_{n+1}(w) - u_n(w) = [g^{n+1}(w) - g^n(w)] - p - (A/p)\log(1 + 1/n),$$

which with (6.5.8) yields

$$\begin{aligned} u_{n+1}(w) - u_n(w) &= A/g^n(w) + \theta(g^n w) - (A/p)\log(1 + 1/n) \\ &= A[1/g^n(w) - 1/np] + \theta(g^n w) + (A/p)[1/n - \log(1 + 1/n)]. \end{aligned}$$

The proof thus reduces to showing that each of the series

$$\sum_n |1/g^n(w) - 1/np|, \qquad \sum_n |\theta(g^n w)|, \qquad \sum_n |\log(1 + 1/n) - 1/n| \tag{6.5.15}$$

converges uniformly on any compact subset $Q$ of $\Pi$.

The convergence of the second series is a direct consequence of (6.5.11), and the convergence of the third series is straightforward, for an application of the Mean Value Theorem to $x \mapsto x - \log(1 + x)$ on $[0, 1/n]$ shows that

$$|\log(1 + 1/n) - 1/n| \leq 1/n^2.$$

Finally, from (6.5.12) with the aid of (6.5.10), we obtain

$$|1/g^n(w) - 1/(np)| \leq (C_5/n^2)\log n;$$

this gives the convergence of the first series in (6.5.15) and so completes the proof of Lemma 6.5.6 and, of course, Theorem 6.5.4. □

We recall that Theorem 6.5.4 has only been proved under the assumption (6.5.1) on the coefficients of $R$. The next step is to prove the corresponding result for functions $f$ which satisfy $f(0) = 0$ and $f'(0) = 1$, but not the condition (6.5.1), and for this we need the following result.

**Theorem 6.5.7.** *Suppose that $f$ is analytic near* 0 *with*

$$f(z) = z + az^{p+1} + O(z^{p+2}), \qquad a \neq 0,$$

*as $z \to 0$. Then $f$ is conjugate near* 0 *to a function*

$$F(z) = z - z^{p+1} + O(z^{2p+1}).$$

With this available, the existence of petals (and the dynamics of $f^n$ on these petals) for analytic maps of the form

$$f(z) = z + az^{p+1} + \cdots,$$

follows immediately. Indeed, by Theorem 6.5.7, such an $f$ is conjugate to some $F$ near the origin, and Theorem 6.5.4 is applicable to $F$; thus we can define the petals for $f$ as the conformal images (under the conjugating map) of the petals for $F$. We shall not give an explicit statement of this result, and it suffices to remark that the conclusions of Theorem 6.5.4 are valid (using Lemma 6.5.2 to justify (d)) in these more general circumstances except for the fact that we no longer have an explicit expression for the petals. We do, however, know that each petal for $f$ subtends an angle $2\pi/p$ at the origin, and that the petals are pairwise disjoint (if taken small enough) for the function conjugating $f$ to $F$ is conformal there.

THE PROOF OF THEOREM 6.5.7. First, by conjugation with a suitable map $z \mapsto \lambda z$, we may assume that $a = 1$. The proof proceeds by induction over a finite number of steps, starting with the given $f$ and conjugating this to $f_1$, $f_2, \ldots$, in turn until the Taylor series has the required form.

Suppose that we have reached a stage (or we are at the initial stage) where

$$f_k(z) = z + z^{p+1} + bz^{p+r+1} + \cdots,$$

$b \neq 0$ and $1 \leq r$. If $r \geq p$ the desired conclusion holds, so we may assume that $1 \leq r < p$. Now define

$$g(z) = z + \beta z^{r+1},$$

where $\beta = b/(p - r)$, and conjugate $f_k$ with $g$ to obtain the function

$$f_{k+1} = g f_k g^{-1}.$$

We can now save a little work by noting that $f_{k+1}$ and $f_k$ have the same number of fixed points at the origin (Lemma 2.6.1), thus $f_{k+1}$ must be given by

$$f_{k+1}(z) = z + \sum_{m=p+1}^{\infty} A_m z^m.$$

We now use the fact that $f_{k+1} g = g f_k$, and express both sides as power series, working modulo addition of functions which have a zero of order at least $p + r + 2$ at the origin (alternatively, adding the $O(z^{p+r+2})$ at the end of every line). With this convention, we have

$$\begin{aligned} g f_k(z) &= (z + z^{p+1} + bz^{p+r+1} + \cdots) + \beta(z + z^{p+1} + bz^{p+r+1} + \cdots)^{r+1} \\ &= z + \beta z^{r+1} + z^{p+1} + [b + \beta(1 + r)]z^{p+r+1} \\ &= z + \beta z^{r+1} + z^{p+1} + \beta(1 + p)z^{p+r+1}, \end{aligned}$$

and

$$f_{k+1} g(z) = (z + \beta z^{r+1}) + \sum_{m=p+1}^{\infty} A_m (z + \beta z^{r+1})^m.$$

Still working modulo functions of order $z^{p+r+2}$ at the origin, we now have

$$\begin{aligned} z^{p+1} + \beta(1 + p)z^{p+r+1} &= \sum_{m=p+1}^{\infty} A_m (z + \beta z^{r+1})^m \\ &= A_{p+1} z^{p+1} + \cdots + A_{p+r+1} z^{p+r+1} \\ &\quad + A_{p+1} \beta(p + 1) z^{p+r+1}, \end{aligned}$$

and from this we see that

$$A_{p+1} = 1, \qquad A_{p+2} = \cdots = A_{p+r} = A_{p+r+1} = 0.$$

It follows that $f_k$ is conjugate (near the origin) to the map

$$f_{k+1}(z) = z + z^{p+1} + O(z^{p+r+2}),$$

and this is the general inductive step.

It may be that the series for $f_{k+1}$ terminates as a polynomial of degree less than $2p + 2$, and if this happens the proof is complete. Otherwise, the argument given above remains valid while $r < p$, and so will be valid up to and including the case $r = p - 1$, when $p + r + 2 = 2p + 1$. We deduce that $f$ is conjugate to some map

$$z \mapsto z + z^{p+1} + O(z^{2p+1}),$$

which itself is conjugate (by a map $z \mapsto \lambda z$) to a function of the form $F$. The proof is complete. □

Theorems 6.5.4 and 6.5.7 describe the action of the iterates of any map of the form

$$f(z) = z + az^{p+1} + \cdots$$

on the $p$ petals of $f$ at the origin. We now consider the case when $f$ is a rational map $R$ and obtain information about the relationship of the petals to the Fatou set of $R$.

**Theorem 6.5.8.** *Let $R$ be a rational map and suppose that*

$$R(z) = z + az^{p+1} + \cdots, \qquad a \neq 0,$$

*near the origin. Let $\Pi_j$ be the petals of $R$ and, for each $j$, let $F_j$ be the component of $F(R)$ that contains $\Pi_j$. Then:*

(a) *$R^n(z) \to 0$ and $\arg R^n(z) \to 2\pi k/p$ on $F_k$; and*
(b) *$F_0, \ldots, F_{p-1}$ are distinct components of $F(R)$.*

The geometric picture, then, is as follows. The rationally indifferent fixed point 0 lies on the boundary of $p$ components $F_k$ of $F(R)$, and each $F_k$ contains a petal $\Pi_k$ which makes an angle $2\pi/p$ at 0. Next, $R$ maps each $F_k$ into itself, and $R^n \to 0$ locally uniformly on each $F_k$ in such a way that for any $z$ in $F_k$, $R^n(z) \to 0$ on a path which is asymptotic to the axis of the petal $\Pi_k$. It is an immediate consequence of this that different petals lie in different components of $F(R)$. In addition, this convergence shows that for any $z$, $R^n(z)$ ultimately lies within $\Pi_k$, so the consecutive inverse images of the petal $\Pi_k$ in $F_k$ expand to fill the entire component $F_k$.

Finally, we note that a rationally indifferent fixed point may also lie on the boundary of components of $F$ other than the $F_j$: indeed, if $P(z) = z + \cdots$ is any non-linear polynomial, then the origin is not only on the boundary of each of the $p$ components $F_j$, but it is also on the boundary of the completely invariant component $F_\infty$ of $F$ which contains $\infty$ (and which is distinct from

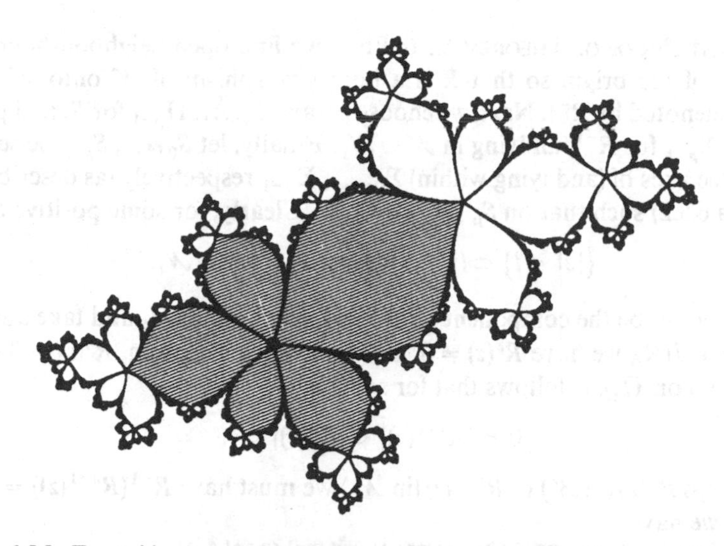

Figure 6.5.8. Fractal image reprinted with permission from *The Beauty of Fractals* by H.-O. Peitgen and P. H. Richter, 1986, Springer-Verlag, Heidelberg, New York.

each $F_j$ because $P^n \to \infty$ on $F_\infty$). As $F_\infty$ is connected, it must therefore contain a thin "tongue" which stretches to the origin *between* the petals: see Figure 6.5.8 (where $p = 5$) and Figure 6.5.9 (where $p = 3$) in which the petals are coloured black.

Theorem 6.5.8 is a fundamental result and we shall give several proofs of it. In each of these the specific shape of the petals is irrelevant so when we refer to Theorem 6.5.4, strictly speaking we mean the (unstated) version of this after it has been modified in the light of Theorem 6.5.7.

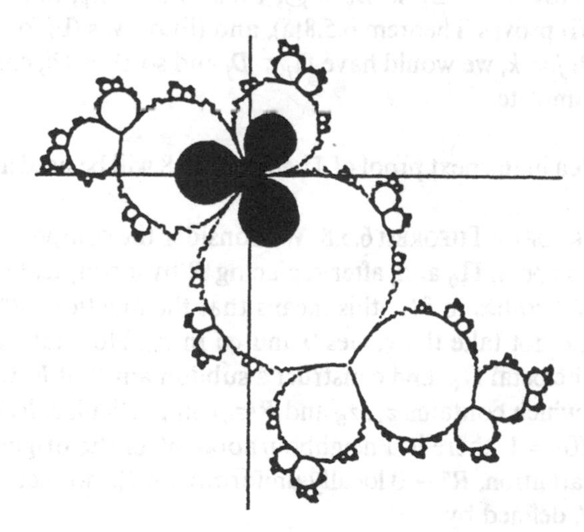

Figure 6.5.9. $z \mapsto e^{2\pi i/3}z + z^2$. Reprinted with permission of the American Mathematical Society.

THE FIRST PROOF OF THEOREM 6.5.6. First, we find open neighbourhoods $\mathcal{N}$ and $\mathcal{N}_1$ of the origin so that $R$ is a homeomorphism of $\mathcal{N}$ onto $\mathcal{N}_1$ with inverse denoted by $R^{-1}$. Next, we choose petals $\Pi_0, \ldots, \Pi_{p-1}$ for $R$, and petals $\Sigma_0, \ldots, \Sigma_{p-1}$ for $R^{-1}$, all lying in $\mathcal{N} \cup \mathcal{N}_1$. Finally, let $S_0, \ldots, S_{p-1}$ be sectors about the axes of (and lying within) $\Sigma_0, \ldots, \Sigma_{p-1}$ respectively (as described in Lemma 6.5.2) such that on $S_j$, $|R^{-1}(z)| < |z|$. Clearly, for some positive $t$,

$$\{|z| < t\} \subset (\bigcup \Pi_j) \cup (\bigcup S_j) \subset \mathcal{N} \cup \mathcal{N}_1.$$

Now let $F_k$ be the component of $F(R)$ which contains $\Pi_k$ and take any $z$ in $F_k$. As $0 \in J(R)$, we have $R^n(z) \neq 0$, and by Vitali's Theorem, $R^n(z) \to 0$ on $F_k$ (as $R^n \to 0$ on $\Pi_k$). It follows that for some $n$,

$$0 < |R^{n+1}(z)| < |R^n(z)| < t.$$

As $R$ maps $R^n(z)$ (in $\mathcal{N}$) to $R^{n+1}(z)$ (in $\mathcal{N}_1$) we must have $R^{-1}(R^{n+1}(z)) = R^n(z)$ and so we have

$$|R^{n+1}(z)| < |R^n(z)| = |R^{-1}(R^{n+1}(z))|.$$

This means that $R^{n+1}(z)$ is not in any $S_j$; thus it is in some $\Pi_j$ and we have proved that

$$F_k = D_0 \cup \cdots \cup D_{p-1},$$

where

$$D_j = \{z \in F_k\text{: for some } n,\ R^n(z) \in \Pi_j\}.$$

Of course, the sets $D_j$ are open subsets of $F_k$, and they are pairwise disjoint because the $\Pi_j$ are disjoint and each is forward invariant. It follows that $F_k$ is the disjoint union of the open sets $D_j$ and as $F_k$ is connected, all but one of the $D_j$ are empty. Now $\Pi_k \subset D_k$ so $D_k \neq \emptyset$, thus (i) $F_k = D_k$, and (ii) for $j \neq k$, $D_j = \emptyset$. Now (i) proves Theorem 6.5.8(a), and (ii) proves (b) for if $F_k$ were to contain $\Pi_j$ with $j \neq k$, we would have $\Pi_j \subset D_j$ and so $D_j \neq \emptyset$, contrary to (ii). The proof is complete. □

The main idea in the next proof of Theorem 6.5.8 will be used again in §7.3.

THE SECOND PROOF OF THEOREM 6.5.8. We consider the component $F_0$ of $F(R)$ that contains the petal $\Pi_0$ and, after replacing $R$ by a conjugate function, we may assume that $\infty$ lies in $\partial F_0$: this means that the functions $R^n$ (which map $F_0$ into itself) do not take the values 0 and $\infty$ in $F_0$. Now take any $z_1$ in $F_0$, and any $z_0$ in the petal $\Pi_0$, and construct a subdomain $V$ of $F_0$ whose closure lies in $F_0$, and which contains $z_1$, $z_0$ and $R(z_0)$: in particular, $R(V)$ meets $V$.

Next, as $R'(0) = 1$, there is a neighbourhood $\mathcal{N}$ of the origin on which $R$ is injective. In addition, $R^n \to 0$ locally uniformly on $F_0$, so there is an integer $N$ such that $W$, defined by

$$W = R^N(V) \cup R^{N+1}(V) \cup R^{N+2}(V) \cup \cdots,$$

is a forward invariant, connected, subdomain of $\mathcal{N} \cap F_0$. Now let $R^N(z_1) = \zeta_0$

(this lies in $\Pi_0 \cap W$) and for $n \geq N$, define the functions $\varphi_n$ on $W$ by

$$\varphi_n(z) = R^n(z)/R^n(\zeta_0).$$

We prove

**Lemma 6.5.9.** *$\{\varphi_n\}$ is normal in $W$.*

Proof. First, $\varphi_n$ does not take the values 0 and $\infty$ in $W$ (because $R^n$ does not). Next, as $W \subset \mathcal{N}$, $R$ is injective on $W$, and as $R(W) \subset W$, so too is each $R^n$: thus each $\varphi_n$ is injective on $W$. Finally, as $\varphi_n(\zeta_0) = 1$, none of the $\varphi_n$ takes any of the values $0, \infty, 1$ in $W - \{\zeta_0\}$, and so certainly, $\{\varphi_n\}$ is normal in $W - \{\zeta_0\}$.

To prove that $\{\zeta_0\}$ is normal in $W$, we need only prove that it is normal near $\zeta_0$. By normality, we can find a sequence $\varphi_{n_j}$ which converges locally uniformly to some function $\varphi$ on $W - \{\zeta_0\}$. Now choose a closed disc centred at $\zeta_0$, and lying in $F_0$, and let $C$ be its bounding circle and $D$ its interior. Now as $n_j \to \infty$,

$$\varphi_n(z) = \frac{1}{2\pi i}\int_C \frac{\varphi_n(w)}{w - z}\,dw \to \frac{1}{2\pi i}\int_C \frac{\varphi(w)}{w - z}\,dw,$$

and this convergence is uniform near $\zeta_0$. The integral on the right is analytic in $D$ with value 1 at $\zeta_0$ and, by definition, it is $\varphi(z)$ when $z \neq \zeta_0$. It follows that $\varphi_{n_j}$ now converges locally uniformly to $\varphi$ throughout $W$, where $\varphi(z_0) = 1$, and the proof of the lemma is complete. □

We continue with the second proof, and we work with the function $\varphi$ used in the proof of the lemma. As the functions $\varphi_{n_j}$ are injective in $W$, Hurwitz's Theorem implies that $\varphi$ is either univalent or constant in $W$. Now for all $n$,

$$\varphi_n(R(z)) = \varphi_n(z)\left(\frac{R^n(Rz)}{R^n(z)}\right) = \varphi_n(z)\left(\frac{R(R^n z) - R(0)}{R^n z - 0}\right),$$

and letting $n \to \infty$ through the sequence $n_j$, we obtain

$$\varphi(Rz) = R'(0)\varphi(z) = \varphi(z).$$

This means that $\varphi$ is not univalent in $W - \{\zeta_0\}$ (for $W$ is forward invariant and $R^n \to 0$): thus $\varphi$ is constant and so for all $z$ in $W$,

$$\varphi(z) = \varphi(\zeta_0) = 1.$$

We deduce that as $n \to \infty$ in $(n_j)$,

$$R^n(z_0)/R^n(\zeta_0) \to 1,$$

and so given any positive $\varepsilon$, for sufficiently large $n$ in $(n_j)$,

$$|R^n(z_0) - R^n(\zeta_0)| < \varepsilon |R^n(z_0)|.$$

This means that the angle subtended at the origin by the line segment $[R^n(z_0), R^n(\zeta_0)]$ tends to zero as $n \to \infty$ in $(n_j)$, and as $R^n(z_0)$ approaches the origin asymptotically along the axis of $\Pi_0$, the same is true of $R^n(\zeta_0)$. This proves Theorem 6.5.8(a), and (b) follows easily as before. □

Our third proof of Theorem 6.5.6 is only sketched, and it is included simply to give more geometric insight into these ideas.

THE SKETCH OF A THIRD PROOF OF THEOREM 6.5.8. We begin by proving (b) in the case $p = 2$ (it will be evident that only trivial modifications are required when $p \geq 3$, and there is nothing to prove when $p = 1$). By conjugation, we may assume that $a = -1$, and as $p = 2$ there are two petals $\Pi_0$ and $\Pi_1$ of $R$ whose axes lie along the positive and negative parts of the real axis respectively. Next, let $\mathcal{N}$ be a disc centred at the origin whose diameter is less than the $\delta$ described in Theorem 2.3.4, and on which $R$ is injective.

We argue by contradiction, so we assume that $\Pi_0$ and $\Pi_1$ lie in the same component, say $F_0$, of $F(R)$ and then construct a closed curve $\gamma$ which lies entirely in $F_0 \cup \{0\}$, and which is as illustrated in Figure 6.5.10, the essential point being that $\gamma$ contains some small real segment $(-\varepsilon, \varepsilon)$. Observe now that $R^n \to 0$ uniformly on $\gamma$: indeed, the convergence is uniform on the segments $(-\varepsilon, 0)$ and $(0, \varepsilon)$ of $\gamma$ (for these lie in the petals), and also on the remainder of $\gamma$ (which is a compact subset of $F_0$ on which $R^n \to 0$).

Because of the uniform convergence of $R^n$ to zero on $\gamma$, we may relabel some $R^m(\gamma)$ and $\gamma$ and so assume:

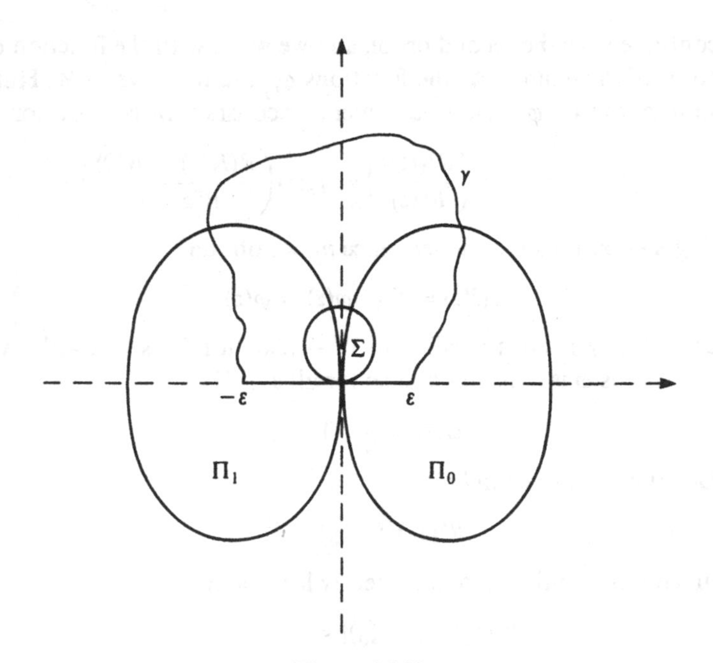


Figure 6.5.10

(i) each closed curve $R^n(\gamma)$, $n \geq 0$, lies in $\mathcal{N}$.

Further, by replacing $\gamma$ (which now lies in $\mathcal{N}$) by a collection of its subarcs, we may assume:

(ii) $\gamma$ is a simple closed curve,

and finally, because of the known nature of the convergence on the petals, a minor modification to our new $\gamma$ enables us to retain the assumption:

(iii) $\gamma$ contains some small real segment $(-\varepsilon, \varepsilon)$.

With these, it follows by induction that every $R^n(\gamma)$ is a simple closed curve in $\mathcal{N}$ with interior $D_n$, say. Also, by Theorem 2.3.4, $R$ maps each $D_n$ into $D_{n+1}$ and as $R^n \to 0$ uniformly on $\gamma$, this shows that $R^n \to 0$ uniformly on both $\gamma$ and its interior $D_0$.

Finally, $R^{-1}$ exists and is injective on some neighbourhood of the origin, and it has two petals whose axes lie along the positive and negative parts of the imaginary axis. It follows from (iii) that if these petals are taken small enough, then one of them, which we denote by $\Sigma$, lies inside $D_0$ (as illustrated in Figure 6.5.10). Now take any $w$ in $\Sigma$. As $R^{-1}$ maps $\Sigma$ into itself, $R^{-n}(w)$ is also in $\Sigma$, and so

$$\sup\{|R^n(z)|: z \in D_0\} \geq |R^n(R^{-n}(w))| = |w|:$$

as this implies that $R^n$ does not converge uniformly to 0 on $D_0$, it is the desired contradiction and (b) follows.

Finally, we use the hyperbolic metric to prove Theorem 6.5.8(a) when $p \geq 2$ (and we only sketch the main ideas). As Theorem 6.5.8(b) holds, it follows that near the origin, the component $F_0$ of $F$ which contains $\Pi_0$ is approximately a wedge $W$ of angle $2\pi/p$ (that is to say, for some small $r$ and small $\varepsilon$, there is some component of $F_0 \cap \{|z| < r\}$ which contains a small sector of angle $2\pi(1-\varepsilon)/p$, and which is contained in a sector of angle $2\pi(1+\varepsilon)/p$). Because of this, the hyperbolic metric $\rho_0$ of $F_0$ is, near the origin, approximately the same as the hyperbolic metric for the wedge $W$ (see, for example, Appendix IV and [20]).

Now take any point $z$ in $F_0$ and any $z_0$ in the petal $\Pi_0$. As $R$ maps $F_0$ onto itself, $R$ does not increase $\rho_0$-distances (this is the general version of the Schwarz–Pick Lemma) so for all $n$,

$$\rho_0(R^n z, R^n z_0) \leq \rho_0(z, z_0) = d,$$

say. Now $R^n(z_0) \to 0$ asymptotically along the axis of the petal, and so we deduce that as $n \to \infty$, the points $R^n(z)$ ultimately approach 0, *staying no more than a $\rho_0$-distance, say 2d, away from the axis of* $\Pi_0$. If $\rho_0$ were to be replaced here by the hyperbolic metric $\rho_1$ of $W$, this would mean that the points $R^n(z)$ approach the origin within a strictly smaller wedge that $W$, and hence ultimately lie within the petal $\Pi_0$, and the fact that the metrics $\rho_0$ and $\rho_1$ are asymptotically the same as the origin is approached means that the same is true for the metric $\rho_0$. We omit the details, and this completes our discussion of Theorem 6.5.8 and its proofs. □

There is still one more result that we must consider before we can give the Petal Theorem in its full generality, namely the extension of Theorem 6.5.8 to the case when $R(0) = 0$ and $R'(0)$ is a root of unity other than 1. We assume, then, that

$$R(z) = az + bz^{p+1} + \cdots, \tag{6.5.16}$$

where $a \neq 1$ but $a = \exp(2\pi ir/q)$ where $r$ and $q$ are coprime (so $a^n = 1$ when $n = q$ but not for any smaller $n$). In this case, $R^q$ is of the form

$$R^q(z) = z + cz^{t+1} + \cdots$$

for some $t$, so $R^q$ has $t$ petals at the origin. This means that there are components $F_1, \ldots, F_t$ of $F(R^q)$, and hence of $F(R)$, each one containing a petal of $R^q$. As $R$ maps each $F_j$ to some component of $F(R)$ which makes an angle $2\pi/t$ at the origin, we see that $R$ must act as a permutation of the set $\{F_1, \ldots, F_t\}$. Further, as $a = \exp(2\pi ir/q)$ with $(r, q) = 1$, we find (from Theorem 6.5.4 applied to $R^q$) that $R$ (viewed as a permutation of the $F_j$) is a composition of, say, $k$ disjoint cycles each of length $q$, and it follows that $t = kq$. Thus, given $R$ in the form (6.5.16), there is an integer $k$ such that $R$ has $kq$ petals at the origin, these dividing naturally into $k$ sets of $q$ petals such that $R$ acts as a cycle of length $q$ on each such set. We shall not give a formal statement of this as it is a special case of our final version of the Petal Theorem which follows very shortly.

Observe that if $R$ is given by (6.5.16), then

$$R^n(z) = a^n z + \cdots,$$

so if $q$ does not divide $n$, then $R^n$ has only one fixed point at the origin. If $q$ does divide $n$, say $n = mq$, then

$$R^q(z) = z + cz^{t+1} + \cdots$$

so

$$\begin{aligned} R^n(z) &= (R^q)^m(z) \\ &= z + mcz^{t+1} + \cdots \end{aligned}$$

and $R^n$ has $t + 1$ $(= kq + 1)$ fixed points at the origin.

We turn now to the final and most comprehensive version of the Petal Theorem. Let $\{\zeta_1, \ldots, \zeta_m\}$ be a rationally indifferent cycle of $R$, and write

$$\zeta_{n+i} = R^n(\zeta_i)$$

for all $n$, $n \geq 1$, so that $\zeta_i = \zeta_j$ if and only if $i \equiv j \bmod m$. Essentially the only difference between this case and the previous one is that now, $R$ maps each component of $F(R)$ that contains a petal at $\zeta_j$ to some other component containing a petal at $\zeta_{j+1}$.

Of course, the previous version applies to the map $R^m$ at each point $\zeta_j$ of the cycle, so we now put $S = R^m$ and suppose that

$$\begin{aligned} S(z) &= R^m(z) \\ &= \zeta_j + \exp(2\pi ir/q)(z - \zeta_j) + b(z - \zeta_j)^{p+1} + \cdots, \end{aligned} \tag{6.5.17}$$

where $(r, q) = 1$. Note that the coefficient $\exp(2\pi ir/q)$ is independent of the choice of $j$ here as it is the multiplier of the cycle. In keeping with the previous case, we also assume that

$$\begin{aligned} S^q(z) &= R^{mq}(z) \\ &= \zeta_j + (z - \zeta_j) + c(z - \zeta_j)^{t+1} + \cdots, \end{aligned} \tag{6.5.18}$$

where (as before, but using $S$ instead of $R$) $t = kq$ for some $k$. Note that $S^q$ acting near $\zeta_j$ is conjugate (by the branch of $R^{-1}$ which maps $\zeta_{j+1}$ to $\zeta_j$) to $RS^qR^{-1}$ $(= S^q)$ acting near $\zeta_{j+1}$, and so $S^q$ has the same number of fixed points at $\zeta_j$ as it has at $\zeta_{j+1}$: this proves that $t$ in (6.5.18), and hence $k$ also, is independent of the choice of $\zeta_j$.

Now $S$ has $qk$ $(= t)$ petals at each $\zeta_j$, hence associated with the cycle $\{\zeta_1, \ldots, \zeta_m\}$ are $mkq$ components $F_1, \ldots, F_{mkq}$ of $F(R)$. The action of $R$ is to map $F_1$, say, with a petal at $\zeta_1$, to $F_2$, say, with a petal at $\zeta_2$, then $F_2$ to $F_3$ with a petal at $\zeta_3$, and so on until $F_{m+1}$ with a petal at $\zeta_{m+1}$ $(= \zeta_1)$. However, $F_{m+1} \neq F_1$ (unless $q = 1$) for $S$ maps $F_1$ to $F_{m+1}$. In this way, $R$ acts as a permutation of $\{F_1, \ldots, F_{mkq}\}$ as the product of $k$ cycles each of length $mq$. Thus we have the next (and final) version of the Petal Theorem.

**Theorem 6.5.10.** *Let $\{\zeta_1, \ldots, \zeta_m\}$ be a rationally indifferent cycle for $R$, and let the multiplier of $R^m$ at each point of the cycle be* $\exp(2\pi ir/q)$, *where* $(r, q) = 1$. *Then there exists an integer $k$, and $mkq$ distinct components $F_1, \ldots, F_{mkq}$ of $F(R)$ such that at each $\zeta_j$, there are exactly $kq$ of these components containing a petal of angle $2\pi/kq$ at $\zeta_j$. Further, $R$ acts as a permutation $\tau$ on $\{F_1, \ldots, F_{mkq}\}$, where $\tau$ is a composition of $k$ disjoint cycles of length $mq$, and a petal based at $\zeta_j$ maps under $R$ to a petal based at $\zeta_{j+1}$.*

More information can be found in §6.10, especially Theorem 6.10.3 and (6.10.3).

We end with a count of the number of fixed points of $R$ in the cycle described in the Petal Theorem: this will be needed in §6.8, and it is convenient to prove it here while the relevant ideas are still fresh in the reader's mind. We recall that $q|n$ means that $q$ divides $n$, while $q\nmid n$ means that it does not.

**Theorem 6.5.11.** *Suppose that the hypotheses of Theorem* 6.5.10 *and also* (6.5.17), (6.5.18) *hold.*

(i) *If $m \nmid n$, then $R^n$ has no fixed points at $\zeta_j$.*
(ii) *If $m|n$ but $mq \nmid n$, then $R^n$ has one fixed point at $\zeta_j$.*
(iii) *If $mq|n$, then $R^n$ has $t + 1$ fixed points at $\zeta_j$.*

PROOF. First, (i) is clear for if $m \nmid n$, then $R^n$ does not fix any $\zeta_j$. Now suppose that $m|n$ and write $n/m = uq + v$, where $0 \leq v < q$.

To prove (ii) we assume that $mq \nmid n$; then $v \neq 0$ so $1 \leq v < q$. As $(r, q) = 1$, this means that $rv/q$ is not an integer so $\exp(2\pi irv/q) \neq 1$. We deduce that

$$\begin{aligned} R^n(z) &= S^{uq+v}(z) \\ &= \zeta_j + \exp(2\pi ir[uq + v]/q)(z - \zeta_j) + \cdots \\ &= \zeta_j + \exp(2\pi irv/q)(z - \zeta_j) + \cdots, \end{aligned}$$

and so (ii) follows.

Finally, if $mq|n$, then $v = 0$ so $n = umq$, in which case

$$\begin{aligned} R^n(z) &= (S^q)^u(z) \\ &= \zeta_j + (z - \zeta_j) + uc(z - \zeta_j)^{t+1} + \cdots \end{aligned}$$

which gives (iii). □

EXERCISE 6.5

1. Investigate (from first principles) the dynamics of the iterates of $z \mapsto z(1 - z^N)$ near the origin.
2. Let $P(z) = -z + z^{p+1}$, where $p$ is a positive integer. How many petals does $P$ have at the origin?
3. Use the method given in the proof of Theorem 6.5.7 to prove that the number of fixed points of an analytic map $f$ at a fixed point is conjugation invariant.
4. Consider the function $g$ defined in the proof of Theorem 6.5.4, and let

$$\Sigma = \{z: |z| > R, |\arg z| < \varphi\},$$

where $\varphi \in (0, \pi/2)$ and $R$ is chosen large enough so that $\Sigma \subset \Pi$. By applying the Cosine Rule to the triangle with vertices 0, $w$ and $w + p$, show that if $w$ is in $\Sigma$ then

$$(|w + p| - |w|)(|w + p| + |w|) \geq 2p|w| \cos \varphi.$$

Deduce that $|w| < |g(w)|$ on $\Sigma$, and interpret this in terms of $f$.

# §6.6. Irrationally Indifferent Cycles in $F$

We recall an earlier definition.

**Definition 6.6.1.** An analytic map $f$ is *linearizable* near a fixed point $\zeta$ if $f$ is conjugate to

$$f_0: z \mapsto \zeta + (z - \zeta)f'(\zeta)$$

in some neighbourhood of $\zeta$: explicitly, if there is some neighbourhood $N$ of

$\zeta$, and some map $g$ which fixes $\zeta$, such that $f$ is analytic in $N$, $g$ is analytic and injective in $N \cup f(N)$, and $gfg^{-1} = f_0$ on $g(N)$.

We have seen that a rational map $R$ is linearizable near each attracting, and each repelling, fixed point. Near an indifferent fixed point $\zeta$, $R$ is or is not linearizable depending *only* on whether $\zeta$ is in $F$ or $J$, and this is our next result.

**Theorem 6.6.2.** *Let $\zeta$ be an indifferent fixed point of a rational map $R$. Then $R$ is linearizable near $\zeta$ if and only if $\zeta$ lies in $F(R)$.*

PROOF. First, we assume that $R$ is linearizable near an indifferent fixed point $\zeta$, so $gRg^{-1} = \varphi$, where $\varphi$ is a Euclidean rotation about $\zeta$, and where $g$ fixes, and is analytic near, $\zeta$. For each sufficiently small disc $D$ centred at $\zeta$, $\varphi(D) = D$, so $R$ maps $g^{-1}(D)$ onto itself. We deduce that $\{R^n\}$ is normal in $g^{-1}(D)$, and so $\zeta$ is in $F$.

We now assume that $\zeta \in F$, and prove that $R$ is linearizable near $\zeta$. We may assume that $\zeta = 0$, so there is a neighbourhood of the origin on which $\{R^n\}$ is equicontinuous and from this, we see that there is some neighbourhood, say $N$, of the origin on which for all $n$,

$$|R^n(z)| = |R^n(z) - R^n(0)| < 1. \tag{6.6.1}$$

Now write $\alpha = R'(0)$, so $|\alpha| = 1$, and for $n \geq 1$, define the functions $T_n$ by

$$T_n(z) = [z + R(z)/\alpha + \cdots + R^{n-1}(z)/\alpha^{n-1}]/n.$$

Note that as $|\alpha| = 1$ and $(R^k)'(0) = \alpha^k$, we have $T_n'(0) = 1$ and

$$|T_n(z)| \leq 1 \tag{6.6.2}$$

on $N$. The functions $T_n$ satisfy

$$\begin{aligned}(n/\alpha)T_n(Rz) + z &= (n+1)T_{n+1}(z)\\ &= nT_n(z) + R^n(z)/\alpha^n,\end{aligned}$$

and using this, $|\alpha| = 1$, and (6.6.1), we see (after dividing through by $n$) that

$$T_n(Rz) - \alpha T_n(z) \to 0 \tag{6.6.3}$$

uniformly on $N$ as $n \to \infty$.

Next, (6.6.2) implies that $\{T_n\}$ is normal in $N$ and it follows that as $n \to \infty$ on some sequence of integers, $T_n$ converges locally uniformly on $N$ to some analytic function $g$ which, by (6.6.3), satisfies

$$g(Rz) = \alpha g(z)$$

there. Because $T_n'(0) = 1$, we also have $g'(0) = 1$ so $g$ is not constant, and this completes the proof. □

With reference to Theorem 6.6.2, we know that there can be indifferent fixed points of $R$ in $J(R)$ (for example, rationally indifferent fixed points), and we shall soon see that there can be indifferent fixed points in $F(R)$. Ignoring (for the moment) this question of existence, we proceed to characterize the indifferent fixed points in the Fatou set by a method which gives an alternative proof of Theorem 6.6.2. We prove

**Theorem 6.6.3.** *Let $R$ be a rational map of degree $d$, $d \geq 2$, and suppose that $\zeta$ is an indifferent fixed point lying in some component $F_0$ of $F(R)$. Then $F_0$ is simply connected, and $R\colon F_0 \to F_0$ is analytically conjugate to a rotation of infinite order of the unit disc $\Delta$.*

Any component $F_0$ of this type is called a *Siegel disc* after C.L. Siegel who, in 1941, was the first to establish their existence (see [89] and [90]). We shall verify their existence here, and for this we need the following two results.

**Theorem 6.6.4.** *Suppose that $|\alpha| = 1$, and that for some positive $M$ and $k$, and all positive integers $n$,*

$$1/|\alpha^n - 1| \leq Mn^k. \tag{6.6.4}$$

*Then any analytic function $f$ with Taylor expansion*

$$f(z) = \alpha z + a_2 z^2 + \cdots$$

*is linearizable near the origin.*

**Theorem 6.6.5.** *Let $S$ be the set of numbers $\alpha$ which satisfy $|\alpha| = 1$ and (6.6.4) for some $M$ and $k$, and all $n$. Then $S$ has measure $2\pi$.*

These imply the existence of Siegel discs, for Theorem 6.6.5 shows that there is some $\alpha$ in $S$, and Theorem 6.6.4 then implies that $P\colon z \mapsto \alpha z + z^2$ is linearizable near the origin. It follows from Theorem 6.6.2 that $0 \in F(P)$ and finally, by Theorem 6.6.3, that the origin is the centre of a Siegel disc for $P$. Much work has been done to find necessary and sufficient conditions on $\alpha$ for $f$ to be linearizable and we refer the reader to [55] for more details.

The remainder of this section consists of the proofs of Theorems 6.6.3, 6.6.4 and 6.6.5: of course, the proof of Theorem 6.6.5 has nothing to do with iteration and involves only measure theory.

The Proof of Theorem 6.6.3. By conjugation we may assume that $\zeta = 0$ so, near the origin,

$$R(z) = az + bz^2 + \cdots,$$

where $|a| = 1$. Further, as rationally indifferent fixed points are in $J(R)$, we see that $a^n \neq 1$ for any positive integer $n$.

As $J$ is infinite, the universal covering space of $F_0$ is the unit disc $\Delta$ (see Appendix IV to this chapter); thus there is an analytic map (the universal

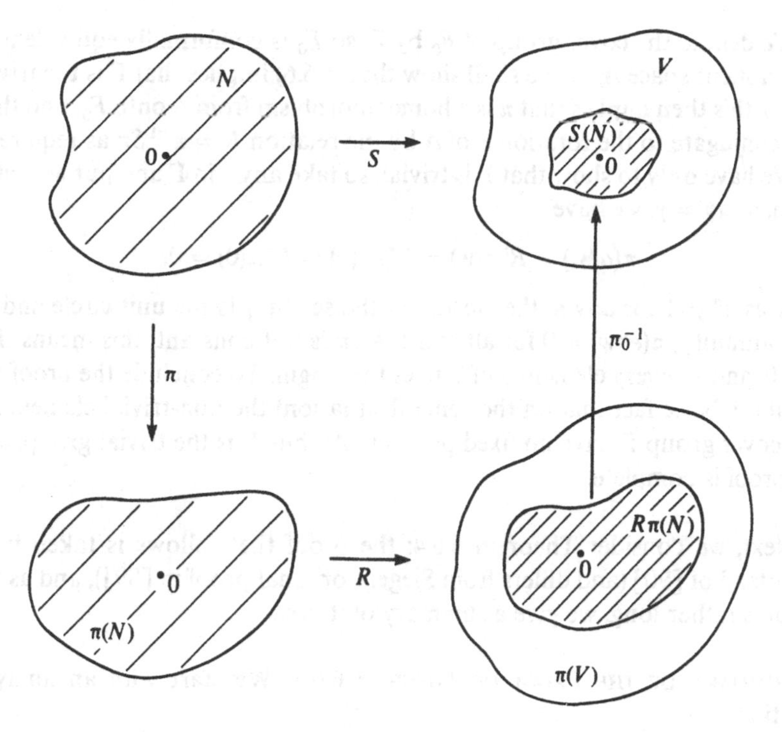


Figure 6.6.1

covering map) $\pi$ of $\Delta$ onto $F_0$ which is locally injective on $\Delta$ with $\pi(0) = 0$. We now choose a neighbourhood $V$ of 0 on which $\pi$ is injective and let $\pi_0$ be the restriction of $\pi$ to $V$ with inverse $\pi_0^{-1}: \pi(V) \to V$. By continuity, there is a small neighbourhood $N$ of the origin which is mapped by $R\pi$ into $\pi(V)$ so we can create the function

$$S = (\pi_0)^{-1} R\pi: N \to \Delta,$$

where, of course, $\pi S = R\pi$ in $N$: see Figure 6.6.1.

The properties of the universal covering map $\pi$ ensure that analytic continuation of $S$ is possible over every curve in $\Delta$, and as $\Delta$ is simply connected, the Monodromy Theorem guarantees that this continuation (which we continue to denote by $S$) is a single valued analytic map of $\Delta$ into itself. Near the origin,

$$\pi S = R\pi, \tag{6.6.5}$$

and so, by analytic continuation, this holds throughout $\Delta$.

We are interested in two consequences of (6.6.5). The first of these is $S'(0) = a$, and as $|a| = 1$, Schwarz's Lemma now tells us that $S(z) = az$. We now combine this with the second consequence of (6.6.5), namely $\pi S^n = R^n \pi$, and so obtain

$$\pi(a^n z) = R^n \pi(z). \tag{6.6.6}$$

We denote the cover group of $F_0$ by $\Gamma$, so $F_0$ is conformally equivalent to the quotient space $\Delta/\Gamma$. We shall show that (6.6.6) implies that $\Gamma$ is the trivial group: this then implies that $\pi$ is a homeomorphism from $\Delta$ onto $F_0$, and then $R$ is conjugate to the rotation $S$ of $\Delta$ by the relation $R = \pi^{-1}S\pi$ as required.

We have only to show that $\Gamma$ is trivial, so take any $\gamma$ in $\Gamma$ and put $w = \gamma(0)$. Then, as $\pi\gamma = \gamma$, we have

$$\pi(a^n w) = R^n\pi(w) = R^n\pi\gamma(0) = R^n\pi(0) = 0.$$

Now as $a^n \neq 1$ for any $n$, the closure of the set $\{a^n\}$ is the unit circle and so by continuity, $\pi(e^{it}w) = 0$ for all real $t$. As $\pi$ is not constant, this means that $w = 0$, and so every element $\gamma$ of $\Gamma$ fixes the origin. To conclude the proof, we recall the basic fact that (in the general situation) the non-trivial elements of any cover group $\Gamma$ have no fixed points in $\Delta$: thus $\Gamma$ is the trivial group, and the proof is complete. □

Next, we consider Theorem 6.6.4: the proof that follows is taken from Chapter 3 of [90] (and differs from Siegel's original proof in [89]), and as the proof is rather long, we give a summary of it first.

A SUMMARY OF THE PROOF OF THEOREM 6.6.4. We start with an analytic function

$$f(z) = \alpha z + a_2 z^2 + \cdots, \tag{6.6.7}$$

where $\alpha \in S$, so there is some positive constant $C$ $(= Mk!)$ such that for all $n$,

$$1/|\alpha^n - 1| \leq (C/k!)n^k. \tag{6.6.8}$$

Our objective is to construct a function

$$\varphi(z) = z + b_2 z^2 + \cdots, \tag{6.6.9}$$

which is analytic, and which satisfies

$$\varphi^{-1}f\varphi(z) = \alpha z,$$

in some neighbourhood of the origin. Given $f$ as in (6.6.7), we create the *formal* power series (6.6.9), and write

$$f(z) = \alpha z + F(z), \qquad \varphi(z) = z + \Phi(z), \tag{6.6.10}$$

so $F$ and $\Phi$ are power series starting with a quadratic term. In terms of these, the relation $f\varphi(z) = \varphi(\alpha z)$ becomes

$$\Phi(\alpha z) - \alpha\Phi(z) = F(\varphi z),$$

which we now consider as an equation in the unknown function $\Phi$. To proceed, we replace this equation by the similar, but simpler, equation

$$\Phi(\alpha z) - \alpha\Phi(z) = F(z), \tag{6.6.11}$$

and then use this to define $\Phi$, and hence $\varphi$, as a formal series about the origin.

In fact, it is easy to see that this series converges on some neighbourhood of the origin, and we are then back in the realm of analytic function theory.

The proof now proceeds by an iterative scheme. Having constructed $\varphi$ from $f$ by means of (6.6.11), we put $g = \varphi^{-1}f\varphi$, and define the generic step of the iterative scheme as $f \mapsto g$. Of course, if $f(z) = \alpha z$, then $F = 0$, $\Phi = 0$ and $\varphi(z) = z$, so the map $z \mapsto \alpha z$ is a fixed point of the process $f \mapsto g$, and this suggests that the proof can be completed by a contraction argument (as in the proof of the Contraction Mapping Theorem).

Unfortunately, if this process is iterated to produce functions $f_n$, where

$$f_0 = f, \qquad f_{n+1} = \varphi^{-1}f_n\varphi,$$

the domains of analyticity of the $f_n$ shrink to a single point as $n \to \infty$, and the proof collapses. To overcome this, we perturb the conjugating map $\varphi$ at the $n$-th stage to produce a new map $\varphi_n$, and so produce sequences $\varphi_n$ and $f_n$, where

$$f_{n+1} = \varphi_n^{-1}f_n\varphi_n.$$

With the appropriate definition for the $\varphi_n$, this overcomes the difficulty described above, and a contraction argument then leads to the existence of a function $\varphi$ satisfying $\varphi^{-1}f\varphi(z) = \alpha z$. We now give the formal proof.

The Proof of Theorem 6.6.4. We start with the analytic function $f$ given by (6.6.7), where $\alpha$ satisfies (6.6.8), and the function $F$ given by (6.6.10). Note that (6.6.8) implies that $\alpha$ is not a root of unity: also, we may assume that $C > 1$.

As described above, we define the formal power series $\varphi$ given in (6.6.9) by the relation (6.6.11), and we note that as (6.6.11) is equivalent to

$$(\alpha^n - \alpha)b_n = a_n, \tag{6.6.12}$$

this does indeed define each $b_n$. Further, the relations (6.6.12) and (6.6.8) show that the radius of convergence of $\varphi$ is not less than that of $f$. Thus $\varphi$ is analytic in any disc centred at the origin in which $f$ is, and near the origin we can define the analytic function $g$ by

$$g(z) = \varphi^{-1}f\varphi(z) = \alpha z + G(z),$$

where $G(z) = O(z^2)$ as $z \to 0$.

The first step in our iterative scheme is $f \mapsto g$ (or, equivalently, $F \mapsto G$) and we must now estimate $|G'(z)|$ in terms of $|F'(z)|$ through the intermediate function $|\Phi'(z)|$. First, we have

**Lemma 6.6.6.** *Suppose that $f$ and $\varphi$ are analytic in $\{|z| < r\}$. Then for $|z| < r$,*

$$|\Phi'(z)| \leq C\|F'\|\left(\frac{r}{r - |z|}\right)^{k+1},$$

*where $\|F'\| = \sup_{|z| \leq r}|F'(z)|$.*

PROOF. First we estimate $a_n$, then $b_n$, and finally $|\Phi'(z)|$. Applying Cauchy's Integral Formula to $h$ ($= F'$), we have

$$|na_n| = \frac{|h^{(n-1)}(0)|}{(n-1)!} \leq \frac{\|F'\|}{r^{n-1}},$$

and this with (6.6.8) and (6.6.12) (and $|\alpha| = 1$) yields

$$|nb_n| \leq \frac{C(n-1)^k \|F'\|}{r^{n-1} k!}.$$

It follows that for $|z| < r$,

$$\begin{aligned}
|\Phi'(z)| &\leq \sum_{m=2}^{\infty} |mb_m| |z|^{m-1} \\
&\leq C\|F'\| \sum_{m=1}^{\infty} \left(\frac{m^k}{k!}\right)\left(\frac{|z|}{r}\right)^m \\
&\leq C\|F'\| \sum_{m=0}^{\infty} \binom{m+k}{k}\left(\frac{|z|}{r}\right)^m \\
&= C\|F'\| \left(\frac{r}{r-|z|}\right)^{k+1}
\end{aligned}$$

as required. □

Our next task is to investigate where $g$ acts and to do this, we suppose that the positive numbers $\delta$, $\theta$ and $r$ satisfy:

(1) $f$ (and hence $\varphi$) is analytic in $\{|z| \leq r\}$;
(2) $\theta < \frac{1}{5}$;
(3) $C\delta < \theta^{k+2}$;
(4) if $|z| \leq r$, then $|F'(z)| \leq \delta$.

Note that as $C > 1$ and $\theta < 1$, we have $0 < \delta < \theta$ and so for $|z| \leq r$,

$$|F(z)| \leq \delta |z| \leq \theta r.$$

To describe where $g$ acts, we introduce the nested sequence of discs

$$D_0 \supset D_1 \supset \cdots \supset D_5,$$

where

$$D_m = \{z\colon |z| < r(1 - m\theta)\}.$$

We then have

**Lemma 6.6.7.** *The maps $\varphi$, $f$ and $\varphi^{-1}$ act according to the scheme*

$$D_4 \xrightarrow{\varphi} D_3 \xrightarrow{f} D_2 \xrightarrow{\varphi^{-1}} D_1$$

PROOF. First, our assumptions (1)–(4) together with Lemma 6.6.6 imply that if $z \in D_1$, then

$$|\Phi'(z)| \le C\delta/\theta^{k+1} < \theta, \tag{6.6.13}$$

and hence that $|\Phi(z)| \le \theta|z|$. It follows that for $z$ in $D_4$,

$$|\varphi(z)| \le |z| + |\Phi(z)| \le |z|(1+\theta) \le r(1-4\theta)(1+\theta) < r(1-3\theta)$$

so $\varphi$ maps $D_4$ into $D_3$.

Next, a similar argument holds for $f$, for if $z \in D_3$, then (as $\delta < \theta$) we have $|F(z)| \le \theta|z|$, and so

$$|f(z)| \le |z| + |F(z)| \le r(1-3\theta)(1+\theta).$$

Finally, suppose that $\varphi(z_1) = \varphi(z_2)$, where $z_1$ and $z_2$ are in $D_1$. Then

$$|z_1 - z_2| = |\Phi(z_2) - \Phi(z_1)| < \theta|z_1 - z_2|$$

and as $\theta < 1$ we have $z_1 = z_2$. This shows that the restriction of $\varphi$ to $D_1$ is a homeomorphism of $D_1$ onto a domain whose boundary lies in $\varphi(\partial D_1)$. However, on $\partial D_1$,

$$|\varphi(z)| \ge |z| - |\Phi(z)| \ge |z|(1-\theta) = r(1-\theta)^2 > r(1-2\theta),$$

so $D_2 \subset \varphi(D_1)$ and the proof of Lemma 6.6.7 is complete. □

Lemma 6.6.7 shows that $g$ maps $D_4$ into $D_1$, and we are now in a position to estimate $G'$. This is given in

**Lemma 6.6.8.** *On* $D_5$, $|G'(z)| \le 2C\delta^2/\theta^{k+2}$.

PROOF. First we write the identity $\varphi g = f\varphi$ in terms of $F$, $\Phi$ and $G$: this gives

$$\alpha z + G(z) + \Phi(\alpha z + G(z)) = \alpha z + \alpha\Phi(z) + F(z + \Phi(z)),$$

and using (6.6.11) to eliminate $\alpha\Phi(z)$, we obtain

$$G(z) = \Phi(\alpha z) - \Phi(\alpha z + G(z)) + F(z + \Phi(z)) - F(z).$$

Next, this and (6.6.13) show that if $z$ is in $D_4$, then

$$\begin{aligned} |G(z)| &\le |G(z)| \sup_{D_4} |\Phi'(z)| + |\Phi(z)| \sup_{D_3} |F'(z)| \\ &\le \theta|G(z)| + \delta|\Phi(z)| \\ &\le |G(z)|/5 + (C\delta^2/\theta^{k+1})|z| \end{aligned}$$

which yields

$$|G(z)| \le 2C\delta^2 r/\theta^{k+1}.$$

Finally, with this, we can estimate $G'(z)$ for $z$ in $D_5$ by using Cauchy's Integral Formula (integrating around the circle centre $z$ and radius $r\theta$) and so obtain the given inequality. □

*To summarize our progress so far*: Given $\alpha$ satisfying (6.6.8), and $\theta$, $\delta$ and $r$ satisfying (1)–(4), we have produced a map

$$f(z) = \alpha z + F(z) \mapsto g(z) = \alpha z + G(z),$$

and we have estimated $|G'(z)|$.

We now return to the iterative scheme suggested earlier, and define sequences $(\varphi_n)$, $(f_n)$ and $(F_n)$ inductively, where

$$f_n(z) = \alpha z + F_n(z).$$

We put $f_0 = f$, and construct $\varphi_n$ from $f_n$ as $\varphi$ was constructed from $f$: after this, we put $f_{n+1} = \varphi_n^{-1} f_n \varphi_n$ (so $F_{n+1}$ is obtained from $F_n$ as $G$ was obtained from $F$). At each stage, this requires a choice of parameters $\theta_n$, $\delta_n$ and $r_n$, and we must exercise some care in this choice: if, for example, $\theta_n = \theta$ and $r_n = r$ for all $n$, then we can only assert that $F_n$ is defined on the disc of radius $r(1 - 5\theta)^n$, and this tends to zero as $n \to \infty$.

Our choice of parameters is as follows. First, we put

$$\theta_n = 1/[10(1 + 2^n)],$$

and

$$r_n = r(1 + 2^{-n})/2.$$

It is convenient to write

$$D_{n,m} = \{z: |z| < r_n(1 - m\theta_n)\},$$

and as $r_{n+1} = r_n(1 - 5\theta_n)$, we have

$$V \subset D_{n,4} \subset D_{n,3} \subset D_{n,0} = D_{n-1,5} \subset D_{n-1,4},$$

where

$$V = \{z: |z| < r/2\}.$$

Next, we select a small positive $\delta$ (the precise requirements will be clear shortly) and then define $\delta_n$ inductively by

$$\delta_{n+1} = 2C\delta_n^2/\theta_n^{k+2}.$$

From this, and assuming that $\delta$ is sufficiently small, we obtain

$$C\delta_n < \theta_n^{k+2}$$

for all $n \geq 0$, and also $\delta_n \to 0$ as $n \to \infty$ (see Exercise 6.6.1).

Clearly, we can also choose $\delta_0$ and $r_0$ ($= r$) sufficiently small so that (1)–(4) hold with $f$, $\delta$, $\theta$ and $r$ replaced by $f_0$ ($= f$), $\delta_0$, $\theta_0$ and $r_0$. More generally, suppose now that (1)–(4) hold with $f$, $\delta$, $\theta$ and $r$ replaced by $f_n$, $\delta_n$, $\theta_n$ and $r_n$, so $f_n$ is analytic in $D_{n,0}$. Then our discussion of the generic step $F \mapsto G$ implies that $f_{n+1}$ is analytic in $D_{n,4}$, and hence on $D_{n+1,0}$. Moreover, as $|F_n'(z)| \leq \delta_n$ on $D_{n,0}$, we also have

$$|F_{n+1}'(z)| \leq 2C\delta_n^2/\theta_n^{k+2} = \delta_{n+1}$$

on $D_{n,5}$, and hence on $D_{n+1,0}$. We conclude that the process can be iterated indefinitely, that each $f_n$ is defined and analytic on $V$, and that $F_n'$, and hence $F_n$ also, converges uniformly to zero on $V$.

Writing

$$\psi_n = \varphi_0 \cdots \varphi_n,$$

we have

$$\psi_n^{-1} f \psi_n(z) = f_{n+1}(z) = \alpha z + F_{n+1}(z),$$

so it only remains to show that $\psi_n$ and $\psi_n^{-1}$ converge uniformly in some neighbourhood of the origin to some non-constant functions $\varphi$ and $\varphi^{-1}$ respectively. Of course, it is sufficient to prove that $\{\psi_n\}$ is a normal family in some neighbourhood of the origin, for if (on a subsequence) $\psi_n \to \varphi$, then

$$\varphi'(0) = \lim \psi_n'(0) = 1$$

(so $\varphi$ is not constant) and $\psi_n^{-1} \to \varphi^{-1}$ (see Exercise 6.6.2). The normality of $\{\psi_n\}$, however, is easily established. From Lemma 6.6.7, $\varphi_n$ maps $D_{n,4}$ into $D_{n,3}$, and hence into $D_{n-1,4}$, and so we find that $\psi_n$ maps $D_{n,4}$ into $D_{0,3}$. It follows that $\{\psi_n\}$ is normal in $V$ and the proof of Theorem 6.6.4 is finally finished. □

We end with the

PROOF OF THEOREM 6.6.5. Let $h$ be the map $x \mapsto \exp(2\pi i x)$ of $\mathbb{R}$ onto $\{|z| = 1\}$. We denote linear measure on $\mathbb{R}$ by $m$, and linear measure on the unit circle by $m_0$, so for any subset $E$ of $[0, 1]$, we have $m_0(h(E)) = 2\pi m(E)$. Next, let $S_0$ be the component of $S$ in $\{|z| = 1\}$, and for each $\delta$ in $(0, 1)$, let

$$E_\delta = \{x \in [0, 1]: \text{for some } m, n \geq 1, |nx - m| < \delta/n^2\}.$$

We shall show that for each positive $\delta$, $m(E_\delta) \leq 4\delta$ and also, $S_0 \subset h(E_\delta)$: then $m_0(S_0) \leq 8\pi\delta$ and letting $\delta \to 0$, we obtain $m_0(S_0) = 0$ as required.

First,

$$E_\delta = \bigcup_{n=1}^{\infty} \bigcup_{m=1}^{\infty} [0, 1] \cap \left(\frac{m}{n} - \frac{\delta}{n^3}, \frac{m}{n} + \frac{\delta}{n^3}\right)$$

$$= \bigcup_{n=1}^{\infty} \bigcup_{m=1}^{n} [0, 1] \cap \left(\frac{m}{n} - \frac{\delta}{n^3}, \frac{m}{n} + \frac{\delta}{n^3}\right)$$

for $m > n$ implies that $m/n - \delta/n^3 \geq 1$. We deduce that

$$m(E_\delta) \leq \sum_{n=1}^{\infty} n\left(\frac{2\delta}{n^3}\right) \leq 4\delta$$

as stated above.

Finally, select any $\delta$ in $(0, 1)$ and suppose that $\alpha$ is in $S_0$. Then for all positive numbers $M$ and $k$, there is some integer $n$ with

$$1/|\alpha^n - 1| > Mn^k,$$

and taking $k = 2$ and $M = \delta^{-1}$, we see that there is some $n$ with

$$|\alpha^n - 1| < \delta/n^2.$$

Now take $x$ in $[0, 1]$ such that $h(x) = \alpha$ and let $m$ be the nearest integer to $nx$: then, as $\sin(y) \geq 2y/\pi$ on $[0, \pi/2]$, we have

$$\begin{aligned} |\alpha^n - 1| &= |e^{n\pi ix} - e^{-n\pi ix}| \\ &= 2|\sin(n\pi x)| \\ &= 2\sin|\pi(nx - m)| \\ &> |nx - m|. \end{aligned}$$

This shows that for these $n$ and $m$, $|nx - m| < \delta/n^2$: thus $x$ is in $E_\delta$, $\alpha$ is in $h(E_\delta)$, and the proof is complete. □

EXERCISE 6.6

1. Suppose that the numbers $\delta_n$ ($n \geq 0$) satisfy
$$\delta_{n+1} \leq AB^n\delta_n^2,$$
where $A > 1$, $B > 1$ and $\delta_0 > 0$. Show (by induction) that
$$AB^{n+1}\delta_n \leq (AB\delta_0)^{2^n},$$
hence if $AB\delta_0 < 1$, then $\delta_n < (AB\delta_0)^n$.
   Using the inequality $10\theta_n > 3^{-n}$, show (in the notation in the text) that for some constants $A$ and $B$, $\delta_n < (AB\delta_0)^n$. Deduce that if $\delta_0$ is chosen sufficiently small, then for all $n$,
$$C\delta_n < \theta_n^{k+2}.$$

2. Suppose that $f$ is analytic and injective near the origin. Show that in suitable circumstances,
$$f^{-1}(w) = \frac{1}{2\pi i}\int_\gamma \frac{zf'(z)}{f(z) - w}\,dz.$$

3. Show that there exists a non-rational function $f$ that is analytic near the origin, for which some $f^n$ is the identity $I$.

4. Use Theorem 6.6.1 to give an alternative proof that a rationally indifferent fixed point is in $J$.

## §6.7. Irrationally Indifferent Cycles in $J$

Our earlier results give us a clear understanding of the dynamics of the iterates $R^n$ in the vicinity of all attracting, repelling, and rationally indifferent cycles, and also irrationally indifferent cycles in $F$, and we shall consider super-attracting cycles in Theorem 6.10.1. The one result in this section shows that the remaining possibility, namely irrationally indifferent cycles in $J$, can occur although the geometry and dynamics of the iterates in a neighbourhood of these cycles seems not to be understood at all. We prove

*Remark.* The reader should compare (6.7.1) with (6.6.4).

**Theorem 6.7.1.** *Suppose that $P(z) = \alpha z + \cdots + z^d$, where $d \geq 2$ and $|\alpha| = 1$ but $\alpha$ is not a root of unity. If*

$$|\alpha^n - 1| \leq (1/n)^{d^n-1} \tag{6.7.1}$$

*for infinitely many $n$, then the origin is an irrationally indifferent fixed point of $P$ that lies in $J(P)$.*

PROOF. First, we show that the periodic points of $P$ accumulate at the origin. Take any $n$, write $N = d^{n-1}$, and let $0, \zeta_1, \ldots, \zeta_N$ be the $N + 1$ fixed points of $P^n$ in $\mathbb{C}$, where

$$0 < |\zeta_1| \leq \cdots \leq |\zeta_N|.$$

Now

$$\begin{aligned} P^n(z) - z &= (\alpha^n - 1)z + \cdots + z^{N+1} \\ &= z(z - \zeta_1)\cdots(z - \zeta_N), \end{aligned}$$

and this with (6.7.1) yields

$$|\zeta_1| \leq |\zeta_1 \cdots \zeta_N|^{1/N} = |\alpha^n - 1|^{1/N} \leq 1/n.$$

We conclude that there are periodic points of $P$ distinct from, but arbitrarily close to, the origin.

Suppose now that the origin is in $F(P)$ so by Theorem 6.6.2 there is some neighbourhood $N$ of the origin, and some disc $D$ centred at the origin, such that $P(N) = N$ and $P: N \to N$ is conjugate to the map $z \mapsto \alpha z$ of $D$ onto itself. Now select any non-zero periodic point $\zeta$ of $P$ in $N$, say of period $m$. Then the corresponding non-zero point $\eta$ in $D$ is of period $m$ for the map $z \mapsto \alpha z$, so $\alpha^m = 1$, contrary to our assumption. We conclude that the origin is in $J(P)$ and the proof is complete. □

Of course, Theorem 6.7.1 does not establish the existence of irrationally indifferent fixed points in $J$, for we have yet to show that numbers $\alpha$ satisfying (6.7.1) do exist. In fact, it is easy to see that such numbers $\alpha$ are dense in the unit circle, and we begin by showing that if $\alpha = \exp(2\pi i\theta)$, then the condition (6.7.1) is closely related to the approximation of $\theta$ by rational numbers.

Now for any positive integers $m$ and $n$, we have

$$\begin{aligned} |\alpha^n - 1| &= |e^{n\pi i\theta} - e^{-n\pi i\theta}| \\ &= 2|\sin(n\pi\theta)| \\ &= 2|\sin \pi(n\theta - m)| \\ &\leq 2\pi|n\theta - m|. \end{aligned}$$

It follows that (6.7.1) holds if for infinitely many $n$ there is some $m$ (depending on $n$) such that

$$|\theta - m/n| \leq 1/n^{N+2}, \tag{6.7.2}$$

for then (with $n > 2\pi$)

$$|\alpha^n - 1| \leq 2\pi n|\theta - m/n| \leq 1/n^N.$$

The problem, then, reduces to showing that the set of $\theta$ satisfying (6.7.2) is dense in $\mathbb{R}$. Now each real number $x$ has a continued fraction expansion and the convergents of the continued fractions are particularly good rational approximations to $x$. By defining $\theta$ in terms of a specified continued fraction expansion, we can construct many $\theta$ satisfying (6.7.2), and it is apparent from the construction that the set of such $\theta$ is dense in $\mathbb{R}$. A self-contained argument for the construction is given in the exericses.

## Exercise 6.7

1. Given positive numbers $a_0, a_1, \ldots$, define the real numbers $[a_0, \ldots, a_n]$ inductive by

$$[a_0] = a_0, \qquad [a_0, \ldots, a_n, a_{n+1}] = [a_0, \ldots, a_{n-1}, a_n + 1/a_{n+1}].$$

Next, define sequences $(p_n)$ and $(q_n)$ inductively by

$$p_n = a_n p_{n-1} + p_{n-2}, \qquad p_0 = a_0, \qquad p_1 = a_0 a_1 + 1;$$

$$q_n = a_n q_{n-1} + q_{n-2}, \qquad q_0 = 1, \qquad q_1 = a_1.$$

(i) Verify that $p_0/q_0 = [a_0]$, and $p_1/q_1 = [a_0, a_1]$.

(ii) Use the definition of $[a_0, \ldots, a_n]$ to prove by induction that for all positive $x$,

$$[a_0, \ldots, a_n, x] = \frac{xp_n + p_{n-1}}{xq_n + q_{n-1}},$$

and deduce that

$$[a_0, \ldots, a_n, a_{n+1}] = p_{n+1}/q_{n+1}.$$

(iii) Suppose that $m$, $n$, $u$, $v$ are positive numbers with $m/n < u/v$. Show that if $a > 0$, then

$$\frac{m}{n} < \frac{am + u}{an + v} < \frac{u}{v}, \qquad \frac{m}{n} < \frac{m + au}{n + av} < \frac{u}{v},$$

(this is trivial, but it shows that $p_{n+2}/q_{n+2}$ lies between $p_n/q_n$ and $p_{n+1}/q_{n+1}$). Deduce that for $m \geq n + 2$, $p_m/q_m$ lies between $p_n/q_n$ and $p_{n+1}/q_{n+1}$.

(iv) Let $\Delta_n = p_n q_{n-1} - p_{n-1} q_n$. Prove that $\Delta_n = -\Delta_{n-1}$, and deduce that $\Delta_n = (-1)^{n+1}$. Deduce that

$$\left|\frac{p_n}{q_n} - \frac{p_{n+1}}{q_{n+1}}\right| < \frac{1}{q_n q_{n+1}}.$$

(v) Now suppose that the $a_j$ are positive integers. Deduce that the $p_j$ and $q_j$ are positive integers. Show also that $q_n \geq q_{n-1} + q_{n-2}$, and deduce that $q_n \geq n + 1$.

(vi) Use (iii), (iv) and (v) to show that $(p_n/q_n)$ is a Cauchy sequence converging to $\theta$, say, and that

$$\left|\frac{p_n}{q_n} - \theta\right| \leq \frac{1}{q_n q_{n+1}}.$$

(vii) Now construct $(a_n)$ inductively as follows. Choose a positive integer $m$, and positive integers $a_0, \ldots, a_m$ arbitrarily. If $a_0, \ldots, a_n$ have been chosen, determine $p_0, \ldots, p_n$ and $q_0, \ldots, q_n$ as above, and then define

$$a_{n+1} = q_n^N, \quad N = d^n - 1.$$

Then

$$q_{n+1} = a_{n+1}q_n + q_{n-1} \geq q_n^{N+1},$$

which, with (vi), yields (6.7.2).

## §6.8. The Proof of the Existence of Periodic Points

We recall Theorem 6.2.1, namely that if a polynomial $P$ of degree $d$ ($d \geq 2$) has no periodic points of period $N$, then $N = 2$ and $P$ is conjugate to $z \mapsto z^2 - z$. In this section, we prove this result and also the corresponding result for rational functions (Theorem 6.2.2).

THE PROOF OF THEOREM 6.2.1. Let $P$ be a polynomial satisfying the hypotheses of Theorem 6.2.1 (so $N \geq 2$), let

$$K = \{z \in \mathbb{C}: P^N(z) = z\},$$

and let

$$M = \{m \in \mathbb{Z}: 1 \leq m < N, m|N\}.$$

Then each $z$ in $K$ is a fixed point of $P^m$ for some $m$ in $M$, and we let $m(z)$ be the minimal such $m$.

The proof depends on establishing the inequalities

$$d^{N-1}(d-1) \leq \sum_{z \in K} [\mu(N, z) - \mu(m(z), z)] \leq N(d-1), \qquad (6.8.1)$$

where $\mu(n, w)$ is the number of fixed points of $P^n$ at $w$. Obviously, (6.8.1) implies that $d^{N-1} \leq N$, and hence that

$$N = 1 + (N-1) \leq 1 + (N-1)(d-1) \leq [1 + (d-1)]^{N-1} = d^{N-1} \leq N.$$

This yields $d = N = 2$, and as we have already proved (in §6.2) that this implies that $P$ is conjugate to $z \mapsto z^2 - z$, it only remains to establish the inequalities (6.8.1).

The lower bound in (6.8.1) is easily obtained and we prove this first. The sum $\sum \mu(N, z)$ over $K$ is simply the total number of fixed points of $P^N$ in $\mathbb{C}$, and so is $d^N$. Next, as each $z$ in $K$ is a fixed point of $P^m$, where $m = m(z)$, we have

$$\sum_{z\in K}\mu(m(z),z)=\sum_{m\in M}\sum_{\substack{m(z)=m\\ z\in K}}\mu(m(z),z)\le\sum_{m\in M}d^m,$$

and so

$$L=d^N-\sum_{m\in M}d^m$$

is a lower bound for the central term in (6.8.1). Note that $L$ gives the lower bound in (6.8.1) when $N=2$, so we may assume that $N\ge 3$. Now $N-1$ and $N$ are coprime (for $N-(N-1)=1$) so if $N-2\ge 1$, then

$$\sum_{m\in M}d^m\le d+\cdots+d^{N-2}\le d^{N-1}.$$

This shows that for all $N$, $N\ge 2$,

$$d^N-d^{N-1}\le L,$$

and this gives the lower bound in (6.8.1).

The verification of the upper bound in (6.8.1) requires more effort. Now each $z$ in $K$ lies in some cycle of length $m(z)$, and we denote these (pairwise disjoint) cycles by $C_1,\ldots,C_q$. Further, we denote that length of $C_j$ by $m_j$, so if $z$ is in $C_j$, then $m(z)=m_j$. As the central term in (6.8.1) is

$$\sum_{j=1}^{q}\sum_{z\in C_j}[\mu(N,z)-\mu(m_j,z)], \tag{6.8.2}$$

we can confine our attention to each cycle separately. Now Corollary 2.6.7 implies that $\mu(N,z)=\mu(m_j,z)$ whenever $z$ lies in a cycle $C_j$ that is not rationally indifferent, so the only nonzero contributions in (6.8.2) arise from the rationally indifferent cycles $C_j$.

Now let $C_j$ be a rationally indifferent cycle so, as in Theorem 6.5.10, there are integers $k_j$, $m_j$, $q_j$ and $p_j$ (corresponding to $k$, $m$, $q$ and $p$ in Theorem 6.5.10), where this $m_j$ is the same as that above (namely, the length of the cycle $C_j$). We want an upper bound of (6.8.2), and we need only concern ourselves with the positive terms, for all others can be replaced by zero (but see Exercise 6.8.1). Now as $\mu(m_j,z)\ge 1$, the term in (6.8.2) is positive only if $\mu(N,z)\ge 2$ and then, from Theorem 6.5.10, $m_jq_j$ divides $N$. It follows that for all $z$ in $C_j$,

$$\mu(N,z)-\mu(m_j,z)\le q_jk_j.$$

Summing over all $z$ in $C_j$, we now find that

$$\sum_{z\in C_j}[\mu(N,z)-\mu(m_j,z)]=q_jk_jm_j\le Nk_j,$$

and so we can take

$$U=N\sum k_j$$

as an upper bound for the central term in (6.8.1).

We need one more fact (which will be verified in Chapter 9, but which is convenient to assume here), namely that if a cycle of components of $F(P)$ contains a petal, then it also contains a critical point of $P$. As there are $k_j$ such cycles of components for the rationally indifferent cycle $C_j$, we see that there are at least $\sum_j k_j$ critical points of $P$ in $\mathbb{C}$; thus $\sum_j k_j \le d - 1$. It follows that we can take the upper bound to be $N(d-1)$ and we have now established (6.8.1). The reference to Chapter 9 is Theorem 9.3.2 (which could be proved now), and we shall not need to use Theorems 6.2.1 and 6.2.2 later. □

Finally, we comment on the

Proof of Theorem 6.2.2. Only minor modifications are needed to the previous proof. Indeed, the argument is exactly the same except for the fact that the polynomial $P$ is replaced by a rational map $R$, and we work in the sphere $\mathbb{C}_\infty$ rather than the plane $\mathbb{C}$. We put

$$K = \{z \in \mathbb{C}_\infty : R^N(z) = z\},$$

and obtain the lower bound

$$L = (d^N + 1) - \sum_{m \in M} (d^m + 1)$$

as a lower bound for the central term in (6.8.1).

Similarly, for the upper bound, we obtain

$$U = N \sum_j k_j \le N(2d - 2),$$

and these lead to the inequality

$$d^N + 1 \le N(2d - 2) + \sum_{m \in M} (d^m + 1). \tag{6.8.3}$$

When $N$ is prime, $M = \{1\}$ and we obtain

$$d^N - d \le 2N(d - 1),$$

or

$$d(1 + d + \cdots + d^{N-2}) \le 2N.$$

For $N = 2$ this yields $d \le 4$, while for $N = 3$, $d = 2$. Assume now that $N \ge 4$. Then $N$ and $N - 1$ are coprime so

$$\sum_{m \in M} d^m \le d + d^2 + \cdots + d^{N-2} \le (d^{N-1} - d)/(d - 1) \le d^{N-1} - d,$$

and this with (6.8.3) yields

$$d^N + 1 \le 2N(d - 1) + (d^{N-1} - d) + N - 2.$$

This is equivalent to

$$d^{N-1}(d - 1) \le (2N - 1)(d - 1) + N - 4,$$

and dividing by $d-1$, we obtain

$$2^{N-1} \leq d^{N-1} \leq (2N-1) + (N-4) = 3N-5,$$

which is false for $N \geq 4$. □

EXERCISE 6.8

1. Use Corollary 2.6.7 to show that each term in (6.8.2) is non-negative.

## §6.9. The Julia Set and Periodic Points

We are now in a position to relate the Julia set to periodic points. First, we prove

**Theorem 6.9.1.** *Let $R$ be a rational map of degree $d$, where $d \geq 2$. Then $J$ is the derived set of the periodic points of $R$.*

We recall that $z$ is in the derived set of $E$ if and only if there are *distinct* points $z_n$ in $E$ such that $z_n \to z$. Later (in §9.6), we shall see that $R$ has at most $2d-2$ non-repelling cycles, and with this, Theorem 6.9.1 implies that $J$ is the derived set of the repelling cycles. As each repelling cycle lies in $J$ (and anticipating the results in §9.6), this yields

**Theorem 6.9.2.** *$J$ is the closure of the repelling periodic points of $R$.*

Next, we give the

PROOF OF THEOREM 6.9.1. We begin by choosing any open set $W$ that meets $J$, and then choosing a point $w$ in $W \cap J$ such that $w$ is not a critical value of $R^2$. Then $R^{-2}\{w\}$ contains at least four points (as $d \geq 2$), and so we can choose three of them, say $w_1$, $w_2$ and $w_3$, which are distinct from $w$. Now construct pairwise disjoint, compact, neighbourhoods $W$, $W_1$, $W_2$ and $W_3$ of $w$, $w_1$, $w_2$ and $w_3$ respectively, such that for each $j$, $R^2$ is a homeomorphism of $W_j$ onto $W$, and let $S_j$: $W \to W_j$ be the inverse of $R^2$: $W_j \to W$; see Figure 6.9.1.

If for all $z$ in $W$, all $j$ in $\{1, 2, 3\}$, and all $n \geq 1$, we have

$$R^n(z) \neq S_j(z),$$

then $\{R^n\}$ is normal in $W$ (Theorem 3.3.6). This cannot be so, however, as $W$ meets $J$: thus there is some $z$ in $W$, some $j$, and some $n$, such that $R^n(z) = S_j(z)$, and hence

$$R^{n+2}(z) = R^2 S_j(z) = z.$$

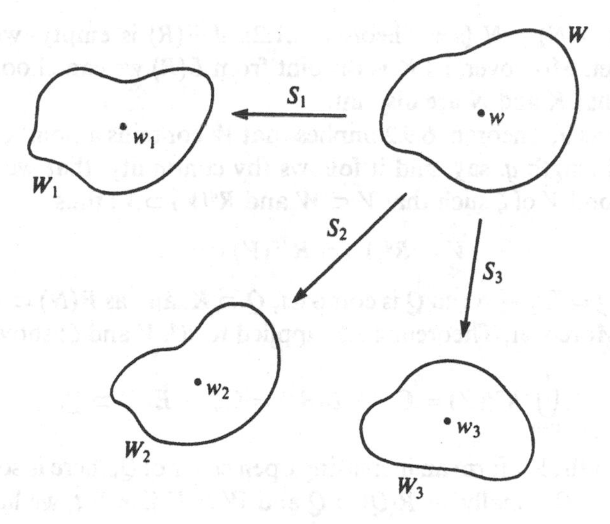


Figure 6.9.1

This shows that $z$ (in $W$) is a periodic point of $R$, and hence that $J$ is contained in the derived set of the periodic points of $R$. □

To prove that the derived set of the periodic points of $R$ is contained in $J$, it is sufficient to prove

**Lemma 6.9.3.** *Any component of $F$ contains at most one periodic point of $R$.*

PROOF. Let $F_0$ be a component of $F$ and suppose that $\alpha$ and $\beta$ are periodic points in $F_0$. By replacing $R$ by a suitable iterate $R^n$, we may assume that both $\alpha$ and $\beta$ are fixed by $R$. If $\alpha$ is a (super)attracting fixed point of $R$: then $R^n \to \alpha$ on $F_0$ and so as $R$ fixes $\beta$, we must have $\alpha = \beta$. The remaining case is when $\alpha$ is an indifferent fixed point of $R$, and then $R: F_0 \to F_0$ is analytically conjugate to a rotation of infinite order of the unit disc (Theorem 6.6.3). In this case, $\alpha$ is the only fixed point of $R$ in $F_0$, so again, $\alpha = \beta$ and the proof is complete. □

We end this section with the following extension of Theorem 4.2.5.

**Theorem 6.9.4.** *Let $R$ be a rational map of degree $d$, where $d \geq 2$, let $W$ be a domain that meets $J$, and let $K$ be a compact set which contains no exceptional points of $R$. Then for all sufficiently large $n$, $R^n(W) \supset K$.*

PROOF. Our knowledge of the exceptional points of $R$ leads easily to the existence of an open set $N$ which contains $E(R)$, which lies in $F(R)$, and which is such that $R(N) \subset N$ (see Theorem 4.1.2): if $E(R)$ is empty, we let $N$ be the empty set. Moreover, as $K$ is disjoint from $E(R)$ we can choose $N$ small enough so that $K$ and $N$ are disjoint.

As $W$ meets $J$, Theorem 6.9.2 implies that $W$ contains a point $\zeta$ of a repelling cycle of length $q$, say, and it follows (by continuity) that we can find a neighbourhood $V$ of $\zeta$ such that $V \subset W$ and $R^q(V) \supset V$: thus

$$V \subset R^q(V) \subset R^{2q}(V) \subset \cdots.$$

Now let $Q = \mathbb{C}_\infty - N$: so $Q$ is compact, $Q \supset K$, and as $R(N) \subset N$, we have $R(Q) \supset Q$. Moreover, Theorem 4.2.5 (applied to $R^q$, $V$ and $\zeta$) shows that

$$\bigcup_{n=0}^{\infty} R^{nq}(V) = \mathbb{C}_\infty - E(R^q) = \mathbb{C}_\infty - E(R) \supset Q.$$

As the sets on the left form an increasing open cover of $Q$, there is some integer $t$ with $R^t(V) \supset Q$. Finally, as $R(Q) \supset Q$ and $W \supset V$, if $m \geq t$, we have

$$R^m(W) \supset R^m(V) = R^{m-t}R^t(V) \supset R^{m-t}(Q) \supset Q \supset K$$

as required. □

## §6.10. Local Conjugacy

Let

$$f(z) = a_1 z + a_2 z^2 + \cdots$$

be a power series, analytic near the origin. In this final section of the chapter, we comment briefly on the possible forms that the local conjugates $gfg^{-1}$ of $f$ can take. We shall assume that $g(0) = 0$, and we shall discuss *analytic conjugacy* (when $g$ is analytic), *topological conjugacy* (when $g$ is a homeomorphism), and *formal conjugacy* (when $g$ is a formal power series).

We have already discussed analytic conjugacy when $|a_1| \neq 0, 1$ (Theorem 6.3.2). The situation when $a_1 = 0$ is straightforward, and the most optimistic guess is that in this case $f$ is conjugate to the first non-trivial term of its Taylor expansion, say to $a_q z^q$ where $q \geq 2$. This is correct, but as $q \geq 2$, the two maps $z \mapsto a_q z^q$ and $z \mapsto z^q$ are conjugate and so we can state this result in the following form.

**Theorem 6.10.1.** *Suppose that*

$$f(z) = b_0 z^q + b_1 z^{q+1} + \cdots$$

*is analytic near the origin, and that $b_0 \neq 0$, $q \geq 2$. Then there exists a unique function $g$, analytic near the origin, such that:*

(a) *$g(0) = 0$, $g'(0) = 1$; and*
(b) *for all $z$ near the origin, $gfg^{-1}(z) = z^q$.*

For the proof we shall need

**Lemma 6.10.2.** *Suppose that $n \geq 2$, let $W = \{w: |w - 1| < \frac{1}{2}\}$, and let $z^{1/n}$ denote the principal branch of the n-th root of z in W. Then*

$$|z^{1/n} - 1| < 2|z - 1|/n < 1/n.$$

PROOF. We simply note that $z^{1/n} - 1$ is the integral of the derivative of $z^{1/n}$ along the segment from 1 to $z$ and make the obvious estimates. □

THE PROOF OF THEOREM 6.10.1. First, we replace $f$ by $\varphi f \varphi^{-1}$, where $\varphi(z) = \lambda z$ and $\lambda^{q-1} = b_0$, so we may assume that $b_0 = 1$. We now define the power series $h(z)$ by

$$f(z) = z^q(1 + h(z)), \tag{6.10.1}$$

so $h(0) = 0$, and there are positive numbers $R$ and $K$ such that

$$|h(z)| \leq K|z|$$

when $|z| \leq R$.

Next, we choose any positive $\delta$ satisfying

$$\delta < \min\{\tfrac{1}{4}, R, 1/(2K)\},$$

and let $D = \{z: |z| < \delta\}$. If $z$ is in $D$, then

$$|f(z)| \leq |z|\delta^{q-1}(1 + K\delta) < |z|/2,$$

so $f$ maps $D$ into itself and on $D$,

$$|f^n(z)| \leq |z|/2^n.$$

For $n \geq 0$ and $z$ in $D$, we have

$$f^{n+1}(z) = f(f^n z) = [f^n(z)]^q[1 + h(f^n z)],$$

where

$$|h(f^n z)| \leq K|f^n(z)| \leq K|z|/2^n < \tfrac{1}{2}.$$

This shows that $1 + h(f^n z)$ lies in the disc $W$ (in Lemma 10.6.2) and so we can find a $q^{n+1}$-th root, say $h_{n+1}(z)$, of $1 + h(f^n z)$ in $D$ with

$$|h_{n+1}(z) - 1| < 2/q^{n+1}.$$

We deduce that the infinite product

$$g(z) = z \prod_{n=1}^{\infty} h_n(z)$$

converges uniformly on $D$, so $g$ is analytic on $D$ with $g(0) = 0$ and $g'(0) = 1$.

Finally,

$$[h_{n+1}(fz)]^{q^{n+1}} = 1 + h(f^{n+1}z) = [h_{n+2}(z)^q]^{q^{n+1}}$$

and so

$$h_{n+1}(fz) = h_{n+2}(z)^q$$

(because their ratio is a root of unity and the two functions are equal when $z = 0$). It follows that if

$$g_n(z) = zh_1(z) \cdots h_n(z),$$

then

$$g_n(fz) = f(z)[h_2(z) \cdots h_{n+1}(z)]^q$$
$$= g_{n+1}(z)^q,$$

and letting $n \to \infty$ we obtain

$$g(fz) = g(z)^q$$

as required. □

We shall now briefly consider the local analytic conjugates of $f$ when $a_1 = 1$. In this case, $f$ has the form

$$f(z) = z(1 + \alpha_p z^p + \alpha_{p+1} z^{p+1} + \cdots) \tag{6.10.2}$$

and by the Residue Theorem, the function $\tau(f)$ defined by

$$\tau(f) = \frac{1}{2\pi i} \int_{|z|=r} \frac{1}{f(z) - z} \, dz$$

is independent of $r$ for $r < r_0$, say. Further, if we expand $(f(z) - z)^{-1}$ in a Laurent series about the origin, we see that $\tau(f)$ is of the form

$$P(\alpha_p, \ldots, \alpha_{2p})/(\alpha_p)^{p+1}$$

for some polynomial $P$: for example, if $p = 1$, then

$$\tau(f) = -\alpha_2/\alpha_1^2.$$

We now prove that the function $\tau$ is invariant under analytic conjugation.

**Theorem 6.10.3.** *If $f$ is of the form* (6.10.2), *and is analytically conjugate to $g$ near the origin, then $\tau(f) = \tau(g)$.*

PROOF. Let $h$ be any conjugating function, that is, $h$ is analytic near the origin with $h(0) = 0$ and $h'(0) \neq 0$. We take a small circle, say $C = \{|z| = 2r\}$, and for $z$ and $w$ in $\{|z| < r\}$, express $h(z)$, $h(w)$ and $h'(z)$ as a Cauchy integral around $C$. An elementary estimation then gives

$$\left| \frac{h(z) - h(w)}{z - w} - h'(z) \right| \leq M|z - w|,$$

say, and hence

$$\left| \frac{1}{z - w} - \frac{h'(z)}{h(z) - h(w)} \right| \leq M \frac{|z - w|}{|h(z) - h(w)|} \leq M_1,$$

say, because the previous inequality shows that the quotient in the middle term is approximately $1/h'(0)$. This means that the two functions

$$\frac{1}{f(z)-z}, \qquad \frac{h'(z)}{h(fz)-h(z)},$$

differ near the origin by some analytic function $g$ and so will have the same integral taken over any small curve surrounding the origin.

Now let $F = hfh^{-1}$, let $\gamma$ be a small circle around the origin and let $\Gamma = h(\gamma)$. Then

$$\begin{aligned}\tau(F) &= \frac{1}{2\pi i}\int_\Gamma \frac{1}{F(z)-z}\,dz\\ &= \frac{1}{2\pi i}\int_\gamma \frac{h'(w)}{F(hw)-h(w)}\,dw\\ &= \frac{1}{2\pi i}\int_\gamma \frac{h'(w)}{h(fw)-h(w)}\,dw\\ &= \frac{1}{2\pi i}\int_\gamma \frac{1}{f(w)-w}\,dw\end{aligned}$$

as required. □

As an application of Theorem 6.10.3, recall that if $f$ is of the form (6.10.2), then $f$ is locally conjugate to a function

$$z \mapsto z(1 + z^p + Az^{2p} + O(z^{2p+1})) \tag{6.10.3}$$

(Theorem 6.5.7), and in this case, a simple calculation gives

$$\tau(f) = -A.$$

Next, we have

**Theorem 6.10.4.** *Suppose that $f$ is given by (6.10.2): then for any integer $q$ with $q > 2p$, $f$ is analytically conjugate to some function*

$$z \mapsto z(1 + z^p + Az^{2p} + O(z^q)).$$

The proof is essentially the same as that for Theorem 6.5.7 and we omit the details. By contrast, we also have

**Theorem 6.10.5.** *Suppose that $f$ has a Taylor expansion*

$$f(z) = a_1 z + a_2 z^2 + \cdots,$$

*where $a_1$ is not a root of unity. Then given any positive integer $N$, there is some function $g$, analytic near and vanishing at the origin, such that near the origin*

$$g^{-1}fg(z) = a_1 z + O(z^N).$$

Proof. It is sufficient to show that any analytic $F$ of the form

$$F(z) = a_1 z + O(z^k)$$

is conjugate near the origin to some map

$$z \mapsto a_1 z + O(z^{k+1}).$$

So, given

$$F(z) = a_1 z + Az^k + \cdots,$$

where $A \neq 0$, put

$$h(z) = z + bz^k, \qquad b = A/(a_1^k - a_1).$$

Then writing

$$g^{-1}Fg(z) = a_1 z + c_2 z^2 + c_3 z^3 + \cdots,$$

we must have

$$a_1(z + bz^k) + A(z + bz^k)^k + \cdots$$
$$= (a_1 z + c_2 z^2 + \cdots) + b(a_1 z + c_2 z^2 + \cdots)^k.$$

Equating coefficients up to and including terms with $z^k$ (and noting that $k \geq 2$), we obtain

$$c_2 = c_3 = \cdots = c_{k-1} = 0$$

and

$$a_1 b + A = c_k + ba_1^k.$$

Substituting the given value of $b$ in this last equation shows that $c_k = 0$, and so

$$g^{-1}Fg(z) = a_1 z + O(z^{k+1})$$

as required. □

We turn now to discuss topological conjugacy and here, of course, the conjugacy classes are much larger. First, observe that the dynamics of the iterates $f^n$ near an attracting fixed point is invariant under topological conjugacy, for if $f^n(z) \to 0$ and $F = hfh^{-1}$, then $F^n(hz) \to 0$ also, but in this case we cannot talk of the multiplier. In fact, given any two functions $f$ and $g$ (not necessarily analytic) which are orientation-preserving homeomorphisms of some neighbourhood of the origin into itself such that $|f(z)| < |z|$ and $|g(z)| < |z|$, then $f$ and $g$ are topologically conjugate [91] (see also [50]).

In the case of rationally indifferent fixed points, we have

**Theorem 6.10.6.** *Suppose that*

$$f(z) = a_1 z + a_2 z^2 + \cdots,$$

*where* $a_1 = \exp(2\pi ip/q)$, $(p, q) = 1$, *and also that no iterate* $f^n$ *is the identity. Then there is some integer* $k$ *such that near the origin,* $f$ *is analytically con-*

*jugate to*

$$z \mapsto a_1 z(1 + z^{kq} + \cdots),$$

*and topologically conjugate to* $z \mapsto z(1 + z^{kq})$.

The reader is referred to [33] for the details.

Finally, we comment briefly on the question of formal conjugacy and we mention just one result which sheds more light on the work in this section.

**Theorem 6.10.7.** *Any formal power series about the origin is formally conjugate to one and only one formal series of the form* $z(1 + z^p + Az^{2p})$.

The proof is by repeated applications of Theorem 6.10.3, and we refer the reader to [101] for more details.

EXERCISE 6.10

1. Show that the maps $x \mapsto 2x$ and $x \mapsto 3x$ of $\mathbb{R}$ into itself are topologically conjugate. [Consider a function $h(x) = ax^d$ for $x > 0$ and $h(x) = -h(-x)$ for $x < 0$.]
2. In the notation of the text, show directly that if $p = 1$, then $\tau(f)$ is invariant under conjugation.
3. Show that the two polynomials $z + z^2$ and $z + z^2 + z^3$ are topologically, but not analytically, nor formally, conjugate near the origin.
4. Let $f(z) = 2z^2 - 1$ and $g(z) = (1 + z^2)/(2z)$. Show that in a neighbourhood of $\infty$, $g^{-1}fg(z) = z^2$.
5. Let $f(z) = \exp[i \log(1 + z)] - 1$, where $\log(1 + z) = 0$ when $z = 0$, so $f'(0) = i$. Show that near the origin, $f^2(z) = -z/(1 + z)$, and $f^4$ is the identity map. [See Theorem 6.10.6, and note that if some iterate of a rational map $R$ is the identity, then $\deg(R) = 1$.]

# Appendix III. Infinite Products

We discuss conditions which imply the convergence and analyticity of an infinite product of analytic functions. Roughly speaking, the convergence of the infinite product is related to that of infinite sums by taking logarithms, but some care is needed in defining the logarithms. We shall prove two (equivalent) results; the first involves logarithms but the second does not.

**Theorem 1.** *Let $D$ be an open disc in the complex plane and suppose that the complex functions $g_1, g_2, \ldots$ are continuous and never zero in $D$, and that $\sum_{n=1}^{\infty} \log g_n(z)$ is locally uniformly convergent on $D$ to $S(z)$. Then as $n \to \infty$,*

$$\prod_{m=1}^{n} g_m(z) \to \exp[S(z)]$$

*locally uniformly on $D$.*

Proof. As $g_n$ does not vanish at any point of the disc $D$, there exists a single-valued, continuous branch of $\log g_n$ in $D$. By assumption, for suitable choices of these branches, the series $\sum_{n=1}^{\infty} \log g_n(z)$ is locally uniformly convergent to $S$ on $D$, so $S$ is continuous on $D$.

Now take any compact subset $Q$ of $D$, so for some number $M$, $|\exp[S(z)]| \leq M$ on $Q$. Next, take any $\varepsilon$ satisfying $0 < \varepsilon < 2M$, and choose $n_0$ such that for all $z$ in $Q$ and all $n \geq n_0$,

$$\left| \sum_{m=1}^{n} \log g_m(z) - S(z) \right| < \varepsilon/2M.$$

In general, if $|\zeta| < \varepsilon/2M$, then $|\zeta| < 1$ and so

$$|\exp(\zeta) - 1| < 2|\zeta|.$$

It follows that if $z \in Q$ and $n \geq n_0$, then

$$\left| \prod_{m=1}^{n} g_m(z) - \exp[S(z)] \right| < (\varepsilon/M)|\exp[S(z)]| < \varepsilon$$

as required. □

**Theorem 2.** *Suppose that the functions $f_1, f_2, \ldots$ are holomorphic in a domain $D$ and that the series*

$$\sum_{n=1}^{\infty} |f_n(z)|$$

*is locally uniformly convergent on $D$. Then*

$$\lim_{n\to\infty} \prod_{m=1}^{n} [1 + f_m(z)]$$

*exists locally uniformly in $D$.*

In the following proof, we shall use $\mathrm{Log}[\zeta]$ to denote the principal branch of the logarithm of $\zeta$ on the complex plane cut along the negative real axis. The inequality

$$|\zeta|/2 < |\mathrm{Log}[1 + \zeta]| < 3|\zeta|/2,$$

is valid for $|\zeta| < \frac{1}{2}$ and is easily proved (although the precise values of the bounds here are irrelevant).

Proof. It is sufficient to establish locally uniform convergence on the interior of any compact disc $K$ within $D$. Now for $n > n(K)$, say, we have $|f_n(z)| < \frac{1}{2}$ on $K$ so, for these $n$, the functions

$$g_n(z) = 1 + f_n(z)$$

satisfy the requirements of Theorem 1 with $D$ as the interior of $K$. Theorem 1 implies that the infinite product taken over $n > n(K)$ converges, and the required result follows from this by including the finite number of factors corresponding to $m = 1, \ldots, n(K)$. □

*Remark.* The results above show that if the factors in a convergent infinite product are holomorphic, then so is the infinite product. Further, the analysis above also shows that if the infinite product vanishes at some point $w$, then a positive, but only finite, number of the factors vanish at $w$.

# Appendix IV. The Universal Covering Surface

The following facts are well known, and together they form the Uniformization Theorem: see [2], [5] or [63] for more details. We call the three domains $\Delta$ (the open unit disc), $\mathbb{C}$ and $\mathbb{C}_\infty$ *canonical domains*, and an analytic automorphism of any one of these is necessarily a Möbius map. Given any subdomain $D$ of the complex sphere (or of any other Riemann surface), there exists a canonical domain $\Sigma$ (the *universal covering surface of* $D$), and an analytic map $\pi$ of $\Sigma$ onto $D$ (the *universal covering map*), such that: (1) each point of $\Sigma$ has a neighbourhood on which $\pi$ is injective; and (2) given any curve $\sigma: [0, 1] \to D$, and any point $w$ over $\sigma(0)$ (that is, with $\pi(w) = \sigma(0)$) there is a unique curve $\sigma_1: [0, 1] \to \Sigma$ such that $\pi\sigma_1 = \sigma$ (so $\sigma_1$ projects under $\pi$ to $\sigma$). We call $\sigma_1$ the *lift* of $\sigma$ *from* $w$.

The *cover group* $\Gamma$ is the group of all (Möbius) automorphisms $\gamma$ of $\Sigma$ such that $\pi\gamma = \pi$, and this has the properties;

(1) $\pi(z) = \pi(w)$ if and only if $z = \gamma(w)$ for some $\gamma$ in $\Gamma$;
(2) the non-trivial elements of $\Gamma$ have no fixed points in $\Sigma$; and
(3) $\Gamma$ is discrete (that is, the $\Gamma$-images of any $z$ do not accumulate in $\Sigma$).

These facts enable us to construct the topological quotient space $\Sigma/\Gamma$, and it is easy to see that $\Sigma/\Gamma$ is a Riemann surface which is conformally equivalent to the original domain $D$. If $D$ is a subdomain of the complex sphere whose complement contains at least three points, then $\Sigma$ must be $\Delta$ and the local hyperbolic geometry on $\Sigma$ projects to the same local geometry on $D$ (for the elements of the cover group are hyperbolic isometries of $\Delta$): for an example of this, see Appendix III in which $D = \mathbb{C}_\infty - \{0, 1, \infty\}$.

Now let $D$ be a domain (or a Riemann surface) whose universal covering surface is the unit disc $\Delta$. We regard $\Delta$ as the hyperbolic plane with metric $ds = 2|dz|/(1 - |z|^2)$, and as the elements of the cover group $\Gamma$ are isometries of this space, it follows that the hyperbolic metric on $\Delta$ projects down to a hyperbolic metric $\rho$ on $D$ (so that the projection map $\pi$ of $\Delta$ onto $D$ is a *local* isometry), and this implies that the intrinsic geometry of $D$ is *non-Euclidean*. This is so even when $D$ is a subdomain of $\mathbb{C}$, and in this case the hyperbolic metric is in sympathy with the geometry of $D$ in a way that the Euclidean metric is not; for example, the boundary of $D$ is an infinite hyperbolic distance away from any point of $D$. If the domains $D_1$ and $D_2$ have hyperbolic metrics $\rho_1$ and $\rho_2$ respectively, then we have the important *Comparison Principle*: *if* $D_1 \subset D_2$, *then* $\rho_2 \leq \rho_1$. This is a consequence of Schwarz's Lemma: see [2] and [20].

Given an analytic map $f: D \to D$, we say that the analytic map $F: \Sigma \to \Sigma$ is a *lift* of $f$ if

$$\pi F = f\pi.$$

Observe that for any lift $F$ and any $\zeta$, $F$ maps $\zeta$ to a point over $f\pi(\zeta)$. Further, if $F$ and $G$ are two lifts of $f$ with $F(\zeta) = G(\zeta)$, then as

$$\pi F(z) = f\pi(z) = \pi G(z),$$

and as $\pi$ is injective near $F(\zeta)$ $(= G(\zeta))$, we see that for all $z$ sufficiently near to $\zeta$, $F(z) = G(z)$ and hence $F = G$. It follows that each lift $F$ maps $\zeta$ to some point $w$ over $f\pi(\zeta)$, and that $F$ is the only lift that maps $\zeta$ to $w$.

We shall now describe how to construct lifts and so obtain the totality of all lifts of $f$. We take $\zeta = 0$ (this is not essential) and take any point $w$ over $f\pi(0)$. Now draw a curve $\sigma$ in $\Sigma$ from 0 to $z$, say, project this to the curve $\pi(\sigma)$ and then apply $f$ to obtain the curve $f\pi(\sigma)$ in $D$ from $f\pi(0)$ to $f\pi(z)$. Next, construct the unique lift $\sigma_1$ of $f\pi(\sigma)$ from $w$ and denote the terminal point of this by $F(z)$. As $\Sigma$ is simply connected, the construction of $F(z)$ depends only on $z$ (and not on $\sigma$), and, by construction, $\pi F(z) = f\pi(z)$. It is easy to see that $F$ is analytic; thus $F$ is a lift of $f$ with $F(0) = w$.

As $\pi\gamma = \pi$ for any $\gamma$ in $\Gamma$, we see that if $F$ is any lift of $f$, then so is $\gamma F$. With this and the argument in the preceding paragraph, we now see that if $F$ is a lift of $f$, then the totality of lifts of $f$ is $\{\gamma F: \gamma \in \Gamma\}$. In particular, $\Gamma$ is the totality of lifts of the identity map of $D$ onto itself.

Next, we consider the nature of the lifts of $f$ when $f$ is an analytic automorphism of $D$ (that is, an analytic bijection of $D$ onto itself). In this case, we select any lift $F$ of $f$. Now $f^{-1}$ is also an automorphism of $D$ and so this too has a lift, say $G$, and as 0 lies over $f^{-1}\pi F(0)$, we may assume that $G$ maps $F(0)$ to 0. As

$$\pi FG = f\pi G = ff^{-1}\pi = \pi$$

and

$$\pi GF = g\pi F = f^{-1}f\pi = \pi.$$

we see that both $FG$ and $GF$ are lifts of the identity map of $D$ onto itself, and hence are elements of $\Gamma$. As $FG$ fixes $F(0)$, and $GF$ fixes 0, we find that $FG$ and $GF$ are the identity maps of $\Sigma$ onto itself and so $F$ is an automorphism of $\Sigma$ with $F^{-1} = G$. In particular, $F$ is a Möbius map.

Now observe that $\pi(F\gamma) = f\pi\gamma = f\pi$, so $F\gamma$ is a lift of $f$. This shows that $F\Gamma \subset \Gamma F$, and using this with $F$ replaced by $F^{-1}$ $(= G)$ we find that $F\Gamma = \Gamma F$.

The *normalizer* of $\Gamma$ is $N(\Gamma)$, where

$$N(\Gamma) = \{\eta : \eta \text{ Möbius}, \eta\Gamma\eta^{-1} = \Gamma\},$$

and so we have seen that *the lift of an automorphism of D lies in the normalizer of* $\Gamma$. It is not difficult to see that every element of $N(\Gamma)$ arises in this way, and moreover, the group Aut($D$) of automorphisms of $D$ is isomorphic to the quotient group $N(\Gamma)/\Gamma$.

# CHAPTER 7
# Forward Invariant Components

In Chapter 8 we shall show that each component of the Fatou set $F$ of $R$ has a forward image that is mapped onto itself by some iterate $R^m$. Anticipating this, we devote this chapter to the classification of the forward invariant components of the Fatou set of a rational map. Although this classification logically follows Chapter 8, by giving it first we emphasize that it does not depend on the more advanced material in Chapter 8.

## §7.1. The Five Possibilities

Our objective is to give a complete analysis of the possible forward invariant components of the Fatou set of a rational map. We shall show that such a component can arise in exactly one of five different ways, and we begin with the terminology used to describe these possibilities: as usual, $R$ is a rational map of degree at least two with Fatou set $F$.

**Definition 7.1.1.** A forward invariant component $F_0$ of $F$ is:

(a) an *attracting component* if it contains an attracting fixed point $\zeta$ of $R$;
(b) a *super-attracting component* if it contains a super-attracting fixed point $\zeta$ of $R$;
(c) a *parabolic component* (or a *Leau domain*) if there is a rationally indifferent fixed point $\zeta$ of $R$ on the boundary of $F_0$, and if $R^n \to \zeta$ on $F_0$;
(d) a *Siegel disc* if $R: F_0 \to F_0$ is analytically conjugate to a Euclidean rotation of the unit disc onto itself;
(e) a *Herman ring* if $R: F_0 \to F_0$ is analytically conjugate to a Euclidean rotation of some annulus onto itself.

Of course, in (d) and (e), the rotations necessarily have infinite order for otherwise, some $R^n$ would be the identity and then $\deg(R) = 1$. We know that attracting, super-attracting and parabolic components exist, and the existence of Siegel discs was established in Chapter 6. The proof that Herman rings exist is more difficult and we comment on this in §7.4.

The central, and most important, result in this chapter is that (a)–(e) are the only possibilities for a forward invariant component of $F$, and this is recorded in

**Theorem 7.1.2.** *A forward invariant component of the Fatou set $F(R)$ is one of the types* (a)–(e) *in Definition* 7.1.1.

Theorem 7.1.2 is simply an amalgamation of several other theorems which occur later in this chapter, but we shall explain the general strategy of the proof now. The five possibilities in Definition 7.1.1 can almost be distinguished from each other by considering those functions which can be expressed as the limit of some subsequence of $R^n$ in $F_0$. In (a) and (b), the only possible limit is the fixed point $\zeta$ and this is in $F_0$; in (c), the only limit function is $\zeta$, and this is in $\partial F_0$; finally, in (d) and (e), non-constant limit functions exist. To complete the description, note that (a) and (b) are distinguished according to whether $\zeta$ is, or is not, a critical point, and (d) and (e) are distinguished either by the connectivity of $F_0$, or by the existence of a fixed point in $F_0$.

EXERCISE 7.1

*Remark.* According to Theorem 5.2.1, any completely invariant component of $F$ is either simply connected or infinitely connected. These six exercises show the existence of completely invariant components of types (a), (b) and (c) which are (i) simply connected, and (ii) infinitely connected.

1. Let $R(z) = z(z - a)/(1 - \bar{a}z)$, where $0 < |a| < 1$. Show that $\Delta$ is a simply connected attracting component of $F(R)$. [As $\{|z| < 1\}$ and $\{|z| > 1\}$ are both forward invariant, they cannot be in the same component of $F$.]

2. Let $R(z) = z/(2 - z^2)$. Show that $F(R)$ has an attracting component of infinite connectivity. [Conjugate the function discussed in §1.8.]

3. Let $R(z) = z^2$. Show that $\Delta$ is a simply connected super-attracting component of $F(R)$.

4. Let $R(z) = 6z(1 - z)$. Draw the graph of $R(x)$ for $0 \leq x \leq 1$, and show that $\bigcap_{n=0}^{\infty} R^{-n}([0, 1])$ is a Cantor set $E$. Prove that $J = E$, and deduce that $R$ has a super-attracting component of infinite connectivity.

5. Let $P(z) = z - z^2$ (see Example 6.5.3). Show that $F(P)$ has a completely invariant, simply connected, parabolic component. Let $F_0$ be the component of $F$ which contains the one petal at the origin. Show that $F_0$ is simply connected ($P$ is a polynomial) and completely invariant ($\deg(P) = 2$), and $F_0$ contains the critical point of $P$.

6. Let $R(z) = z/(1 + z - z^2)$. Find $\mathrm{Im}[R(z)]$ and deduce that each of
$$\{x + iy: y > 0\}, \quad \{x + iy: y < 0\}, \quad \mathbb{R} \cup \{\infty\},$$
is completely invariant under $R$. Show that $0 < R(x) < x < 1$ on $(0, 1)$; deduce that $R^n \to 0$ on $(0, 1)$, and hence (by considering the periodic points of $R$ in $J$) that $J \cap (0, 1) = \varnothing$. Show now that $F$ is completely invariant, infinitely connected and of type (c).

## §7.2. Limit Functions

We now consider the class of functions which arise as locally uniform limits of subsequences of $(R^n)$ in $F_0$. First (and regardless of whether $F_0$ is forward invariant or not) we have

**Definition 7.2.1.** A function $\varphi$ is a *limit function* on a component $F_0$ of $F(R)$ if there is some subsequence of $(R^n)$ which converges locally uniformly to $\varphi$ on $F_0$. The class of limit functions on $F_0$ is denoted by $\mathscr{F}(F_0)$.

As $\{R^n: n \geq 1\}$ is normal in $F_0$, $\mathscr{F}(F_0)$ is always non-empty, and each map in $\mathscr{F}(F_0)$ is analytic in $F_0$. If $F_0$ is forward invariant, and if $\varphi$ is a limit function in $F_0$, then $\varphi(F_0)$ lies in the closure of $F_0$: in particular, if $\varphi$ is constant, with value $\zeta$ say, then $\zeta \in F_0 \cup \partial F_0$. Moreover, if $z$ is in $F_0$, then so is $R(z)$, and so on some sequence of integers $n$ tending to $\infty$,

$$R(\zeta) = R\left(\lim_{n\to\infty} R^n(z)\right) = \lim_{n\to\infty} R^n(R(z)) = \varphi(R(z)) = \zeta.$$

This proves

**Lemma 7.2.2.** *If $F_0$ is forward invariant, and if there exists a constant limit function with value $\zeta$, then $\zeta$ is a fixed point of $R$.*

Of course, if the component $F_0$ is (super)-attracting with fixed point $\zeta$, say, then $R^n \to \zeta$ locally uniformly on $F_0$ and so $\mathscr{F}(F_0)$ contains a single function, namely the constant function with value $\zeta$. The same is true (by definition) if $F_0$ is a parabolic component as in Definition 7.1.1(c): note that it is necessary to assume that $R^n \to \zeta$ in this case for a rationally indifferent fixed point can lie on the boundary of a super-attracting component of $F$ (for example, when $R$ is a polynomial).

The next two results enable us to separate (a), (b) and (c) from (d) and (e) but before giving these, we recall that an *automorphism* of a domain $D$ is an analytic bijection of $D$ onto itself. The group of automorphisms of $D$ is denoted by $\mathrm{Aut}(D)$.

**Theorem 7.2.3.** *Suppose that $F_0$ is forward invariant, and that every function in $\mathscr{F}(F_0)$ is constant. Then $\mathscr{F}(F_0)$ contains exactly one function, with value $\zeta$, say, where $R(\zeta) = \zeta$ and $R^n \to \zeta$ locally uniformly on $F_0$.*

**Theorem 7.2.4.** *Suppose that $F_0$ is forward invariant, and that $\mathscr{F}(F_0)$ contains some non-constant function. Then $R$ is in* $\text{Aut}(F_0)$, *and the identity map $I$ is in* $\mathscr{F}(F_0)$.

Proof of Theorem 7.2.3. In conjunction with Lemma 7.2.2, the hypotheses imply that $\mathscr{F}(F_0)$ cntains at most $\deg(R) + 1$ functions, and each is constant. Suppose first that some limit function has the value $\zeta$, where $\zeta \in F_0$: then $R(\zeta) = \zeta$, and for some sequence $n_j$, $R^{n_j} \to \zeta$ locally uniformly on $F_0$. Let $N$ be an open Euclidean disc with centre $\zeta$, and whose closure lies in $F_0$. By assumption, for some integer $m$ of the form $n_j$, $R^m(\bar{N})$ is strictly contained in $N$, and so, by the Schwarz Lemma, $|(R^m)'(\zeta)| < 1$. However, $R$ fixes $\zeta$, so

$$|R'(\zeta)|^m = |(R^m)'(\zeta)| < 1$$

and $\zeta$ is a (super)attracting fixed point of $R$. It follows that the full sequence of iterates $R^n$, $n \geq 1$, converges locally uniformly to $\zeta$ on $F_0$, and in this case there is only one limit function, namely the constant map with value $\zeta$.

It remains to consider the case when $\mathscr{F}(F_0)$ consists of a finite number of constant limit functions, the value of each being a fixed point of $R$ on $\partial F_0$. Take any compact subset $K$ of $F_0$ and, enlarging $K$ if necessary, we may assume that $K$ is connected, and contains some pair of points $w$ and $R(w)$. It follows that $R(K)$ meets $K$, $R^2(K)$ meets $R(K)$, and so on, and so for any $n_0$,

$$\bigcup_{n=n_0}^{\infty} R^n(K)$$

is connected.

There are only a finite number of fixed points of $R$ on $\partial F_0$, so let these be $\zeta_1, \ldots, \zeta_r$, and take any mutually disjoint open neighbourhoods $V_j$ of $\zeta_j$ respectively. If there is a strictly increasing sequence of integers $k_j$ such that each $R^{k_j}(K)$ meets the complement of $\bigcup_{j=1}^r V_j$, then no subsequence of $(R^{k_j})$ can converge locally uniformly to any of $\zeta_1, \ldots, \zeta_r$ (which it must); thus for some $n_0$,

$$\bigcup_{n=n_0}^{\infty} R^n(K) \subset \bigcup_{j=1}^{r} V_j.$$

As the union of the $R^n(K)$ is connected, it must lie in one $V_j$, say in $V_1$, and it follows that $\zeta_1$ is the only limit function in $F_0$. Moreover, the fact that $R^n(K) \subset V_1$ for all sufficiently large $n$, means that $R^n \to \zeta_1$ uniformly on $K$, and hence locally uniformly on $F_0$. □

Proof of Theorem 7.2.4. By assumption, there are non-constant analytic limit functions in $\mathscr{F}(F_0)$, and we begin by proving that if $\varphi$ is any one of these, then

$$\varphi(F_0) \subset F_0. \tag{7.2.1}$$

First, there is some sequence of integers $n_j$ such that $R^{n_j} \to \varphi$ locally uniformly on $F_0$. Now take any $w$ in $F_0$. As $\varphi$ is not constant, the zeros of the map $z \mapsto \varphi(z) - \varphi(w)$ are isolated, so let $C$ be the boundary of some closed disc centred at $w$ and lying in $F_0$, and which is such that $\varphi \neq \varphi(w)$ on $C$. Then (by the uniform convergence on $C$), for all sufficiently large $j$, and all $z$ on $C$,

$$|R^{n_j}(z) - \varphi(z)| < \inf_C |\varphi(z) - \varphi(w)|.$$

By Rouché's Theorem, the two functions $\varphi(z) - \varphi(w)$ and $R^{n_j}(z) - \varphi(w)$ have the same number of zeros inside $C$. As the first function vanishes at $w$, we deduce that $\varphi(w)$ lies in $R^{n_j}(F_0)$ $(=F_0)$ and so (7.2.1) holds.

Consider now the non-constant function $\varphi$ and the sequence $n_j$ as above. By passing to a subsequence of the $n_j$ and relabelling, we may assume that

$$m_j = n_j - n_{j-1} \to +\infty$$

as $j \to \infty$. Now $\{R^{m_j}\}$ is normal in $F_0$ so there is some function $\psi$ on $F_0$ such that $R^{m_j} \to \psi$ locally uniformly on $F_0$ as $j \to \infty$ in some set $N$ of positive integers.

Now take any $z$ in $F_0$: then $R^{n_j}(z) \to \varphi(z)$ where, by (7.2.1), $\varphi(z)$ is in $F_0$. As $j \to \infty$ in $N$, so $R^{m_j}$ converges uniformly to $\psi$ on some compact neighbourhood of $\varphi(z)$ and hence, letting $j \to \infty$ in $N$, we have

$$\psi\varphi(z) = \lim R^{m_j}(R^{n_{j-1}}(z)) = \lim R^{n_j}(z) = \varphi(z) \tag{7.2.2}$$

(see Exercise 7.2.1). As $\varphi$ is not constant, $\psi$ must be the identity map $I$, and this proves that $I$ is in $\mathscr{F}(F_0)$.

It is now easy to see that $R$ is in $\mathrm{Aut}(F_0)$. Indeed, because $F_0$ is forward invariant, $R$ must map $F_0$ onto itself. Further, the injectivity of $R$ is trivial for if $R(z) = R(w)$, then

$$R^{m_j}(z) = R^{m_j-1}(Rz) = R^{m_j-1}(Rw) = R^{m_j}(w)$$

and letting $j \to \infty$ in $N$, we obtain $z = I(z) = I(w) = w$. This completes the proof of Theorem 7.2.4. □

The same ideas enable one to show (in the proof of Theorem 7.2.4) that if $\varphi$ is a non-constant limit function, then $\varphi \in \mathrm{Aut}(F_0)$. Indeed, $R \in \mathrm{Aut}(F_0)$ and so has an inverse, say $S\colon F_0 \to F_0$. Now $\{S^n\}$ is a normal family in $F_0$ (because the values taken by $S^n$ do not lie in $J$), thus given any sequence $n_j$ with $R^{n_j} \to \varphi$ locally uniformly in $F_0$, we can take a subsequence $m_j$ of $n_j$ such that $S^{m_j}$ converges locally uniformly to some function $\psi$ on $F_0$. Now as $\varphi$ maps $F_0$ into itself, it follows that

$$\psi\varphi(z) = \lim S^{m_j}R^{m_j}(z) = z, \tag{7.2.3}$$

and hence that $\varphi$ is injective on $F_0$. Further, (7.2.3) shows that $\psi$ is not constant, and exactly as in the proof of (7.2.1), we see that $\psi$ maps $F_0$ into itself. With this, we can reverse the roles of $\varphi$ and $\psi$ in (7.2.3) and deduce that $\varphi\psi(z) = z$ on $F_0$. It follows that $\varphi$ is a bijection of $F_0$ onto itself and so is in $\mathrm{Aut}(F_0)$. For other results, see Exercises 7.2.2 and 7.2.3.

EXERCISE 7.2

1. Let $D$ be a domain and suppose that $f_n \to f$ and $g_n \to g$ locally uniformly on $D$ (with all maps analytic). Suppose also that $g(D) \subset D$. Prove that $f_n g_n \to fg$ on $D$.

2. In the context of the proof of Theorem 7.2.4, show that if $\varphi_1$ and $\varphi_2$ are non-constant functions in $\mathscr{F}(F_0)$, then $\varphi_1 \varphi_2$ is in $\mathscr{F}(F_0)$. Show also that $\varphi_1 \varphi_2 = \varphi_2 \varphi_1$.

3. In the context of the proof of Theorem 7.2.4, show that if $\varphi$ is a non-constant function in $\mathscr{F}(F_0)$, then $\varphi^{-1}$, which exists in $\mathrm{Aut}(F_0)$, is also in $\mathscr{F}(F_0)$. [*Hint.* Take $R^{n_j} \to \varphi$ and $R^{m_j} \to I$ and, by passing to subsequences, assume that $m_j - n_j$ strictly increases to $+\infty$. Now as $j \to \infty$ in some suitable sequence, $R^{m_j} \to I$, $R^{m_j - n_j} \to \psi$ and $R^{n_j} \to \varphi$.]

4. Use Theorem 7.1.2 to show that if $\mathscr{F}(F_0)$ contains non-constant functions, then $F(R)$ has infinitely many components.

## §7.3. Parabolic Domains

Suppose now that $F_0$ is a forward invariant component of $F(R)$ and for the moment, regard $\partial F_0$ as the ideal boundary point of the one-point compactification of $F_0$: explicitly,

$$R^n \to \partial F_0 \tag{7.3.1}$$

means that for each compact subset $K$ of $F_0$, $R^n(K)$ is disjoint from $K$ for all but a finite set of $n$. If this is so, then any limit function $\varphi$ on $F_0$ satisfies $\varphi(F_0) \subset \partial F_0$, and as $\partial F_0$ has an empty interior (simply because $F_0$ is a domain), we deduce (from the Open Mapping Theorem) that $\varphi$ is necessarily constant. Thus if (7.3.1) holds, then by Theorem 7.2.3, there is some fixed point $\zeta$ of $R$ in $\partial F_0$ such that $R^n \to \zeta$ locally uniformly on $F_0$. Our next result contains these conclusions and adds one other important fact.

**Theorem 7.3.1.** *Suppose that $F_0$ is a forward invariant component of $F(R)$, and that $R^n \to \partial F_0$ as $n \to \infty$. Then there is some rationally indifferent fixed point $\zeta$ of $R$ in $\partial F_0$ such that:*

(a) *$R^n \to \zeta$ locally uniformly on $F_0$ as $n \to \infty$; and*
(b) $R'(\zeta) = 1$.

PROOF. The remarks preceding the theorem show that there is a fixed point $\zeta$ in $\partial F_0$ such that (a) holds; thus $\zeta$ is uniquely determined by the action of $R$ on $F_0$ and we only have to show that $R'(\zeta) = 1$. Obviously, $|R'(\zeta)| \geq 1$ for otherwise, $\zeta$ would be a (super)-attracting fixed point of $R$ and so would lie in $F(R)$. It is also easy to see that $|R'(\zeta)| \leq 1$, for if not, $\zeta$ would be a repelling fixed point of $R$. However, we showed in §1.1 that $R^n(z)$ can only converge to a repelling fixed point $\zeta$ if $R^n(z) = \zeta$ for some $n$, and as $z \in F$ and $\zeta \in J$, this cannot be so. It follows that

$$|R'(\zeta)| = 1, \tag{7.3.2}$$

and we shall use this to prove that $R'(\zeta) = 1$. We remark that we expect $R'(\zeta) = 1$ to hold because by (7.3.2), $R$ acts like a rotation about $\zeta$ on a small neighbourhood of $\zeta$, and in order that $R(F_0) = F_0$, this rotation must be trivial.

The first part of our argument is contained in the second proof of Theorem 6.5.8 and we shall only outline the details of this part. By conjugation, we may assume that $\zeta = 0$ and that $\infty \in \partial F_0$. Next, put $\lambda = R'(0)$ and note that as $|\lambda| = 1$, $R$ is injective in some neighbourhood $\mathcal{N}$ of the origin. Exactly as in the second proof of Theorem 6.5.8, we can now construct a forward invariant subdomain $W$ of $F_0 \cap \mathcal{N}$.

Now take any point $\zeta_0$ in $W$, and for $n \geq 1$ define the functions $\varphi_n$ on $W$ by

$$\varphi_n(z) = R^n(z)/R^n(\zeta_0).$$

By Lemma 6.5.9, $\{\varphi_n\}$ is normal in $W$ and so some sequence $\varphi_{n_j}$ converges locally uniformly to some function $\varphi$ on $W$. Further, by the remarks following the proof of Lemma 6.5.9,

$$\varphi(Rz) = \lambda\varphi(z): \tag{7.3.3}$$

at this point in §6.5 we knew that $\lambda = 1$, but here, we only know that $|\lambda| = 1$.

We now complete the proof of Theorem 7.3.1 without further reference to §6.5. First, a non-constant locally uniform limit of injective analytic maps is injective (Hurwitz's Theorem); thus either $\varphi$ is constant in $W$, or it is injective in $W$. If $\varphi$ is constant in $W$, then its value is 1 (its value at $\zeta_0$) and from (7.3.3), we see that $\lambda = 1$ as required.

If $\varphi$ is not constant on $W$, then it has an inverse $\varphi^{-1}$ which maps $\varphi(W)$ onto $W$. However, from (7.3.3) we obtain

$$\varphi(R^n\zeta_0) = \lambda^m\varphi(\zeta_0) = \lambda^n.$$

Now from (7.3.2), $|\lambda| = 1$, so there is an increasing sequence of integers $m_j$ such that $\lambda^{m_j} \to 1$ (Exercise 7.3.1). It follows that $\varphi(R^{m_j}\zeta_0) \to 1$, and as the open set $\varphi(W)$ contains the point 1 ($= \varphi(\zeta_0)$), we see that $\varphi(R^{m_j}\zeta_0) \in \varphi(W)$ for all sufficiently large $j$. For these $j$, then, we have

$$R^{m_j}(\zeta_0) = \varphi^{-1}(\lambda^{m_j}) \to \varphi^{-1}(1) = \zeta_0,$$

and this is false as we know that $R^n \to 0$ ($= \zeta$) on $W$. We deduce that $\varphi$ is necessarily constant and hence that $\lambda = 1$. □

*Remark.* Suppose that $F_0$ is forward invariant and that every limit function in $\mathcal{F}(F_0)$ is constant. Then for some fixed point $\zeta$, $R^n \to \zeta$ in $F_0$ and $\zeta \in F_0 \cup \partial F_0$. If $\zeta \in F_0$, then $F_0$ is a (super)attracting component of $F(R)$ (see the proof of Theorem 7.2.3), while if $\zeta \in \partial F_0$, then by Theorems 7.2.3 and 7.3.1, $F_0$ is a parabolic component of $F(R)$. Thus with reference to the proof of Theorem

7.1.2, we have shown that if every limit function in $\mathscr{F}(F_0)$ is constant, then $F_0$ is one of the types (a), (b) and (c) in Definition 7.1.1.

EXERCISE 7.3

1. Suppose that $|\lambda| = 1$. Show that either $\lambda^m = 1$ for some $m$ or $\{\lambda^n: n \geq 1\}$ is dense in the unit circle. Deduce that there is a strictly increasing sequence of integers $m_j$ such that $\lambda^{m_j} \to 1$.

## §7.4. Siegel Discs and Herman Rings

In this section we prove

**Theorem 7.4.1.** *Suppose that $F_0$ is a forward invariant component of $F(R)$, where $\deg(R) \geq 2$, and that $\mathscr{F}(F_0)$ contains non-constant functions. Then $F_0$ is either a Siegel disc or a Herman ring.*

*Remark.* With this, we will have completed the proof of Theorem 7.1.2.

The idea of the proof of Theorem 7.4.1 is as follows. As $J$ has at least three points, the universal covering space of $F_0$ is the unit disc $\Delta$, and $F_0$ is conformally equivalent to $\Delta/\Gamma$, where $\Gamma$ is the cover group (of Möbius transformations) acting on $\Delta$. Now the conformal automorphisms $R^n$ of $F_0$ accumulate at the identity map (see Theorem 7.2.4), and if we rewrite this information in terms of the group $\Gamma$ and its normalizer, we find that $\Gamma$ is abelian. However, the only abelian cover groups are cyclic, and if we now list all possible cyclic groups and compute the corresponding quotient spaces, we find that $\Delta/\Gamma$, and hence $F_0$, is simply or doubly connected. It follows that $F_0$ is conformally equivalent to either a disc or an annulus and the rest of the argument is straightforward. We now give the details.

PROOF OF THEOREM 7.4.1. As $J$ is infinite, the universal covering space of $F_0$ is the unit disc $\Delta$ (see Appendix IV to Chapter 6). We denote the cover group by $\Gamma$ and the universal covering map of $\Delta$ onto $F_0$ by $\pi$. By replacing $R$ by a conjugate, we may assume that the origin lies in $F_0$, and that $\pi$ is chosen so that $\pi(0) = 0$. Next, we select a neighbourhood $V$ of the origin such that the restriction $\pi_0$ of $\pi$ to $V$ is injective there.

Now by Theorem 7.2.4, $\mathscr{F}(F_0)$ contains the identity map $I$ so $R^n \to I$ locally uniformly in $F_0$ as $n \to \infty$ through some sequence of integers. Working only with integers $n$ in this sequence, and taking $n$ sufficiently large, we find that $R^n(0) \in \pi_0(V)$. It follows that for these $n$ we can define functions $S_n$ by

$$S_n = \pi_0^{-1} R^n \pi,$$

and each of these is analytic in some neighbourhood of the origin. However, the properties of the universal covering map guarantee that each $S_n$ has an analytic continuation to $\Delta$, and as $\Delta$ is simply connected, $S_n$ is single valued on $\Delta$: in fact, $S_n$ is a lift of $R^n$ so $R^n\pi = \pi S_n$ throughout $\Delta$ (see Appendix IV). By definition, $S_n(0) \in V$ and so as $n \to \infty$ (on the given sequence),

$$S_n(0) = \pi_0^{-1}R^n\pi(0) \to 0;$$

thus there are two cases to consider, namely:

*Case* 1: for some $n$, $S_n(0) = 0$; and
*Case* 2: for all $n$, $S_n(0) \neq 0$.

First, in Case 1 (and for this $n$) we have

$$R^n(0) = R^n\pi(0) = \pi S_n(0) = \pi(0) = 0,$$

so 0 is a fixed point of $R^n$. Now 0 is not a repelling fixed point of $R^n$ for it lies in $F(R^n)$, nor is it an attracting fixed point of $R^n$ for if it were, then by considering the iterates of $R^n$ acting on a set $\{z, R(z), \ldots, R^{n-1}(z)\}$, we see that the full sequence of iterates $R^m$, $m \geq 1$, converges to 0 (contrary to the existence of non-constant limit functions in $F_0$). It follows that 0 is an indifferent fixed point of $R^n$ and so from Theorem 6.6.3, $F_0$ is a Siegel disc for $R^n$. In particular, $F_0$ is simply connected and so $R$ (which, by Theorem 7.2.4, is an analytic automorphism of $F_0$) is conjugate (by the Riemann Mapping Theorem) to a Möbius automorphism $M$, say, of the unit disc $\Delta$. It follows that $M^n$ has a fixed point say $\xi$ in $\Delta$ (corresponding to the fixed point 0 of $R$ in $F_0$) and hence (see Exercise 7.4.1) that $M$ also fixes $\xi$. This implies that $R$ fixes 0, and also that 0 is an indifferent fixed point of $R$ (for if not, it would not be an indifferent fixed point of $R^n$). Applying Theorem 6.6.3 again, we see that $F_0$ is a Siegel disc for $R$.

For an alternative argument to this last part, we observe that as 0 is a fixed point of $R^n$, it must be a fixed point of $R$ for $F_0$ is forward invariant and any component of $F$ can contain at most one periodic point (Lemma 6.9.3).

It remains to consider Case 2. We recall that as $n \to \infty$ on some suitable sequence, $R^n \to I$ locally uniformly on $F_0$. It follows that for some suitably small neighbourhood $\mathcal{N}$ of the origin, $R^n(\mathcal{N}) \subset \pi_0(V)$ and so $S_n \to I$ uniformly on $\mathcal{N}$. As $\{S_n\}$ is normal in $\Delta$ (for each $S_n$ maps $\Delta$ into itself), Vitali's Theorem now implies that $S_n \to I$ locally uniformly in $\Delta$. Further, as the $S_n$ are automorphisms of $\Delta$ (see Appendix IV), we easily see that $S_n^{-1} \to I$ locally uniformly on $\Delta$.

Now take any element $\gamma$ in the cover group $\Gamma$ and consider the sequence $S_n^{-1}\gamma S_n$. First, as the $S_n$ are in $N(\Gamma)$ (see Appendix IV), these elements lie in $\Gamma$ and second, because $S_n$ and $S_n^{-1}$ converge locally uniformly to $I$ on $\Delta$, $S_n^{-1}\gamma S_n \to \gamma$. As the $\Gamma$-orbit of a point in $\Delta$ cannot accumulate in $\Delta$, we deduce that for all sufficiently large $n$, say for $n > n(\gamma)$, $S_n^{-1}\gamma S_n = \gamma$ and so $S_n$ commutes with $\gamma$. Using the same argument for any other element $\rho$ of $\Gamma$, we now see that for any pair of elements $\gamma$ and $\rho$ of $\Gamma$, $S_n$ commutes with both $\rho$ and $\gamma$ when $n$ is sufficiently large.

It is convenient now to replace $\Delta$ (as the universal covering space of $F_0$) by the upper half-plane $H$: the new cover group (acting on $H$) is some conjugate group $h\Gamma h^{-1}$, where $h$ is Möbius and $h(\Delta) = H$, but for simplicity we continue to use the notation $\Gamma$ and $S_n$, where now the $S_n$ are automorphisms of $H$ (and $S_n$ is still a lift of $R^n$). By choosing $h$ suitably, we may assume that our chosen element $\gamma$ of $\Gamma$ is one of the maps

$$z \mapsto z + 1, \qquad z \mapsto kz,$$

where $k > 1$.

Now recall that for large $n$, $S_n$ commutes with $\gamma$ and $\rho$. A trivial exercise shows that the only Möbius maps which preserve $H$ and which commute with a map $z \mapsto z + t$, $t$ real, are the maps $z \mapsto z + s$, $s$ real. Thus if $\gamma(z) = z + 1$, then $S_n(z) = z + t$ ($t$ real) and hence $\rho(z) = z + s$ ($s$ real). This means that $\Gamma$ is a discrete group of real translations and so up to conjugacy is generated by $z \mapsto z + 1$. In this case the quotient map is $z \mapsto \exp(2\pi iz)$, and then $H/\Gamma$ is the punctured disc

$$\Delta^* = \{z: 0 < |z| < 1\}.$$

It follows that there is an analytic bijection, say $\varphi$, of $\Delta^*$ onto $F_0$. Now $\varphi$ extends to an analytic bijection of $\Delta$ onto $F_0 \cup \{\xi\}$, say, where $\xi$ is an isolated point of $\partial F_0$. This cannot be so, however, for it implies that $\xi$ is an isolated point of $J(R)$: thus $\gamma$ is not the map $z \mapsto z + 1$.

We know now that $\gamma(z) = kz$, where $k > 1$, and the argument follows essentially the same lines as above (see Exercise 7.4.2 for an alternative approach). The only Möbius maps which commute with $\gamma$ are the maps $z \mapsto tz$, where $t > 0$, and the maps $z \mapsto \mu/z$. Now maps of the second type are of order two, so if $S_n$ is of this type then $R^2$ is the identity on $F_0$ and this cannot be so. It follows that each $S_n$ is of the form $z \mapsto tz$ as above. Now for large $n$, the element $\rho$ of $\Gamma$ commutes with $S_n$, so $\rho$ is of one of the maps $z \mapsto sz$, where $s > 0$ and $s \neq 1$, or $z \mapsto \mu/z$. We know that $\rho$ has no fixed points in $H$ so $\rho$ must be of the form $z \mapsto sz$. Finally, as $\rho$ was any element of $\Gamma$, we deduce that $\Gamma$ is a discrete subgroup of $\{z \mapsto tz: t > 0\}$ and so is a cyclic group generated by some map $z \mapsto kz$.

It is now evident that $H/\Gamma$ is a double connected region (it is obtained topologically by identifying the edges of

$$\{z \in H: 1 < |z| < k\}$$

under $z \mapsto kz$, and as it is conformally equivalent to $F_0$, we deduce that $F_0$ is doubly connected (see Exercise 7.4.3 for an alternative approach to this part). To complete the proof of Theorem 7.4.1, we need only establish the following result.

**Lemma 7.4.2.** *Let*

$$A = \{z: 1/r < |z| < r\},$$

*where $r > 1$. Then the Euclidean rotations of $A$ are the only automorphisms of $A$ that are of infinite order.*

PROOF OF THEOREM 7.4.2. Let $f$ be an automorphism of $A$. We use Exercise 7.4.3 so $k > 1$ and $A = q(H)$. Using the same ideas as in the earlier parts of the proof, we can lift the map $f$ to a Möbius map $F$ of $H$ onto itself with the property $qF = fq$. This shows that for any two points $z$ and $w$ in $H$ with $q(z) = q(w)$, we have $F(z) = \gamma^m F(w)$ for some integer $m$, and it follows from this that for each $z$ in $H$ there is an integer $m(z)$ such that

$$F(kz) = k^{m(z)}F(z).$$

Clearly, the function $m(z)$ varies continuously with $z$, and as $H$ is connected, $m(z)$ is constant on $H$, say with value $m$: thus for all $z$ in $H$, $F(kz) = k^m F(z)$, and so, by induction,

$$F(k^n z) = k^{mn}F(z).$$

Recalling that $F$ is Möbius and letting $n \to \infty$, we find that $F(\infty)$ is either $\infty$ (if $m > 0$), or 0 (if $m < 0$): note that $m \neq 0$, else $F$ is constant with value $F(\infty)$. By letting $n \to -\infty$, we find that $F(0)$ is either 0 or $\infty$, so we find that either $F$ fixes both 0 and $\infty$, or it interchanges them. This means that $F$ is of one of the forms $z \mapsto az$, $z \mapsto b/z$, where $a > 0$ and $b < 0$, and a computation using $qF = fq$ shows that $f$ is of one of the forms $z \mapsto e^{i\theta}z$ or $z \mapsto e^{i\theta}/z$, where $\theta$ is real. As $f$ is of infinite order it is a rotation and this completes the proofs of Lemma 7.4.2, Theorem 7.4.1 and Theorem 7.1.2. □

We end this section with some brief remarks on the existence of Herman rings. Let

$$R(z) = \lambda z^2\left(\frac{1 + \bar{\alpha}z}{z + \alpha}\right), \tag{7.4.1}$$

where $|\lambda| = 1$ and $0 < |\alpha| < 1$. If $|\alpha|$ is sufficiently small, then $R$ is a homeomorphism of the unit circle $\partial\Delta$ onto itself (see Exercise 7.4.4). Now it is known that for suitable choices of $\lambda$ and $\alpha$, $R: V \to V$ in analytically conjugate to a rotation (necessarily of infinite order) of $\partial\Delta$ and moreover, this conjugacy extends to an $R$-invariant neighbourhood $V$ of $\partial\Delta$ such that $R: V \to V$ is analytically conjugate to a rotation of an annulus (see [6] and [52]). Clearly, $\{R^n: n \geq 1\}$ is normal in $V$, and if $F_0$ is the component of $F(R)$ which contains $V$, then $\mathcal{F}(F_0)$ contains non-constant limit functions. We deduce that for these choices of $\lambda$ and $\alpha$, $F_0$ is either a Siegel disc or a Herman ring. As both 0 and $\infty$ are super-attracting fixed points of $R$, $F_0$ is not simply connected and so $F_0$ must be a Herman ring.

For an alternative construction of Herman rings, and quantitative information about them see [88]; this construction uses quasiconformal maps and we consider these and part of the material in [88] in Chapter 8 and §9.8.

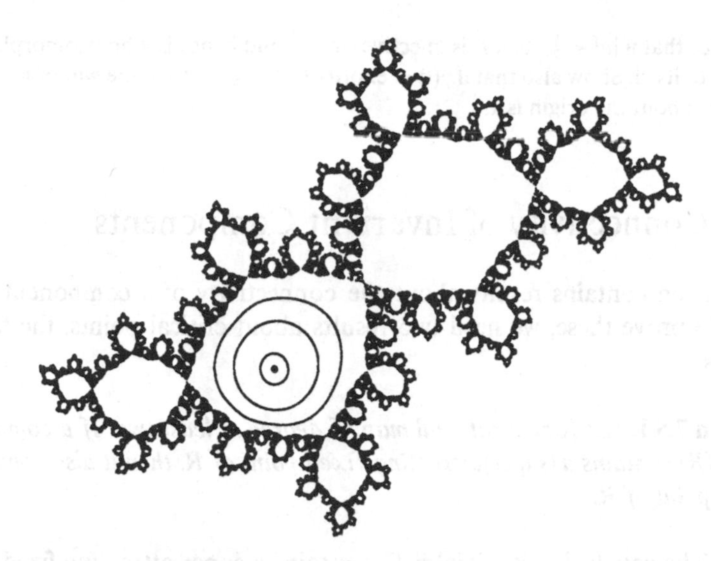

Figure 7.4.1. $z \mapsto \lambda z + z^2, \lambda = \exp(\pi i(\sqrt{5} - 1))$. Fractal image reprinted with permission from *The Beauty of Fractals* by H.-O. Peitgen and P.H. Richter, 1986, Springer-Verlag Heidelberg New York.

For computer-generated illustrations of Siegel discs see Figure 7.4.1, [80], and for Herman rings, see [100]. For further information, see also [36], [37], [54] and [70].

EXERCISE 7.4

1. Let $M$ be any Möbius transformation and let $\Sigma(M)$ be the set of fixed points of $M$. Show that either for some $n$, $M^n$ is the identity, or for all $n$, $\Sigma(M^n) = \Sigma(M)$. [*Hint*: Without loss of generality, $\Sigma(M) = \{0, \infty\}$ or $\Sigma(M) = \{\infty\}$.]

2. Using $\Sigma(S)$ and $\Sigma(T)$ as in Question 1, show that the Möbius maps $S$ and $T$ commute if and only if $T$ maps $\Sigma(S)$ onto itself, and $S$ maps $\Sigma(T)$ onto itself. Use this to argue (in the proof of Theorem 7.4.1) that $\Gamma$ is a cyclic group generated by some $z \mapsto kz$.

3. Suppose that $k > 1$, let $\gamma(z) = kz$, and let $\Gamma$ be the group generated by $\gamma$. Show that for a suitable choice of constants $r$ and $\lambda$, the analytic function $q: H \to \mathbb{C}$ by
$$q(z) = r \exp(\lambda \log z),$$
maps $H$ onto $\{z: 1/r < |z| < r\}$, and satisfies $q(z) = q(w)$ if and only if $w = \gamma^m(z)$ for some integer $m$. Deduce that the quotient space $H/\Gamma$ (and hence $F_0$ also) is conformally equivalent to the annulus $q(H)$.

4. Suppose that $R$ is given by (7.4.1), where $|\lambda| = 1$ and $0 < |a| < 1$. Show that if $R(z) = R(w)$ and $z \neq w$, then
$$zw + \alpha(z + w) + \bar{\alpha}zw(z + w) + |\alpha|^2(z^2 + zw + w^2) = 0.$$

Deduce that if $|a| < \frac{1}{7}$, then $R$ is injective on $\partial\Delta$, and hence is a homeomorphism of $\partial\Delta$ onto itself. Show also that if $\gamma(t) = \exp(it)$, $0 \leq t \leq 2\pi$, then the winding number of $R(\gamma)$ about the origin is 1.

## §7.5. Connectivity of Invariant Components

This section contains results about the connectivity of a component $F_0$ of $F(R)$. To prove these, we need two results about critical points, the first of which is

**Theorem 7.5.1.** *Let $R$ be a rational map of degree at least two. If a component $F_0$ of $F(R)$ contains a (super)attracting fixed point of $R$, then it also contains a critical point of $R$.*

PROOF. The conclusion is trivial if $F_0$ contains a super-attracting fixed point for such a point is a critical point. We assume, then, that $F_0$ contains an attracting fixed point $\zeta$, and by conjugation, we may assume that $\zeta \neq \infty$. Thus $|R'(\zeta)| < 1$ and we can construct a Euclidean disc

$$V = \{z: |z - \zeta| < r\},$$

in $F_0$ such that $R$ maps the closure of $V$ into $V$.

The proof proceeds by contradiction, so we assume now that $F_0$ does not contain any critical points of $R$. Let $U_n$ be the component of $(R^n)^{-1}(V)$ which contains $\zeta$. Clearly, $U_n$ lies in $F_0$, and the Riemann–Hurwitz formula together with our assumption about critical points yields

$$1 \geq X(U_n) = X(U_n) + \delta(U_n) = mX(V) = m \geq 1.$$

This shows that $m = X(U_n) = 1$ and so $R$ is a homeomorphism of the simply connected domain $U_n$ onto the disc $V$.

Next, $U_n \subset U_{n+1}$ (for $U_n$ is connected and contains $\zeta$, and it is mapped by $R^{n+1}$ into $R(V)$) and so

$$U_1 \subset U_2 \subset U_3 \subset \cdots.$$

We need one more fact concerning the $U_n$. As $\zeta$ is an attracting fixed point, $R^n \to \zeta$ locally uniformly on $F_0$. Now take any $z$ in $F_0$, and join $z$ to $\zeta$ by a curve $\sigma$ in $F_0$. As $\sigma$ is compact, we see that for some $n$, $R^n(\sigma)$ lies in $V$ and so $\sigma$, and hence $z$, lies in $U_n$. This shows that

$$F_0 = \bigcup_{n=1}^{\infty} U_n,$$

and so $F_0$ is simply connected (Proposition 5.1.7). The Monodromy Theorem, or

$$1 + 0 = X(F_0) + \delta(F_0) = mX(F_0) = m \geq 1,$$

now shows that $R$ is a homeomorphism of the simply connected domain $F_0$ onto itself. Applying the Riemann Mapping Theorem, we deduce that the map $R$ of $F_0$ onto itself is analytically conjugate to an automorphism $S$ of the unit disc which fixes the origin and as such, $S$ must be a Euclidean rotation of $\Delta$. However, this implies that

$$1 = |S'(0)| = |R'(\zeta)| < 1,$$

a contradiction, and the proof is complete. □

The second result about critical points is the corresponding result for Petal domains, namely

**Theorem 7.5.2.** *Let $R$ be a rational map of degree at least two, let $\zeta$ be a rationally indifferent fixed point lying on the boundary of some forward invariant component $F_0$ of $F(R)$, and suppose that $R^n \to \zeta$ locally uniformly in $F_0$. Then $F_0$ contains a critical point of $R$.*

Although we shall not prove this until Chapter 9 (which is devoted to a study of critical points), it is convenient (but not necessary) to use it in the proof of the next result. For a proof, readers may now consult Theorem 9.3.2 and its proof, but for those who would prefer not to do this, we have also included an outline of a proof of the next result which does not use Theorem 7.5.2. We can now start to derive results about the connectivity of forward invariant components of $F(R)$. First, we have

**Theorem 7.5.3.** *Let $R$ be a rational map of degree at least two. Then any forward invariant component of $F(R)$ is simply, doubly, or infinitely, connected.*

PROOF. As $F_0$ is forward invariant, it is of one of the five types listed in Definition 7.1.1 (this is Theorem 7.1.2). If $F_0$ is a Siegel disc or a Herman ring, then $F_0$ is simply or doubly connected as required. We may assume, then, that $F_0$ is one of the types (a), (b) or (c) and so, by Theorems 7.5.1 and 7.5.2, $F_0$ contains a critical point of $R$.

We may also assume that $F_0$ has finite connectivity (else the conclusion holds) and as $F_0$ is forward invariant, the restriction of $R$ to $F_0$ is an $m$-fold covering map of $F_0$ onto itself for some positive integer $m$. With these facts, we have the basic relation

$$\chi(F_0) + \delta(F_0) = m\chi(F_0),$$

with all terms finite. We deduce that

$$(m-1)\chi(F_0) = \delta(F_0) > 0,$$

and so $m \geq 2$ and $\chi(F_0) > 0$. Thus in these cases, $\chi(F_0) = 1$ and so $F_0$ is simply connected. This completes the proof subject to proving Theorem 7.5.2. □

Theorem 7.5.3 and its proof contains the following results.

**Corollary 7.5.4.** *If F is connected, then it is either simply or infinitely connected.*

**Corollary 7.5.5.** *If a component of $F(R)$ contains a fixed point of R, then it is either simply or infinitely connected.*

**Corollary 7.5.6.** *A forward invariant component of R is doubly connected if and only if it is a Herman ring.*

If a forward invariant component $F_0$ is of one of the types (a), (b) or (c), then according to Theorem 7.5.3 and Corollary 7.5.6, it must be simply or infinitely connected: in fact, all six possibilities can occur as is shown in Exercise 7.1.

We end this section with a sketch of a proof of Theorem 7.5.3 that does not use Theorem 7.5.2. This argument is closer in spirit to the proof of Theorem 7.5.1, it requires more background knowledge, and we omit the details.

We assume that the connectivity of $F_0$ is finite and at least three, that is, $-\infty < \chi(F_0) < 0$, and seek a contradiction. We also use the same notation as in the proof of Theorem 7.4.1: thus $F_0$ is conformally equivalent to the quotient space $\Delta/\Gamma$, where the unit disc $\Delta$ is the universal cover of $F_0$ and $\Gamma$ is the cove group. Now as before,

$$0 \geq (m-1)\chi(F_0) = \delta(F_0) \geq 0,$$

and so $m = 1$: thus $R$ is in $\text{Aut}(F_0)$.

Now the lifts of all maps in $\text{Aut}(F_0)$ constitute the normalizer $N(\Gamma)$ of $\Gamma$, and $\text{Aut}(F_0)$ is isomorphic to the quotient group $N(\Gamma)/\Gamma$ (see Appendix IV to Chapter 6). Further, it can be shown that as the connectivity $F_0$ is at least three, the group $\Gamma$ is not abelian and as a consequence of this, $N(\Gamma)$ is discrete (if $\Gamma$ is abelian, then $N(\Gamma)$ is not discrete: see Exercise 7.5.1). There is a standard proof (which involves arguments about the areas of fundamental regions of $\Gamma$ and $N(\Gamma)$) that if, in the general case, $\Delta/\Gamma$ is compact, then the quotient $N(\Gamma)/\Gamma$ is a finite group. In our case, $\Delta/\Gamma$ is not compact (it is homeomorphic to $F_0$), but as $F_0$ has finite connectivity, $\Delta/\Gamma$ is a compact surface from which a finite number of discs have been removed. With this fact available, the argument that yields the finiteness of $N(\Gamma)/\Gamma$ in the compact case can now be modified so as to produce the same conclusion (the modification here is to analyse the action of the groups on the Nielsen convex hull, which intersects the fundamental regions in a set of finite area, rather than the full disc). In any event, in this way we can show that $N(\Gamma)/\Gamma$, and hence $\text{Aut}(F_0)$, is a finite group. However, $R$ is in $\text{Aut}(F_0)$, and it cannot have finite order; thus we have the required contradiction.

Clearly, to give all of the details of this argument would take us too far afield; we emphasize, however, that the results in Chapter 9 are proved independently of Theorem 7.5.1, so the first proof of this result is much the simpler of the two proofs.

EXERCISE 7.5

1. Let $\gamma(z) = z + 1$, let $\Gamma$ be the group generated by $\gamma$, and let $N(\Gamma)$ be the normalizer of $\Gamma$ (that is, the group of all Möbius maps $g$ which preserve $H$ and which satisfy $g\Gamma = \Gamma g$). Show that all real translations are in $N(\Gamma)$, so $N(\Gamma)$ is not discrete. Show that $z \mapsto mz$ is in $N(\Gamma)$ when $m$ is an integer. Find $N(\Gamma)$.

2. Construct, and solve, the analogous question to Question 1 when $\gamma(z) = kz$, where $k > 0$ and $k \neq 1$.

3. Let $F_0$ be a forward invariant component of the Fatou set of a rational map $R$ of degree at least two. Show that $R$ is an automorphism of $F_0$ if and only if $F_0$ is a Siegel disc or a Herman ring.

CHAPTER 8

# The No Wandering Domains Theorem

This chapter is devoted to the proof of Sullivan's important theorem, namely that for any component $\Omega$ of the Fatou set of $R$, the sequence $\Omega$, $R(\Omega)$, $R^2(\Omega), \ldots$ is eventually periodic.

## §8.1. The No Wandering Domains Theorem

Consider the sequence

$$\Omega, R(\Omega), R^2(\Omega), \ldots, R^n(\Omega), \ldots \tag{8.1.1}$$

of successive images of a component $\Omega$ of the Fatou set of a rational map $R$. This chapter is devoted to the proof that this sequence is eventually periodic (the only alternative being that the regions in (8.1.1) are pairwise disjoint). We begin with the terminology which enables us to refer briefly to each possibility and, as usual, $R$ is a rational map with $\deg(R) \geq 2$.

**Definition 8.1.1.** A component $\Omega$ of the Fatou set $F(R)$ is:

(a) *periodic* if for some positive integer $n$, $R^n(\Omega) = \Omega$;
(b) *eventually periodic* if for some positive integer $m$, $R^m(\Omega)$ is periodic; and
(c) *wandering* if the sets $R^n(\Omega)$, $n \geq 0$, are pairwise disjoint.

A wandering component of $F(R)$ is called a *wandering domain* of $R$. Fatou asked whether or not such domains can exist and in 1983, [93], Sullivan proved that they cannot (for an alternative proof, see [21]).

**Theorem 8.1.2.** *Every component of the Fatou set of a rational map is eventually periodic.*

Sullivan's result is probably the most significant advance made in this subject in recent times and we shall devote this entire chapter to its proof. It implies, for example, that the long-term dynamics of a rational map $R$ on its Fatou set $F$ is the action of $R$ on the periodic components of $F$, and as a periodic component of $F$ is forward invariant under some $R^m$, we have already analysed this situation in Chapter 7. Further important applications arise by combining it with the results on critical points in Chapter 9, for periodic components imply the existence of critical points (see Theorems 7.5.1 and 7.5.2), and from the Riemann–Hurwitz relation we know that there are at most $2d - 2$ critical points. The rest of this chapter is concerned only with the proof of Theorem 8.1.2, and the reader who is primarily interested in applications can now pass safely on to Chapter 9.

Although we shall not consider entire functions in this text, we remark that wandering domains can exist for these: see [11], [12], [13], [15] and [41].

## §8.2. A Preliminary Result

The proof of Theorem 8.1.2 will be by contradiction so we suppose now that $R$ does have a wandering domain. In this section we shall derive a consequence of this assumption which enables us to simplify Sullivan's original proof by restricting ourselves to simply connected components of $F(R)$: of course, this only contributes towards the required contradiction and it will have no significance at all once Theorem 8.1.2 has been proved. We prove

**Lemma 8.2.1.** *Suppose that $R$ has a wandering domain. Then for some component $W$ of $F(R)$, the components*

$$W, R(W), R^2(W), \ldots, R^n(W), \ldots \tag{8.2.1}$$

*of $F(R)$ are pairwise disjoint, simply connected, and contain no critical points of $R$. In particular, each is mapped by $R$ homeomorphically onto the next.*

The heart of the argument lies in showing that the existence of a wandering domain implies the existence of a simply connected wandering domain. We use $\operatorname{diam}[E]$ to denote the diameter of a set $E$ with respect to the spherical metric, and as a first step, we prove

**Lemma 8.2.2.** *Suppose that $W$ is a wandering domain. Then for any compact subset $K$ of $W$, $\operatorname{diam}[R^n(K)] \to 0$ as $n \to \infty$.*

Proof. Suppose that this is false: then there is some compact subset $K$ of $W$, some positive $\varepsilon$, and some increasing sequence $n_j$ of integers such that for $j = 1, 2, \ldots$,

$$\operatorname{diam}[R^{n_j}(K)] \geq \varepsilon. \tag{8.2.2}$$

As $\{R^n\}$ is normal in $W$, there is a subsequence of $R^{n_j}$ which converges locally uniformly on $W$ to some analytic function $g$: for convenience, we relabel this subsequence and so assume that $R^{n_j}$ itself has this property. Now if $g$ is constant, with value $\alpha$ say, on $W$, then $R^{n_j}$ converges uniformly to $\alpha$ on $K$ and so, for large $j$, $R^{n_j}(K)$ lies in an $\varepsilon/3$-neighbourhood of $\alpha$. This contradicts (8.2.2) and we conclude that $R^{n_j}$ converges to a non-constant $g$ locally uniformly on $W$.

We now take a point $\zeta$ in $W$ where $g'(\zeta) \neq 0$ and draw a small circle $C$ with centre $\zeta$ which with its interior $D$ lies in $W$, and which is such that $g(z) \neq g(\zeta)$ when $z$ is on $C$. Then for $j \geq j_0$, say,

$$|R^{n_j}(z) - g(z)| < \inf_{w \in C} |g(w) - g(\zeta)| < |g(z) - g(\zeta)|$$

on $C$ and so by Rouché's Theorem, $R^{n_j}(D)$ contains the point $g(\zeta)$. This contradicts the fact that $W$ is a wandering domain and so completes the proof. □

We can now give the

Proof of Lemma 8.2.1. First, $R$ has a wandering domain, $\Omega$ say, so only a finite number of the sets $R^n(\Omega)$ contain critical points of $R$. Taking $N$ to be sufficiently large, we let $W = R^N(\Omega)$ and $W_n = R^n(W)$, so $W$ is a wandering domain and no $W_j$ contains any critical points of $R$. As a consequence of this, we have

$$\chi(W_n) = m_n \chi(W_{n+1}), \tag{8.2.3}$$

where $R$ is an $m_n$-fold map of $W_n$ onto $W_{n+1}$ (Theorem 5.5.4). Now it is only necessary to prove that $W$ $(= W_0)$ is simply connected for then, from (8.2.3), $m_0 \chi(W_1) = 1$; hence $W_1$ is also simply connected, $m_0 = 1$ and the argument then proceeds by induction in the same way to derive the simple connectivity of $W_n$ for all subsequent $n$.

To complete the proof, then, we show that $W$ is simply connected and again we proceed by contradiction. Given that $W$ is not simply connected, there is a simple closed curve, say $\gamma$, in $W$ which is not homotopically trivial in $W$, and we define $\gamma_n = R^n(\gamma)$. Now

$$R^n \colon W \to R^n(W)$$

is a smooth unlimited covering map (for by construction, there are no critical points of any of the factors $R$ in the composition $R^n$), and this ensures that $\gamma_n$ is not homotopically trivial in $R^n(W)$, for if it were, the Monodromy Theorem would enable us to take the deformation of $\gamma_n$ into a point of $R^n(W)$ and lift it to a deformation of $\gamma$ into a point of $W$.

We now recall Theorem 2.3.4 and the $\delta$ defined in that result. By Lemma 8.2.2 (with $K = \gamma$) there is some $m$ such that for $n \geq m$,

$$\operatorname{diam}[\gamma_n] \leq \delta.$$

Now according to Theorem 2.3.4, if $n \geq m$ then $R$ maps the union of $\gamma_n$ and its interior components into the union of $\gamma_{n+1}$ and its interior components. Using Theorem 3.3.5, we deduce that $\{R^n: n \geq 1\}$ is normal in each interior component of $\gamma_m$ and so each such component lies in $F$. It now follows that the union of $\gamma_m$ and all its interior components is a connected, compact subset of $F(R)$ whose complement is (by Proposition 5.1.3) simply connected. We deduce that $\gamma_m$ is homotopic to a point in $F(R)$ (see Exercise 8.2.1), and hence also in $W_m$, contrary to our earlier observation. It follows that $W$ is simply connected and the proof is complete. □

*Remark.* When (in §8.6) we proceed to the proof of Theorem 8.1.2 by contradiction, we shall assume that the (hypothetical) wandering domain has those properties given in Lemma 8.2.1.

Exercise 8.2

1. Let $D$ be a simply connected domain on the complex sphere, let $K$ be its (compact, connected) complement and suppose that $D$ contains a compact set $J$. Prove that there is a Jordan curve $\sigma$ in $D$ which separates $K$ from $J$. [Assume that $\infty \in J$, put a square grid on the plane so that no square meets both $K$ and $J$ and consider the outer boundary of the squares that meet $K$.]

   By taking $K$ to be the union of $\gamma_m$ and its interior components, and $J$ to be $J(R)$, show (in the proof of Lemma 8.2.1) that $\gamma_m$ is homotopically trivial in $F(R)$.

2. Let $W$ be any component of $F(R)$ and let $W_j$ be the components of the completely invariant set $[W]$ generated by $W$. Show that if one $W_j$ is a wandering domain of $R$, then so is every $W_j$.

# §8.3. Conformal Structures

We shall need certain facts about quasiconformal maps and in order to avoid too great a disruption to the proof of Theorem 8.1.2, we shall adopt a rather axiomatic view and confine ourselves to a brief discussion of the few standard results that we need. The reader is referred to [1], [63] and [64] for the proofs and the background material.

We begin with a generalization of the Cauchy–Riemann equations. Given any function $f$ with continuous partial derivatives in a domain $D$, we introduce the two differential operators

$$\frac{\partial f}{\partial \bar{z}} = \frac{1}{2}\left(\frac{\partial f}{\partial x} + i\frac{\partial f}{\partial y}\right), \qquad \frac{\partial f}{\partial z} = \frac{1}{2i}\left(\frac{\partial f}{\partial x} - i\frac{\partial f}{\partial y}\right).$$

Observe that if $f$ is analytic, then

$$\frac{\partial f}{\partial \bar{z}} = 0, \qquad \frac{\partial f}{\partial z} = f'(z),$$

the first of these being the Cauchy–Riemann equations. Our generalization of these equations is the *Beltrami equation*

$$\frac{\partial f}{\partial \bar{z}} = \mu \frac{\partial f}{\partial z}, \tag{8.3.1}$$

where $\mu$ is some suitable complex-valued function on $D$, and the basic idea is that if $\mu = 0$ throughout $D$ then any sufficiently smooth solution $f$ of (8.3.1) is analytic in $D$, while in the general situation, $\mu$ is taken as a measure of the deviation of a solution $f$ from conformality. Of course, given a function $f$ on $D$, we can use (8.3.1) to define $\mu$; this $\mu$ (when it exists) is called the complex dilatation of $f$ and is denoted by $\mu_f$.

An essential feature of the theory is that it is developed for a very large class of $\mu$; most notably, $\mu$ is not required to be continuous and as a result we have a tool of great flexibility. The functions $\mu: D \to \mathbb{C}$ are required to be Lebesgue measurable and to satisfy

$$\|\mu\|_\infty < 1 \tag{8.3.2}$$

in $D$, so $|\mu| \leq \|\mu\|_\infty$ almost everywhere (that is, except on a set of zero area), and we shall call any such function $\mu$ a *Beltrami coefficient* on $D$.

Given a domain $D$ and a Beltrami coefficient $\mu$ on $D$, we say that a homeomorphism $f$ on $D$ is *quasiconformal* with *complex dilatation* $\mu$ on $D$ if $f$ is an $L^2$-solution of (8.3.1) in $D$ (see [63], p. 24). The reader need not worry about the technical nature of the definition as we shall not mention these ideas again; however, some explanatory remarks may be helpful. First, a quasiconformal map $f$ is differentiable (in the sense that its derivative is a linear map from $\mathbb{R}^2$ to itself) almost everywhere (a.e.) in $D$, and we only require (8.3.1) to be satisfied a.e. in $D$. Second, the intuitive implication of (8.3.1) is that for almost all $z$, the function $f$ maps an infinitesimal circle $C$ centred at $z$ to an infinitesimal ellipse $E$ centred at $f(z)$, where both the eccentricity and the orientation of $E$ and are determined by $\mu(z)$. The significance of (8.3.2) is that it controls this distortion induced by $f$, and it implies that the eccentricity of $E$ is bounded above throughout $D$.

We turn now to the basic existence and uniqueness theorem for solutions of (8.3.1); see Theorems 4.2 and 4.4 in [63].

**Theorem 8.3.1.** *Let $\mu$ be a Beltrami coefficient on a domain $D$ on the complex sphere. Then*:

(i) *there exists a quasiconformal map with complex dilatation $\mu$ on $D$*; *and*
(ii) *if $\varphi$ and $\psi$ are two such maps, then $\varphi\psi^{-1}$ is analytic.*

Taking $\mu$ to be identically zero and $\psi$ to be the identity, we obtain

**Corollary 8.3.2.** *If $\varphi$ is quasiconformal with dilatation* 0 *on $D$, then it is conformal on $D$.*

*Remark.* Given a simply connected domain $D$ (conformally equivalent to the unit disc $\Delta$), and a Beltrami coefficient $\mu$ on $D$, then there is a quasiconformal map $\varphi$ of $D$ onto $\Delta$ with dilatation $\mu$ almost everywhere in $D$: this is the *Measurable Riemann Mapping Theorem* for the case when $\mu$ is identically zero is the classical Riemann Mapping Theorem.

Theorem 8.3.1 leads naturally to the idea that each $\mu$ on $D$ creates a Riemann surface, namely the space $D$ together with a conformal structure on $D$ determined by $\mu$. For a given $D$ and $\mu$, this conformal structure is defined by the atlas $\mathscr{A}(\mu)$ (the totality of charts) consisting of the family of all maps $\varphi: D \to \mathbb{C}_\infty$ which are quasiconformal with dilatation $\mu$ on $D$. Theorem 8.3.1(i) guarantees that each point of $D$ is in the domain of some such $\varphi$, and Theorem 8.3.1(ii) guarantees that the transition map between two different co-ordinate charts is analytic; thus $\mathscr{A}(\mu)$ defines a conformal structure on $D$ and so converts $D$ into a Riemann surface. We shall denote this Riemann surface by $D[\mu]$, and we call the structure the *$\mu$-conformal structure* on $D$. Of course, if $\mu \equiv 0$, then $D[\mu]$ is simply $D$ with the usual structure as a subdomain of $\mathbb{C}_\infty$ and we usually use $D$ rather than $D[0]$; however, where there may be some gain in clarity, we shall use $D[0]$.

Given the Riemann surface $D[\mu]$, we say that the map

$$f: D \to \mathbb{C}_\infty,$$

is *$\mu$-analytic* if it is an analytic map between the Riemann surfaces $D[\mu]$ and $\mathbb{C}_\infty$ (with the usual structure): if, in addition, $f$ is a homeomorphism, we say that it is *$\mu$-conformal* on $D$. Of course, we retain the usual usage of *analytic* and *conformal* when $\mu$ is identically zero. Note that if $f$ is $\mu$-analytic on $D$, and if $g$ is analytic on $f(D)$, then (as the composition of analytic maps is analytic) $gf$ is $\mu$-analytic on $D$. Quite generally, a homeomorphism $\varphi$ from an open subset of a Riemann surface $\mathscr{R}$ to $\mathbb{C}_\infty$ is analytic if and only if it is a chart in the maximal atlas for $\mathscr{R}$; thus we record the different points of view in

**Lemma 8.3.3.** *Let $\varphi$ be a homeomorphism defined on $D$. Then the following are equivalent:*

(i) *$\varphi$ is $\mu$-conformal on $D$;*
(ii) *$\varphi$ is a chart for $D[\mu]$;*
(iii) *$\varphi: D[\mu] \to \mathbb{C}_\infty[0]$ is analytic.*

Our last task in this section is to formulate the transfer of a conformal structure from one domain to another in terms of complex dilatations. If $g$ is a bijection of a Riemann surface $X$ onto a set $Y$, then $g$ can be used to transfer the conformal structure on $X$ to $Y$ in a natural, trivial, and unique, way so that $g: X \to Y$ is analytic. We need to express this transfer of structure in terms of Beltrami coefficients and we prove

**Lemma 8.3.4.** *Suppose that $\mu$ and $v$ are Beltrami coefficients on the domains $U$ and $V$ respectively and let $g$ be an analytic map of $U$ onto $V$. Then:*

(i) $g: U[\mu] \to V[v]$ *is analytic;*

*if and only if*

(ii) $v(gz) = [g'(z)/\overline{g'(z)}]\mu(z)$ *a.e. in* $U$.

Proof. First, if $\psi$ is $v$-conformal on $V$, then the Chain Rule for partial derivatives (which exist almost everywhere) leads directly to the fact that the complex dilatation of the composition $\psi g$ is $\mu_1$, where

$$\mu_1(z) = [\overline{g'(z)}/g'(z)]v(gz)$$

(see Exercise 8.3.1). Further, by the definition of analyticity for maps between Riemann surfaces, (i) holds if and only if the composite map

$$\psi g: U[\mu] \xrightarrow{g} V[v] \xrightarrow{\psi} \mathbb{C}_\infty[0]$$

is analytic.

If (i) holds, then $\psi g$ is analytic, and hence is $\mu$-conformal in a neighbourhood of each $z$ in $U$ except at an isolated set of points. We deduce that $\mu = \mu_1$ a.e. in $U$, and this is (ii). If (ii) holds, then $\psi g$ is $\mu$-conformal in a neighbourhood of each $z$ in $U$ except in a set $\{z_j\}$ of isolated points. It follows that $\psi g$ is analytic in the complement of $\{z_j\}$ and hence (as $g$ is continuous in $U$) in $U$ itself. □

We shall use Lemma 8.3.4 to transfer structure in two different ways. First, if $g$ is an analytic homeomorphism from $U$ onto $V$, and if $\mu$ is given on $U$, then we can define $v$ on $V$ by (ii) and, as a result, (i) will hold: thus we have transferred the structure from $U$ to $V$. If we no longer assume that $g$ is injective, this process may not be valid as (ii) may provide conflicting definitions of $v$ at some point of $V$. However, in this case, if $v$ is given, we can use (ii) to define $\mu$ and again force (i) to be true: thus we have transferred the structure from $V$ to $U$. Thus, quite generally, we can pull back a conformal structure against the action of an analytic function so that the function remains analytic with respect to the two new structures. We shall use both of these ideas in the next section.

## Exercise 8.3

1. Suppose that $f$ is $\mu$-conformal, that $g$ is $v$-conformal, and that the composition $g \circ f$ ($= gf$) is defined. Writing $\partial f$ and $\bar{\partial} f$ for $\partial f/\partial z$ and $\partial f/\partial \bar{z}$ respectively, prove that

$$\partial(g \circ f) = (\partial g \circ f)\partial f + (\bar{\partial} g \circ f)\partial\bar{f},$$
$$\bar{\partial}(g \circ f) = (\partial g \circ f)\bar{\partial} f + (\bar{\partial} g \circ f)\bar{\partial}\bar{f}.$$

Show also that

$$\partial \bar{f} = \overline{\bar{\partial} f}, \qquad \bar{\partial} \bar{f} = \overline{\partial f},$$

and deduce that if $f$ is analytic, then

$$\mu_{gf}(z) = \mu_g(fz)[\overline{f'(z)}/f'(z)].$$

## §8.4. Quasiconformal Conjugates of Rational Maps

In this section we focus attention on the conjugate $\varphi R \varphi^{-1}$ of a rational map $R$ by some $\mu$-conformal map $\varphi$ of the complex sphere onto itself. In general, $\varphi R \varphi^{-1}$ will not be analytic and our next result enables us to recognize when it is. We have

**Lemma 8.4.1.** *Suppose that $R$ is a rational map, and that $\varphi$ is a $\mu$-conformal map of the complex sphere onto itself. Then $\varphi R \varphi^{-1}$ is rational if and only if*

$$\mu(Rz) = [R'(z)/\overline{R'(z)}]\mu(z) \tag{8.4.1}$$

*a.e. on the sphere, and when this is so,* $\deg(\varphi R \varphi^{-1}) = \deg(R)$.

We emphasize that (8.4.1) guarantees that we can conjugate a rational map by a *non-analytic* function $\varphi$ and yet still retain analyticity.

Proof. By virtue of Lemma 8.3.4, condition (8.4.1) is equivalent to the statement that

$$R\colon \mathbb{C}_\infty[\mu] \to \mathbb{C}_\infty[\mu]$$

is analytic and this, in turn, is equivalent to the statement that $\varphi R \varphi^{-1}$ is an analytic map of the complex sphere onto itself with the usual structure (for $\varphi$ is a chart of the Riemann surface $\mathbb{C}_\infty[\mu]$). This proves that that (8.4.1) holds if and only if $\varphi R \varphi^{-1}$ is analytic throughout the sphere, and hence rational. Moreover, $\varphi R \varphi^{-1}$ must have the same degree as $R$ because $\varphi$ is a homeomorphism of the sphere onto itself and the degree of $\varphi R \varphi^{-1}$ is given by the cardinality of $(\varphi R \varphi^{-1})^{-1}\{w\}$ for most points $w$. ☐

We shall now incorporate this result into the situation in which a wandering domain exists, and in doing so, we take a significant step forward in our proof of Theorem 8.1.2. We prove

**Lemma 8.4.2.** *Let $R$ be a rational map, and suppose that $W$ satisfies the conditions of Lemma* 8.2.1. *Then any Beltrami coefficient $\mu$ on $W$ extends to a Beltrami coefficient on the complex sphere with the property that if $\varphi$ is any $\mu$-conformal map of the sphere onto itself, then $\varphi R \varphi^{-1}$ is a rational map with the same degree as $R$.*

PROOF. We begin with a Beltrami coefficient $\mu$ on $W$ and our objective is to extend $\mu$ to the sphere so that the extension satisfies (8.4.1) a.e. on the sphere. As the sphere is the disjoint union of the completely invariant set $[W]$ and its (completely invariant) complement, say $K$, we can extend $\mu$ to $[W]$ and to $K$ separately and then verify that (8.4.1) holds on each of these two sets. We define $\mu$ to be identically zero on $K$ so (8.4.1) is certainly satisfied there.

The extension from $W$ to $[W]$ is carried out as follows. First, using Lemma 8.2.1 and the ideas expressed at the end of the preceding section, we define $\mu$ on the (pairwise disjoint) sets $W_n (= R^n(W))$, $n \geq 0$, by ensuring that the map

$$R: W_n[\mu] \to W_{n+1}[\mu], \qquad W_n = R^n(W),$$

is analytic. Next, we define $\mu$ on successive inverse images of the $W_n$ by the requirement that $R$ provides an analytic map between any component $\Omega$ and its image under $R$ (with the $\mu$-structures): this process determines $\mu$ uniquely on $[W]$ and is characterized by the fact that for every component $\Omega$ of $[W]$, the map

$$R: \Omega[\mu] \to R(\Omega)[\mu]$$

is analytic. As a consequence of this, the functional relation (8.4.1) holds a.e. on $[W]$ (this is simply Lemma 8.3.4) and this proves Lemma 8.4.2. □

## §8.5. Boundary Values of Conjugate Maps

Our proof of Theorem 8.1.2 requires a comparison between the boundary values of two functions which are conjugate to each other by a conformal map, so we devote this section to this topic. Let $\mu$ be a Beltrami coefficient on the unit disc $\Delta$, let $g$ be a conformal map of $\Delta$ onto a simply connected domain $W$ in $\mathbb{C}$, and use $g$ to transfer $\mu$ to a Beltrami coefficient $\nu$ on $W$ (as in Lemma 8.3.4). Suppose now that $\Phi$ is a $\nu$-conformal map of $W$ onto itself, and that $\Phi$ extends to a homeomorphism of the closure of $W$ onto itself such that $\Phi = I$ (the identity map) on $\partial W$. Then the map

$$\varphi = g^{-1}\Phi g$$

is a $\mu$-conformal map of $\Delta$ onto itself, and as such, it extends to a homeomorphism of the closed disc onto itself ([1], p. 47). Our objective is to study the behaviour of $\varphi$ on $\partial\Delta$: one approach to this problem is through the theory of prime ends, but we shall take a different route.

When comparing maps which are conjugate by a conformal map, it is natural to try to exploit conformal invariants and one such invariant is the hyperbolic metric. For any homeomorphism $\Phi$ of a simply connected domain $\Omega$ onto itself, we define the (hyperbolic) *displacement function* of $\Phi$ by

$$z \mapsto \rho(\Phi z, z),$$

where $\rho$ is the hyperbolic metric of $\Omega$ (which exists unless $\Omega$ is conformally equivalent to $\mathbb{C}$ or $\mathbb{C}_\infty$).

Returning to the discussion above, the conformal equivalence $g$ is an isometry with respect to the hyperbolic metrics in $\Delta$ and $W$ and so the property of having a bounded displacement is invariant under conjugation by $g$: thus if $\Phi$ has a bounded displacement on $W$, then $\varphi$ has a bounded displacement on $\Delta$ (and conversely). This, however, implies that $\varphi = I$ on $\partial\Delta$ for it implies that $\varphi(z)$ lies in the hyperbolic disc $\Delta(z, d)$ with centre $z$ and some (fixed) radius $d$, and as $z$ converges to any $\zeta$ on $\partial\Delta$, the Euclidean radius of $\Delta(z, d)$ tends to zero (see Exercise 8.5.1).

We now seek a criterion for $\Phi$ to have a bounded displacement function in $W$, and for our purposes, it will be sufficient to assume that $v$ is a Beltrami coefficient on the entire sphere, and that $\Phi$ is a $v$-conformal map of the sphere onto itself. With this additional information, we can readily obtain a criterion for $\Phi$ to have a bounded displacement function with respect to the chordal metric on the sphere, and we can then convert this to a criterion for $\Phi$ to have a bounded hyperbolic displacement on $W$. First, we have

**Lemma 8.5.1.** *For each positive $\varepsilon$ there is a positive $\delta$ such that if the Beltrami coefficient $v$ on $\mathbb{C}_\infty$ satisfies $\|v\| < \delta$, then $\sigma(Fz, z) < \varepsilon$ in $\mathbb{C}_\infty$ for every function $F$ that is a $v$-conformal map of $\mathbb{C}_\infty$ onto itself which fixes* 0, 1 *and* $\infty$.

PROOF. This is a non-sequential version of Theorem 4.6, [63]. If the conclusion is false there is a positive $\varepsilon$, a sequence of $v_n$-conformal maps $F_n$, and a sequence $z_n$, such that:

(i) $F_n$ fixes 0, 1 and $\infty$;
(ii) $\|v_n\| < 1/n$; and
(iii) $\sigma(F_n z_n, z_n) \geq \varepsilon$.

By Theorem 4.6, [63], (i) and (ii) imply that $F_n \to I$ uniformly on $\mathbb{C}_\infty$, and this contradicts (iii). □

Now let $\delta_0$ be the value of $\delta$ corresponding to $\varepsilon = \frac{1}{8}$ in Lemma 8.5.1. Then we have

**Lemma 8.5.2.** *Let $\rho$ be the hyperbolic metric of a simply connected subdomain $W$ of $\mathbb{C}_\infty$. If the Beltrami coefficient $v$ on $\mathbb{C}_\infty$ satisfies $\|v\| < \delta_0$, then $\rho(\Phi z, z) <$* log 2 *in $W$ for every function $\Phi$ which is a $v$-conformal map of $\mathbb{C}_\infty$ onto itself, and which satisfies $\Phi(W) = W$, and $\Phi = I$ on $\partial W$.*

PROOF. As the statement of Lemma 8.5.2 is invariant under conjugation by a Möbius map, we may assume that $\infty \in \partial W$. Now suppose that $W$, $\rho$, $v$ and $\Phi$ satisfy the hypotheses of Lemma 8.5.2 with $\|v\| < \delta_0$.

We take any $\zeta$ in $W$, let $r(\zeta)$ be the (Euclidean) distance between $\zeta$ and $\partial W$, and suppose that this is attained at $\alpha$ in $\partial W$. As $\partial W$ is a connected set containing $\alpha$ and $\infty$, there is some point $\beta$ in $\partial W$ such that

$$|\beta - \alpha| = |\zeta - \alpha| = r(\zeta),$$

and we define the Euclidean similarity $h$ by

$$h(z) = \frac{z - \alpha}{\beta - \alpha}.$$

The function $F$ defined by $F = h\Phi h^{-1}$ is a $\nu$-conformal map of $\mathbb{C}_\infty$ onto itself, and it fixes 0, 1 and $\infty$ (because $\Phi$ fixes $\alpha$, $\beta$ and $\infty$); thus by Lemma 8.5.1, $\sigma(Fz, z) < \frac{1}{8}$ on $\mathbb{C}_\infty$. Putting $z = h(\zeta)$, we obtain

$$\sigma(h(\Phi\zeta), h\zeta) < \tfrac{1}{8} < \sigma(1, 2),$$

and as $|h(\zeta)| = 1$, this implies that $|h(\Phi\zeta)| \leq 2$, and hence that

$$\begin{aligned} 2|h(\Phi\zeta) - h\zeta| &= \sigma(h\Phi(\zeta), h\zeta)[(1 + |h(\Phi\zeta)|^2)(1 + |h(\zeta)|^2)]^{1/2} \\ &< \tfrac{1}{2}. \end{aligned}$$

Using the definition of $h$, this now yields

$$|\Phi(\zeta) - \zeta| < r(\zeta)/4.$$

Now the disc $\{z\colon |z - \zeta| < r(\zeta)\}$ lies in $W$, so if $\rho_0$ denotes the hyperbolic metric of this disc then, by the Comparison Principle for hyperbolic metric,

$$\rho(\Phi\zeta, \zeta) \leq \rho_0(\Phi\zeta, \zeta) \leq \log\left(\frac{1 + \frac{1}{4}}{1 - \frac{1}{4}}\right) \leq \log 2$$

as required. □

In conclusion, if $\varphi\colon \Delta \to \Delta$ and $\Phi\colon W \to W$ are as given above with $\varphi = g^{-1}\Phi g$, and if $\Phi$ satisfies the hypotheses of Lemma 8.5.2, then $\varphi = I$ on $\partial\Delta$.

EXERCISE 8.5

1. Suppose that the displacement function of $\varphi\colon \Delta \to \Delta$ (as in the text) is bounded above by $d$. Use the formula

$$\sinh^2\left[\frac{\rho(z, w)}{2}\right] = \frac{4|z - w|^2}{(1 - |z|^2)(1 - |w|^2)}$$

(valid for all $z$ and $w$ in $\Delta$) to show that for some $M$,

$$|z - \varphi(z)|^2 \leq M(1 - |z|),$$

and deduce that $\varphi$ extends continuously to the identity on $\partial\Delta$.

## §8.6. The Proof of Theorem 8.1.2

In this, the final section of the chapter, we complete the proof of Theorem 8.1.2. We assume that $R$ has a wandering domain, so there exists a component $W$ of $F(R)$ with the properties expressed in Lemma 8.2.1. As $W$ is simply connected (and $J$ is infinite), there is a conformal equivalence $g$ of $\Delta$ onto $W$. Now let $\mu$ be any Beltrami coefficient on $\Delta$. Using $g$, we can transfer $\mu$ to $\nu$ on $W$ (as in Lemma 8.3.4), and then extend $\nu$ to $\mathbb{C}_\infty$ as in Lemma 8.4.2. Solving the Beltrami equation with coefficient $\nu$ throughout the sphere, we obtain a $\nu$-conformal map $\varphi$ of the sphere onto itself such that $\varphi R \varphi^{-1}$ is rational (and of the same degree as $R$), and by combining $\varphi$ with a Möbius map we may assume that $\varphi$ fixes 0, 1 and $\infty$ and so is uniquely determined by $\mu$. In this way we have created the composite map

$$\mu \mapsto \nu \mapsto \varphi \mapsto \varphi R \varphi^{-1} \tag{8.6.1}$$

of the space of Beltrami coefficients on $\Delta$ to the space of rational maps of degree $\deg(R)$.

Roughly speaking, the space of Beltrami coefficients on $\Delta$ is infinite dimensional while the space of rational maps of degree $\deg(R)$ is finite dimensional, so the map $\mu \mapsto \varphi R \varphi^{-1}$ must map a large subspace of Beltrami differentials onto a single rational function $S$. We shall justify this later, but to maintain the flow of ideas, we formulate the result we need as a lemma and continue with the proof of Theorem 8.1.2. For the moment, then, we assume the validity of

**Lemma 8.6.1.** *Suppose that $\eta_0 > 0$. For each $t$ in $[0, 1]$ we can construct a Beltrami coefficient $\mu_t$ in $\Delta$ such that:*

(a) $\|\mu_t\|_\infty < \eta_0$; *and*
(b) (8.6.1) *maps each $\mu_t$ to the same rational function $S$.*

*Further, this construction can be made so that for each $z$, the map $t \mapsto \varphi_t(z)$ is continuous on $[0, 1]$, where $\varphi_t$ is the image of $\mu_t$ under the first two maps in* (8.5.1).

A consequence of this is that for all $t$ in $[0, 1]$,

$$\varphi_t R \varphi_t^{-1} = S = \varphi_0 R \varphi_0^{-1},$$

so if we put

$$\Phi_t = \varphi_0^{-1} \varphi_t,$$

we find that:

(1) $\Phi_0(z) = z$ on $\mathbb{C}_\infty$;
(2) for each $t$, the map $z \mapsto \Phi_t(z)$ commutes with $R$;
(3) for each $t$, the map $z \mapsto \Phi_t(z)$ is a homeomorphism of $\mathbb{C}_\infty$ onto itself;
(4) for each $z$, the map $t \mapsto \Phi_t(z)$ is continuous on $[0, 1]$.

These properties lead to two striking facts.

**Lemma 8.6.2.** *For each $t$ in $[0, 1]$, $\Phi_t = I$ on $J(R)$. Further, $\Phi_t$ maps each component of $F(R)$ onto itself.*

PROOF. For $p = 1, 2, \ldots$, let $F_p$ be the set of fixed points of $R^p$. By virtue of (2), if $z$ is in $F_p$, then so is $\Phi_t(z)$; thus $\Phi_t$ maps each finite set $F_p$ into itself. Now by (4), for each $z$ in $F_p$, the map $t \mapsto \Phi_t(z)$ is a continuous map of $[0, 1]$ into the discrete set $F_t$; thus $\Phi_t(z)$ is independent of $t$ and so, by (1), for all $t$ in $[0, 1]$ and all $z$ in $F_p$, $\Phi_t(z) = z$. We deduce that for each $t$, $\Phi_t$ is the identity map on the union of the $F_p$, and hence by (3), also on the closure of this union. As the closure of the periodic points of $R$ contains $J$ (Theorem 4.2.6), the restriction of $\Phi_t$ to $J$ is the identity.

Because of this, and because $\Phi_t$ is a homeomorphism of the sphere onto itself, each $\Phi_t$ must permute the components of $F(R)$. Now take any $z$ in any component $F_0$ of $F(R)$. By virtue of (1) and (4), the image of $[0, 1]$ under the map $t \mapsto \Phi_t(z)$ is a curve which lies in $F$ and which starts at $z$. It follows that this curve lies entirely in $F_0$, and so for each $t$, $\Phi_t$ must map the arbitrary component $F_0$ into itself. This completes the proof of Lemma 8.6.2. □

As $\Phi_t$ maps each component $F_0$ of $F(R)$ onto itself, the homeomorphisms $\varphi_t$ must map $W$ onto the simply connected domain $W_0$, where $W_0$ $(= \varphi_0(W))$ is independent of $t$. It follows that there is a conformal equivalence $h$ of $W_0$ onto $\Delta$, and we have now reached the situation summarized in Figure 8.6.1.

Now in the construction of $\nu$ (on $\mathbb{C}_\infty$) from $\mu$ (on $\Delta$) in (8.6.1) we have $\|\nu\|_\infty = \|\mu\|_\infty$ (see Lemmas 8.3.4 and 8.4.1), so, by virtue of Lemma 8.6.1(a), $\|\nu_t\|_\infty < \eta_0$, where, of course, $\mu_t$ maps to $\nu_t$ in (8.6.1). By taking $\eta_0$ small enough, we can ensure that the complex dilatation of $\Phi_t$ $(= \varphi_0^{-1}\varphi_t)$ has $L^\infty$-norm less than $\delta_0$ as given in Lemma 8.5.2. By this lemma, and the remarks following

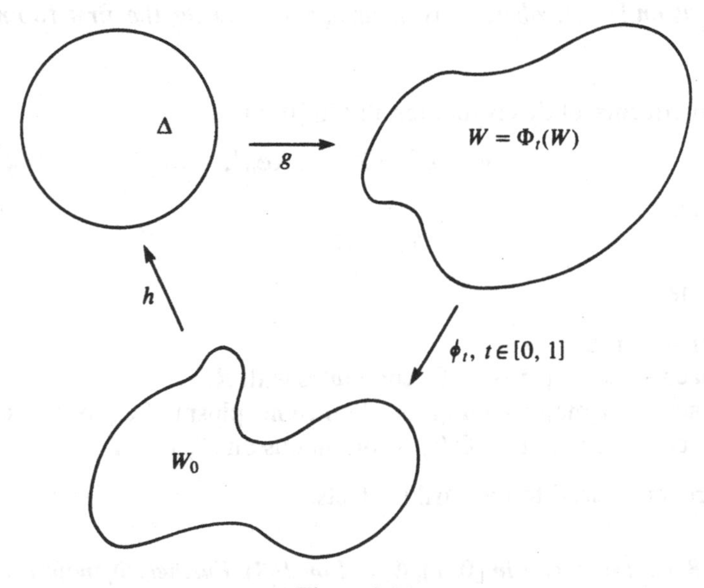


Figure 8.6.1

it, we can now conclude that the map $g^{-1}\Phi_t g$ of $\Delta$ onto itself has bounded displacement function and so extends to the identity on $\partial\Delta$.

Now observe that each of the maps in the composition

$$\Delta[\mu_t] \xrightarrow{g} W[v_t] \xrightarrow{\varphi_t} W_0[0] \xrightarrow{h} \Delta[0]$$

is analytic, whence $\psi_t$ $(= h\varphi_t g)$ is a $\mu_t$-conformal map of $\Delta$ onto itself and so extends to a homeomorphism of the closed disc onto itself. Moreover, in the open disc

$$\psi_0^{-1}\psi_t = g^{-1}\Phi_t g,$$

thus on $\partial\Delta$, $\psi_0^{-1}\psi_t = I$, and so $\psi_0 = \psi_t$.

Now let $\Psi_t$ be any $\mu_t$-conformal map of $\Delta$ onto itself. By Theorem 8.3.1, $\psi_t = M_t\Psi_t$ for some Möbius automorphism $M_t$ of $\Delta$, and so we deduce that

$$M_t\Psi_t = M_0\Psi_0$$

on $\partial\Delta$. In proving Lemma 8.6.1, we shall construct maps $\Psi_t$ so that $\Psi_t = \Psi_0$ on some open arc of $\partial\Delta$, and this will imply that $M_t = M_0$, and hence that $\Psi_t = \Psi_0$. However, it will be clear from our construction that this is not so, and this will be the contradiction that completes the proof of Theorem 8.1.2. It remains, then, to prove Lemma 8.6.1 and to construct the maps $\Psi_t$ with the properties used above.

First, we take an integer $N$ satisfying $N > 4d + 2$ (the space of rational maps of degree $d$, $d = \deg(R)$, having $4d + 2$ real degrees of freedom). We shall now construct, for each vector $T$ lying in the cube $[0, \varepsilon_1]^N$ (where $\varepsilon_1$ is some small number yet to be chosen), a Beltrami coefficient $\mu_T$ on $\Delta$ and a $\mu_T$-conformal map $\Psi_T$ of $\Delta$ onto itself. In fact, we shall construct the maps $\Psi_T$ first, and then define $\mu_T$ to be their complex dilatations.

Divide the interval $[0, 2\pi]$ into $2N$ equal and consecutive arcs $\sigma_1, \tau_1, \ldots, \tau_N, \sigma_N$, and for each of the arcs $\sigma_j$ we construct a $C^\infty$-function $\omega_j$ on $[0, 2\pi]$ such that:

(i) $\omega_j(x) > 0$ on the interior of $\sigma_j$, while $\omega_j(x) = 0$ otherwise; and
(ii) $|\omega_j'(x)| < \frac{1}{2}$ for all $x$ and $j$

(see Exercise 8.6.1). Now define the map

$$\Psi_T(z) = z \exp\left[ i \sum_{j=1}^{N} t_j \omega_j(\theta) \right],$$

where $z = r\exp(i\theta)$, of $\Delta$ onto itself, and let the Beltrami coefficient $\mu_T$ on $\Delta$ be the complex dilatation of $\Psi_T$.

The complex dilatation $\mu_T$ can easily be computed in terms of the partial derivatives $\partial\Psi_T/\partial r$ and $\partial\Psi_T/\partial\theta$ (Exercise 8.6.2) and when this has been done we find that

$$\mu_T(re^{i\theta}) = \frac{e^{2i\theta}\sum_{j=1}^{N} t_j\omega_j'(\theta)}{2 + \sum_{j=1}^{N} t_j\omega_j'(\theta)}, \tag{8.6.2}$$

and hence for all $T$,

$$|\mu_T(re^{i\theta})| \leq \frac{\sum_{j=1}^{N} |t_j\omega_j'(\theta)|}{2 - \sum_{j=1}^{N} |t_j\omega_j'(\theta)|} \leq \frac{N\varepsilon_1}{2 - N\varepsilon_1}.$$

Note that:

(a) if $\arg z$ lies in any of the intervals $\tau_j$, then $\Psi_T(z) = z$;
(b) distinct values of $T$ lead to distinct functions $\Psi_T$; and
(c) given any $\eta$, $\|\mu_T\|_\infty < \eta$ if $\varepsilon_1$ is small enough.

The first two of these are the properties required of the function $\Psi_t$ at the end of the proof of Theorem 8.1.2, and by choosing $\varepsilon_1$ small enough, (c) guarantees that we can satisfy Lemma 8.6.1(a) for all $\mathcal{M}_T$. We fix this value of $\varepsilon_1$, and it remains to complete the proof of Lemma 8.6.1.

The Proof of Lemma 8.6.1. Given the Beltrami coefficient $\mu_T$, let the composition (8.6.1) be

$$T \to \mu_T \to \nu_T \to \varphi_T \to R_T = \varphi_T R \varphi_T^{-1}. \tag{8.6.3}$$

The idea of the proof is to factor the map $T \mapsto R_T$ through the intermediate stage of a vector whose components are the zeros and poles of $R_T$, and then to show that the first factor, and hence the composition, is constant on some curve.

We may replace $R$ by any conjugate with respect to a Möbius map, so we may assume that $R$ is such that:

(1) $R$ has distinct zeros $a_1, \ldots, a_d$ in $\mathbb{C}$;
(2) $R$ has distinct poles $b_1, \ldots, b_d$ in $\mathbb{C}$; and
(3) $R(0) = 1$.

Now recall that $\varphi_T$ in (8.6.3) fixes 0, 1 and $\infty$. It follows that $R_T$ is the unique rational function with zeros $\varphi_T(a_j)$, with poles $\varphi_T(b_j)$, and with $R_T(0) = 1$, and so we only have to show that the map

$$\Pi(T) = (\varphi_T(a_1), \ldots, \varphi_T(a_d), \varphi_T(b_1), \ldots, \varphi_T(b_d))$$

of $(0, \varepsilon_1)^N$ into $\mathbb{C}^{2d}$ is constant on some curve.

Now in the fundamental paper [4], Ahlfors and Bers examined the general question of how a solution $z \mapsto \varphi_t(z)$ of the Beltrami equation with coefficient $\mu_t(z)$ varies when the parameter $t$ varies, and they showed that (in an appropriate sense) for each $z$, the map $t \mapsto \varphi_t(z)$ varies as smoothly as the map $t \mapsto \mu_t(z)$. We shall not enter into the details of this here, and we refer the interested reader to the original paper for the precise formulation of this result. For us, the important outcome is that because the map $T \mapsto \mu_T(z)$ is $C^\infty$, the same is true of $T \mapsto \nu_T(z)$ (in the sense of [4]), and hence each of the maps

$$t \mapsto \varphi_T(a_j), \qquad t \mapsto \varphi_T(b_j),$$

is $C^\infty$. We conclude that $\Pi$ is a $C^\infty$-map of the open cube $\mathscr{C}$, $\mathscr{C} = (0, \varepsilon_1)^N$, into $\mathbb{C}^{2d}$.

The set of points where the rank of $\Pi'$ is maximal, say $k$, is an open subset $\mathscr{C}_0$ of $\mathscr{C}$ (for the rank of $\Pi'$ is upper semi-continuous) and so the restriction of $\Pi$ to $\mathscr{C}_0$ is a map of constant rank. It follows from the Implicit Function Theorem (see, for example, [23], p. 79) that the inverse image of a point is a sub-manifold of $\mathscr{C}_0$ of dimension $N - k$ (which is positive) and so there is some curve in $\mathscr{C}_0$ on which $\Pi$ is constant. The proof of the lemma, and of Theorem 8.1.2, is now finished. □

EXERCISE 8.6

1. Given an interval $[a, b]$, define the function $\omega$ on $\mathbb{R}$ by putting $\omega(x) = A(x-a)^2(x-b)^2$ on $[a, b]$ and $\omega(x) = 0$ otherwise, where $A$ is chosen so that $|\omega'(x)| \leq \frac{1}{2}$ on $[a, b]$. Show that $\omega$ is $C^1$, and that for each $t$ in $[0, 1]$, the map $x \mapsto x + tg(x)$ is a strictly increasing map of $[a, b]$ onto itself.

   By giving an alternative definition on $[a, b]$, show that there is a $C^\infty$-function $\omega$ with these properties.

2. Express the complex dilatation of a $C^1$-function in terms of the usual partial derivatives $\partial/\partial r$ and $\partial/\partial\theta$, and hence verify (8.6.2).

CHAPTER 9

# Critical Points

We give evidence to support the assertion that the forward orbits of the critical points of a rational map determine the general features of the global dynamics of the map.

## §9.1. Introductory Remarks

The point $z_0$ is a critical point of a rational map $R$ if $R$ is not injective in any neighbourhood of $z_0$, and the influence of these points on the dynamics of $R$ stems largely from the quantitative constraint imposed by the Riemann–Hurwitz relation, namely that $R$ has at most $2d - 2$ critical points. We denote the set of critical points of $R$ by $C$ (or by $C(R)$ when there is need to specify $R$ explicitly), and we use $C^+$, or $C^+(R)$, to denote the forward images of $C$; thus

$$C^+ = \bigcup_{n=0}^{\infty} R^n(C).$$

Now let $d = \deg(R)$. If $w_0$ is not a critical value of $R^n$, then there are $d^n$ distinct branches, say $S_j$, $j = 1, \ldots, d^n$, of $(R^n)^{-1}$ defined in some neighbourhood of $w_0$. If $V$ is a domain which does not contain any critical values of $R^n$, and if $U$ is a component of $(R^n)^{-1}(V)$, then $R^n$ is a unbranched covering map of $U$ onto $V$: is simply connected, there are $d^n$ such components $U$ and $R^n$ is a homeomorphism of each onto $V$. With these facts in mind, we recall from Theorem 2.7.3 that the set of critical values of $R^n$ is $\bigcup_{k=1}^{n} R^k(C)$.

As $\infty$ is a critical point of a polynomial $P$ of degree $d$, where $d \geq 2$, it is often convenient to confine our attention to the $d - 1$ *finite critical points* of $P$, namely those in $\mathbb{C}$.

## §9.2. The Normality of Inverse Maps

We prove the following result about branches $S_j$ of $(R^n)^{-1}$.

**Theorem 9.2.1.** *Let R be a rational map with* $\deg(R) \geq 2$, *and suppose that the family* $\{S_n: n \geq 1\}$ *is such that each* $S_n$ *is a single-valued analytic branch of some* $(R^m)^{-1}$ *in a domain D. Then* $\{S_n: n \geq 1\}$ *is normal in D.*

*Remark.* Here, $S_n$ is not necessarily a branch of $(R^n)^{-1}$; the family may contain many branches of each $(R^m)^{-1}$.

PROOF. Take two disjoint cycles $A$ and $B$ of $R$, each containing at least three points (see §6.2). Now $S_n$ cannot map any point $\zeta$ of $D - A$ to $A$ (else for some $m$, $\zeta = R^m S_n(\zeta) \in A$), so by Theorem 3.3.5, $\{S_n\}$ is normal in $D - A$. The same argument holds for $B$, so $\{S_n\}$ is normal in the union, $D$, of $D - A$ and $D - B$. □

The following extension of Theorem 9.2.1 will be useful.

**Lemma 9.2.2.** *Suppose that D and* $\{S_n\}$ *are as in Theorem* 9.2.1. *If D meets* $J(R)$, *then any locally uniform limit* $\varphi$ *of a subsequence of* $(S_n)$ *is constant.*

PROOF. For each $n$, let $m(n)$ be such that $S_n$ is a branch of $(R^{m(n)})^{-1}$ in $D$. Suppose now that $S_n \to \varphi$ locally uniformly in $D$ as $n \to \infty$ in some infinite set $N_1$ of positive integers, and assume that $\varphi$ not constant: we seek a contradiction. Now each $S_n$ is univalent in $D$ (for $S_n$ has a left inverse $R^{m(n)}$) and as $\varphi$ is not constant, Hurwitz's Theorem implies that $\varphi$ is univalent in $D$.

Now take any $\zeta$ in $D \cap J$ and draw a small circle $\gamma$ about $\zeta$ which (with its interior) lies in $D$. As $S_n \to \varphi$ uniformly on $\gamma$, it follows from the Argument Principle that for all sufficiently large $n$ in $N_1$, $S_n(D)$ contains some neighbourhood $W$ of $\varphi(\zeta)$. Now Theorem 4.2.5 is applicable because as $S_n(\zeta) \to \varphi(\zeta)$, $\varphi(\zeta)$ is also in $J$; thus for all $k \geq k_0$, say, we have $R^k(W) \supset J$. We can now take $n$ in $N_2$ such that $m(n) \geq k_0$; then

$$J \subset R^{m(n)}(W) \subset R^{m(n)} S_n(D) = D.$$

The hypotheses of Lemma 9.2.2 apply equally well to any subdomain of $D$ so, starting with a subdomain $D_0$ of $D$ which meets, but which does not contain, $J$, the above argument shows that $J \subset D_0$ which is false. Thus $\varphi$ is constant on $D_0$, and hence also on $D$. □

EXERCISE 9.2

1. In the notation of Theorem 9.2.1, suppose that $D \subset F(R)$, and give a direct proof of the normality of $\{S_n\}$ in $D$.

## §9.3. Critical Points and Periodic Domains

We now provide evidence that the dynamics of a rational map $R$ is influenced by the orbits of its critical points by obtaining specific information about the location of these orbits in relation to the periodic components of $F$ and the cycles of $R$. Roughly speaking, we shall show that each attracting cycle, each rationally indifferent cycle, the boundary of each Siegel disc, and each boundary component of each Herman ring, attracts an infinite forward orbit of some critical point of $R$. As $R$ has at most $2d - 2$ critical points, this leads directly to quantitative information about the dynamics of $R$.

We recall that the *immediate basin* of a (super)attracting cycle $\{\zeta_1, \ldots, \zeta_q\}$ is the union of those components of $F$ which contain some $\zeta_j$, and with this we have

**Theorem 9.3.1.** *Let $R$ be a rational map of degree at least two. Then the immediate basin of each (super)attracting cycle of $R$ contains a critical point of $R$.*

Theorem 9.3.1 extends in a natural way to the rationally indifferent cycles of $R$, where the *immediate basin* of such a cycle is a cycle of components of $F$, each of which contains some petal at some point of the cycle (see §6.5 and also Exercise 9.3.1). Note that with this definition, a rationally indifferent cycle may have several basins (for example, $z + z^{p+1}$ has $p$ immediate basins): nevertheless, we have

**Theorem 9.3.2.** *Each immediate basin of a rationally indifferent cycle of $R$ contains a critical point of $R$.*

Of course, the critical point mentioned in Theorem 9.3.2 necessarily has an infinite forward orbit. The critical point mentioned in Theorem 9.3.1 has an infinite forward image if the component is attracting; if the component is super-attracting, then either it has an infinite forward image, or it lies in the cycle (and so is periodic). Note, however, that a super-attracting cycle always contains a critical point.

The next result concerns the location of the forward orbits of critical points in relation to the Siegel discs and Hermann rings (for more information, see [37] and [54]).

**Theorem 9.3.3.** *Let $\{\Omega_1, \ldots, \Omega_q\}$ be a cycle of Siegel discs, or of Herman rings, of a rational map $R$. Then the closure of $C^+(R)$ contains $\bigcup \partial\Omega_j$.*

Together, Theorems 9.3.1, 9.3.2 and 9.3.3 lead to many interesting consequences and we shall discuss these in the next section. To complete the picture, we now consider the irrationally indifferent cycles in $J$, and although these are not related in any obvious way to the components of the Fatou set, we can still prove

**Theorem 9.3.4.** *Let $R$ be a rational map $R$ of degree at least two. They every irrationally indifferent cycle of $R$ in $J$ lies in the derived set of $C^+(R)$.*

The remainder of this section contains the proofs of these results. First, we remark that we may confine our attention to attracting cycles in the proof of Theorem 9.3.1, for a super-attracting cycle contains critical points. Moreover, it is sufficient to prove the result for an attracting fixed point, for suppose that $\{\zeta_1, \ldots, \zeta_q\}$ is an attracting cycle, and let $F_j$ be the component of $F$ which contains $\zeta_j$. If we assume that the conclusion of Theorem 9.3.1 holds in the case of an attracting fixed point, we can conclude that $R^n$ has a critical point in each $F_j$, and hence that $R$ must have a critical point in some $F_j$ (as otherwise, $R^q: F_1 \to F_1$ would have no critical points). As the basin of the cycle is $\bigcup F_j$, the more general result follows. In fact, we have already proved Theorem 9.3.1 in the case of an attracting fixed point (it is Theorem 7.5.1), but we now give an alternative, and shorter, proof.

Proof of Theorem 9.3.1. We need only consider the case of a fixed point $\zeta$, so let $F_\zeta$ be the component of $F(R)$ which contains $\zeta$. As $\Delta$ is the universal cover of $F_\zeta$, the map $R: F_\zeta \to F_\zeta$ can be lifted to a single-valued analytic map $\Phi$, say, of $\Delta$ into itself, and we may assume that $\pi(0) = \zeta$ and $\Phi(0) = 0$.

Assume now that $R$ has no critical points in $F_\zeta$. Then the branch of $R^{-1}$ which fixes $\zeta$ can be continued analytically over every curve in $F_\zeta$, and this implies that given any curve $\gamma$ in $\Delta$, there is a curve $\Gamma$, namely $\pi^{-1}R^{-1}\pi(\gamma)$, which maps by $\Phi$ onto $\gamma$. In particular, $\Phi(\Delta) = \Delta$. Now as $R$ has no critical points in $F_\zeta$, we see that $\Phi$ has no critical points in $\Delta$ (for the cover map $\pi$ is a local homeomorphism); thus $\Phi$ is a smooth covering map of $\Delta$ onto itself and so is a homeomorphism, and hence a Möbius map of $\Delta$ onto itself. As $\pi\Phi = R\pi$, we obtain

$$|\Phi'(0)| = |R'(\zeta)| < 1,$$

and this cannot be so as the Möbius map $\Phi$ is a hyperbolic isometry and hence $|\Phi(z)| = |z|$ for all $z$. □

There is another way of expressing this last part of the argument. Using the earlier proof (of the equivalent Theorem 7.5.1), we see that $F_\zeta$ is simply connected, and then $\pi$ is a conformal equivalence of $\Delta$ onto $F_\zeta$ which conjugates $\Phi$ with $R$. As $\Phi$ is a hyperbolic isometry of $\Delta$, and $\pi: \Delta \to F_\zeta$ is an isometry (between the respective hyperbolic metrics), we see that $R$ is an isometry of $F_\zeta$ onto itself. This, however, conflicts with the fact that $R$ contracts distances near $\zeta$. Note that this argument does not require us to examine the hyperbolic metric at $\zeta$, and it is this form that we shall use for the corresponding proof in the case of an indifferent fixed point.

We come now to the proof of Theorem 9.3.2, and this is an amalgam of the two proofs of Theorem 9.3.1. As before, we need only consider the case of a rationally indifferent fixed point $\zeta$, and after again replacing $R$ by some iterate $R^n$, we may also assume that the basin in Theorem 9.3.2 is a forward invariant component $F_\zeta$ which contains a petal at $\zeta$.

THE PROOF OF THEOREM 9.3.2. The idea is that the rationally indifferent case is the limit of the attracting case as we move the attracting fixed point out to the Julia set. We assume that $R$ has no critical points in $F_\zeta$, and we proceed in essentially the same way as in the proof of Theorem 7.5.1, this time taking $V$ to be a petal at $\zeta$. We know (see Chapter 6) that $R$ maps $V$ into itself, and that $F_\zeta$ is the union of the backward images $R^{-n}(V)$ of $V$. As $V$ is simply connected, the domains $R^{-n}(V)$ provide an increasing sequence of simply connected domains which expand to fill $F_\zeta$, thus $F_\zeta$ is simply connected. This means that $R: F_\zeta \to F_\zeta$ is analytically conjugate to a Möbius map $\Phi$ of the unit disc $\Delta$ onto itself, and as $\Phi$ is a hyperbolic isometry of $\Delta$, we find that $R$ is a hyperbolic isometry of $F_\zeta$ onto itself (for the hyperbolic metrics are transferred without change by the conjugating map $\pi: \Delta \to F_\zeta$). The required contradiction is now obtained from the following lemma, for this implies that $R$ is not an isometry of $F_\zeta$. □

**Lemma 9.3.5.** *Suppose that $\zeta$ is a rationally indifferent fixed point of $R$, that $F_\zeta$ is a forward invariant component of $F(R)$ which contains a petal at $\zeta$, and that $\rho_1$ is the hyperbolic metric of $F_\zeta$. Then for any $z$ in $F_\zeta$, $\rho_1(R^n z, R^{n+1} z) \to 0$ as $n \to \infty$.*

PROOF. A close examination of the proof of the Petal Theorem (Theorem 6.5.2) shows that there is a subdomain $V_0$ of the petal $V$ which is mapped conformally by some map $\sigma$ onto a half-plane

$$H = \{x + iy: x > x_0\},$$

and which is such that $R: V_0 \to V_0$ is conjugate under $\sigma$ to a map $g: H \to H$ which satisfies

$$g^n(z) = np + o(n)$$

as $n \to \infty$. Now let $\rho_0$ be the hyperbolic metric on $V_0$. As $F_\zeta$ contains $V_0$, the Comparison Principle for hyperbolic metrics implies that $\rho_1 \leq \rho_0$, so it is enough to prove the result for $\rho_0$ rather than $\rho_1$ (see Appendix IV).

Now let $\tau$ be the hyperbolic metric on $H$. As the conjugating map $\sigma$ between $(V_0, \rho_0)$ and $(H, \tau)$ is an isometry, we now only prove the corresponding result for the action of $g$ on $H$. This, however, is entirely straightforward and is left to the reader (see Exercise 9.3.2). The proof of Theorem 9.3.2 is now complete. □

Next, we give the

PROOF OF THEOREM 9.3.3. As before, we may replace $R$ by some iterate $R^m$ and so assume that $R$ maps $\Omega$ onto itself, where $\Omega$ is either a Siegel disc or a Herman ring. In these circumstances, $R$ is a conformal bijection of $\Omega$ onto itself (see Chapter 7) and so has an inverse $S\colon \Omega \to \Omega$, with $S^n$ denoting both the $n$-th iterate of $S$ and also the inverse of $R^n\colon \Omega \to \Omega$.

We shall proceed by contradiction, so we suppose that there is a point $\zeta$ on $\partial\Omega$, and a disc, say

$$D = \{z\colon |z - \zeta| < 3\varepsilon\},$$

whose closure does not meet $C^+(R)$. Now take a point $w$ in $\Omega \cap D$, and let $U_n$ be the component of $R^{-n}(D)$ which contains $S^n(w)$. By our assumption about $C^+(R)$ (and by arguments that we have already used in this chapter), $R^n\colon U_n \to D$ is a smooth covering map with $D$ simply connected, so $U_n$ is simply connected and $R$ is a bijection of $U_n$ onto $D$ with inverse $S_n$, say. Of course, $S_n = S^n$ near $w$, so $S_n$ gives an analytic continuation of $S^n$ (defined near $w$) to $D$.

Next, the map $R$ of $\Omega$ onto itself is conjugate to a rotation of infinite order of a disc or annulus; thus the same is true of $S$ and so there is a set $N$, say, of integers such that $S^n \to I$ locally uniformly on $\Omega$ as $n \to \infty$ in $N$. As $S_n = S^n$ near $w$, we find that $S_n \to I$ uniformly in some neighbourhood of $w$. Now by Theorem 9.2.1, $\{S_n\colon n \in N\}$ is normal in $D$ and this in conjunction with Vitali's Theorem (Theorem 3.3.3) shows that $S_n \to I$ locally uniformly in $D$ as $n \to \infty$ in $N$.

Now let

$$U = \{z\colon |z - \zeta| < \varepsilon\}$$

and

$$V = \{z\colon |z - \zeta| < 2\varepsilon\}:$$

see Figure 9.3.1. A simple argument based on Rouché's Theorem shows that if $n$ is sufficiently large and in $N$, then $S_n(V) \supset U$. Applying the map $R^n\colon U_n \to D$

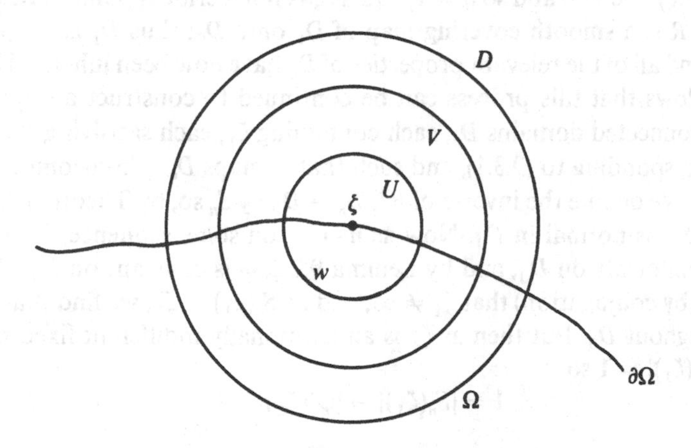


Figure 9.3.1

(the inverse of $S_n$), we obtain

$$R^n(U) \subset R^n S_n(V) = V$$

for these $n$ and so, from Theorem 4.2.5, $J \subset V$. Clearly, we may assume that $D$ was originally chosen small enough so that it does not contain $J$: this is the required contradiction and the proof is complete. □

*Remark.* For further information about the location of the forward orbits of critical points in relation to the boundaries of Siegel discs and Herman rings we refer the reader to [37], [54] and [55].

We end this section with the

PROOF OF THEOREM 9.3.4. We first prove the result for an irrationally indifferent fixed point $\zeta_1$ of $R$, and then extend this to cover the case of a cycle. We proceed by contradiction, so we suppose that there is a domain $D_0$ which contains $\zeta_1$ and is such that

$$C^+(R) \cap (D_0 - \{\zeta_1\}) = \varnothing. \tag{9.3.1}$$

We may take $D_0$ to be simply connected, so let $D_1$ be the component of $R^{-1}(D_0)$ that contains $\zeta_1$, write

$$R^{-1}(\zeta_1) \cap D_1 = \{\zeta_1, \ldots, \zeta_q\},$$

say, where $q \geq 1$, and consider the map

$$R: D_1 - \{\zeta_1, \ldots, \zeta_q\} \to D_0 - \{\zeta_1\}. \tag{9.3.2}$$

If $z$ lies in $D_1 - \{\zeta_1, \ldots, \zeta_q\}$, then $z$ cannot be a critical point of $R$ (else $C^+(R)$ would meet $D_0 - \{\zeta_1\}$ at $R(z)$), so the map (9.3.2) is a smooth covering map of $D_1 - \{\zeta_1, \ldots, \zeta_q\}$ onto $D_0 - \{\zeta\}$. The Riemann–Hurwitz formula shows that $(1 - q) + 0 = 0$ and so $q = 1$, and as $\zeta$ is not a critical point of $R$, we now see that $R$ is a smooth covering map of $D_1$ onto $D_0$; thus $D_1$ is simply connected and all of the relevant properties of $D_0$ have now been inherited by $D_1$.

It follows that this process can be continued to construct a sequence of simply connected domains $D_n$, each containing $\zeta_1$, each satisfying the condition corresponding to (9.3.1), and such that $R$ maps $D_{n+1}$ homeomorphically onto $D_n$. We denote the inverse of $R^n$: $D_n \to D_0$ by $S_n$ so, by Theorem 9.2.1, the family $\{S_n\}$ is normal in $D_0$. Now as $n \to \infty$ on some sequence, $S_n \to \varphi$, say, locally uniformly on $D_0$, and by Lemma 9.2.2, $\varphi$ is constant on $D_0$. We may assume (by conjugation) that $\zeta_1 \neq \infty$, and as $S_n(\zeta_1) = \zeta_1$, we find that $\varphi(z) = \zeta_1$ throughout $D_0$. But then as $\zeta_1$ is an irrationally indifferent fixed point of $R$, $|(R^n)'(\zeta_1)| = 1$ so

$$1 = |S_n'(\zeta_1)| \to |\varphi'(\zeta_1)| = 0,$$

and this is the contradiction we are seeking.

Now suppose that $\{\zeta_1, \ldots, \zeta_N\}$ is an irrationally indifferent cycle for $R$, and put $S = R^N$. We have just shown that there is a critical point $\eta$, say, of $S$, and a sequence $n_j$ such that the distinct points $S^{n_j}(\eta)$ converge to $\zeta_1$. As one of $\eta$, $R(\eta), \ldots, R^{N-1}(\eta)$ is a critical point of $R$, we conclude that $\zeta_1$, and hence each $\zeta_j$, is in the derived set of the forward orbit of some critical point of $R$. Thc proof is now complete. □

EXERCISE 9.3

1. Let $P(z) = z(1 + z^2)$. Show that $F(P)$ has at least three components, namely two petal domains at the origin and an unbounded component. Show also that the axes of the petals lie on the imaginary axis. Prove that $P$ maps each of the sets

$$\{iy: 0, y < 1\}, \qquad \{-iy: 0 < y < 1\}$$

into itself, and that $P^n \to 0$ on each.

Find the two finite critical points of $P$ and use this $P$ to illustrate Theorem 9.3.2.

2. Let $H = \{x + iy: y > 0\}$ and let $\tau$ be the hyperbolic metric on $H$ (so the $\tau$-distance is obtained from the differential $ds = |dz|/y$). Show that if $z$ and $w$ lie in $\{x+iy: y > y_0\}$, then $\tau(z, w) \leq |z - w|/y_0$. Deduce that if $z_n = in + o(n)$ as $n \to \infty$, then $\tau(z_n, z_{n+1}) \to 0$ as $n \to \infty$.

# §9.4. Applications

Here, we give various applications of the results about critical points obtained in the previous section, and we shall not bother to refer to these results explicitly. First, a rather obvious consequence of these results is obtained by noting that distinct cycles have disjoint basins. As $R$ has at most $2d - 2$ critical points, we obtain the following result which is all the more impressive if we bear in mind that $R$ has infinitely many cycles.

**Theorem 9.4.1.** *Let $R$ be a rational map of degree $d$, where $d \geq 2$. Then the combined number of (super)attracting and rationally indifferent cycles is at most $2d - 2$.*

We shall return to a sharper version of this result in §9.6, but even now we can extract more information in specific cases. For example, if $R$ is a quadratic polynomials, then there are only two critical points (one at $\infty$ and one in $\mathbb{C}$) and so there are at most two periodic cycles of components of types (a), (b), (c) in Definition 7.1.1 (one of which is the completely invariant component $F_\infty$ which contains $\infty$). If the finite critical point is attracted towards $\infty$, then there cannot be another such component so, in this case, $F(R) = F_\infty$, and hence $F(R)$ is connected.

Returning to Theorem 9.4.1, we see that a rational map of degree $d$ can have at most $2d - 2$ (super)attracting cycles. In fact, this upper bound is sharp for given any $d$ with $d \geq 2$, there is a rational map $R$ of degree $d$ with exactly $2d - 2$ (super)attracting cycles (see [88]). In the other direction, $R$ need not have any (super)attracting cycles at all. Of course, this must be the case for any map $R$ whose Fatou set is empty, but as this is rather artificial, we give a more interesting example.

**Example 9.4.2.** We claim that the map

$$R(z) = \frac{3z^2 + 1}{z^2 + 3}$$

has no (super)attracting cycles (of any period), and to verify this we show that the forward orbit of each critical point of $R$ accumulates at a point in $J$ and then appeal to Theorem 9.3.1. Now $R(z) = h(z^2)$, where $h$ is a Möbius map of the unit disc onto itself, so the two critical points of $R$ are 0 and $\infty$ and we must find the forward images of these points.

Note that $R$ has all three of its fixed points at the point 1, and as this is a rationally indifferent fixed point, it lies in $J$. Further, an easy computation shows that:

(i) $x < R(x) < 1$ on $(0, 1)$; and
(ii) $1 < R(x) < x$ on $(1, +\infty)$;

and from these we see that $R^n \to 1$ on $(0, +\infty)$ because $R^n$ can only converge to a fixed point of $R$. As $R(0) = \frac{1}{3}$ and $R(\infty) = 3$, we find that the forward orbit of each critical point accumulates only at $J$ and so $R$ has no (super)attracting cycles.

In fact, in this example $J$ is the unit circle and $R^n \to 1$ on both components of $F(R)$. To see this, observe that the unit circle is completely invariant under both $h$ and $z \mapsto z^2$, and hence under $R$, so it only remains to show that $J$ is the entire circle. Now the fixed point 1 is a rationally indifferent fixed point, and as

$$R(1 + z) = 1 + z - z^3/4 + O(z^4),$$

we see that $F(R)$ has exactly two (disjoint) components that contain petals based at 1 (see Theorem 6.5.8). As $J \subset \partial\Delta$, these components can only be the inside and the outside of the unit circle, with the entire circle as $J(R)$ to separate them.

Our next application was first mentioned in Chapter 5 (see Theorem 5.6.1 and the remarks following it).

**Theorem 9.4.3.** *If the Fatou set $F(R)$ of $R$ has two completely invariant components, then these are the only components of $F(R)$.*

PROOF. By assumption, there exist two completely invariant components, say $F_1$ and $F_2$. Now neither $F_1$ nor $F_2$ can be a Siegel disc or a Hermann ring (for if so, $R$ would be an injective map of the $F_j$ onto itself and so, by complete invariance, we would have $\deg(R) = 1$). It follows that each $F_j$ is an attractive, a super-attractive, or a parabolic component of $F(R)$, and in each of these cases, the forward orbit of any point in $F_j$ converges to a fixed point $\zeta_j$ in the closure of $F_j$.

Now every component of $F$ is simply connected (Theorems 5.2.1 and 5.6.1) and in particular, $F_1$ and $F_2$ are. By considering the map $R$ of $F_j$ onto itself, and using Theorem 5.5.4 together with the simple connectivity and complete invariance of each $F_j$, we obtain

$$\delta(F_1) = d - 1 = \delta(F_2),$$

where $\deg(R) = d$. This, together with fact that $R$ has at most $2d - 2$ critical points, implies that the forward orbit of every critical point lies in $F_1 \cup F_2$ and accumulates only at $\zeta_1$ or $\zeta_2$.

Suppose now that $F$ had other components: then (by Sullivan's Theorem, §8.1) it would necessarily contain some cycle of components disjoint from the invariant components $F_1$ and $F_2$. Now by Theorems 9.3.1, 9.3.2 and 9.3.3, such a cycle would either have to meet some forward orbit of critical points (and it cannot because $F_1$ and $F_2$ contain all of the critical points of $R$), or it would have to contain Siegel discs or Herman rings. In this latter case, there would be infinitely many accumulation points of forward orbits of critical points, and as these forward orbits accumulate only at $\zeta_1$ and $\zeta_2$ this case cannot occur either. We have now eliminated all possibilities, so $F(R)$ has no other components. □

Finally, we prove Theorem 4.3.1 which was only stated in §4.3, and which we now restate as

**Theorem 9.4.4.** *If every critical point of $R$ is pre-periodic, then $J(R) = \mathbb{C}_\infty$.*

PROOF. We suppose that every critical point of $R$ is pre-periodic, and that $F$ is non-empty, and we seek a contradiction. As $F$ is non-empty then, by Sullivan's Theorem, there is some component of $F$ which is periodic under the action of $R$. Now such a cycle of components is associated with a super-attracting cycle, an attracting cycle, a rationally indifferent cycle, a cycle of Siegel discs, or a cycle of Herman rings. In the first case there is some periodic critical point, while in the remaining cases, there is a critical point with an infinite forward orbit. By assumption, none of these cases can arise so $F$ must be empty and $J = \mathbb{C}_\infty$.

Recall that Theorem 9.4.4 shows that the rational map

$$R(z) = (z - 2)^2/z^2$$

has an empty Fatou set (see the remarks preceding Theorem 4.3.2). We shall give another, more direct, discussion (which does not use Sullivan's deep result) of this example in §11.9. In fact, Theorem 9.4.4 can also be proved without appealing to Sullivan's result (see Exercise 9.4.3). □

EXERCISE 9.4

1. Show that the rational map $R$ discussed in Example 9.4.2 is conjugate to the map $z \mapsto z + 1/z$ (see §1.7 and Exercise 1.7.4).

2. Investigate the rational maps defined by $z \mapsto h(z^2)$, where $h(z) = (az + c)/(cz + a)$, $0 < c < a$ (see Example 9.4.2).

3. Let $R$ be a rational map with $\deg(R) \geq 2$. Use Sullivan's Theorem to show that if $\varphi$ is a constant limit function in some component of $F(R)$, then the value of $\varphi$ lies in the closure of $C^+(R)$ (for the definition of limit functions, see §7.2).

   Now prove this result without using Sullivan's Theorem. [Suppose that $R^n \to 0$ locally uniformly on some component $F_0$ of $F(R)$ as $n \to \infty$ in some set $N$ of positive integers, and assume that $D = \{|z| < \varepsilon\}$ is disjoint from $C^+(R)$. Take $z_0$ in $F_0$ and for each $n$ in $N$ with $R^n(z_0) \in D$, construct a single-valued analytic branch $S_n$ of $(R^n)^{-1}$ in $D$ with $S_n R^n(z_0) = z_0$. Show that $\{S_n\}$ is normal in $D$, and assume that $S_n \to \Phi$ as $n \to \infty$ on some sequence. Show that

$$z_0 = S_n R^n(z_0) \to \Phi(0)$$

   as $n \to \infty$ in $N_1$. Now take another point, say, $z_1$ in $F_0$: show that $z_1 = \Phi(0)$, and hence that $z_0 = z_1$, a contradiction.]

## §9.5. The Fatou Set of a Polynomial

Let $P$ be a non-linear polynomial. We know that every bounded component of $F(P)$ is simply connected, and that the unbounded component $F_\infty$ of $F(P)$ is either simply connected or infinitely connected (Theorem 5.2.3). Our first result shows that the connectivity of $F_\infty$ is determined by the dynamics of the finite critical points of $P$.

**Theorem 9.5.1.** *Let $P$ be a polynomial with* $\deg(P) \geq 2$. *Then the following are equivalent*:

(a) *$F_\infty$ is simply connected;*
(b) *$J$ is connected;*
(c) *there are no finite critical points of $P$ in $F_\infty$.*

PROOF. We already know that (a) and (b) are equivalent (Theorem 5.2.1). First, we prove that (a) implies (c), so assume that $F_\infty$ is simple connected; thus $\chi(F_\infty) = 1$. Applying the Riemann–Hurwitz relation to the map $P$ of $F_\infty$ onto itself, we obtain

$$1 + \delta(F_\infty) = d,$$

and as $P$ has deficiency $d-1$ at $\infty$, there cannot be any finite critical points of $P$ in $F_\infty$: thus (c) follows.

It remains to prove that (c) implies (a), so assume now that are no finite critical points in $F_\infty$. From this and the complete invariance of $F_\infty$, we see that $P^n$ has no finite critical points in $F_\infty$ (see Theorem 2.7.3). Now find a disc $D$ centred at $\infty$ and such that

$$P(D) \subset D \subset F_\infty,$$

and define $D_0 = D$, and $D_n = P^{-n}(D)$. Then, as before, each $D_n$ is a domain containing $\infty$, and the $D_n$ satisfy

$$D = D_0 \subset D_1 \subset D_2 \subset \cdots.$$

Applying the Riemann–Hurwitz relation to the map $P^n$ of $D_n$ onto $D$, and noting that there are no finite critical points of $P^n$ in $F_\infty$, we find that

$$\chi(D_n) + (d^n - 1) = \chi(D_n) + \delta(D_n) = d^n\chi(D).$$

Now $D$ is simply connected, so $\chi(D_n) = 1$ and $D_n$ is simply connected. As $F_\infty$ is the union of the increasing sequence of simply connected domains $D_n$ it, too, is simply connected and the proof is complete. □

Theorem 9.5.1 implies that for the quadratic polynomial $P: z \mapsto z^2 + c$, the connectivity of $F_\infty$ is completely determined by the behaviour of the sequence $P^n(0)$, and we shall examine this situation in detail in §9.10. Returning to the general case, and recalling that $F_\infty$ is completely invariant, Theorem 9.5.1 leads immediately to the following two corollaries.

**Corollary 9.5.2.** *If either*:

(a) *every finite critical point of $P$ lies in $J$*; *or*
(b) *every finite critical point of $P$ has a finite forward orbit, then $F_\infty$ is simply connected.*

**Corollary 9.5.3.** *If every finite critical point of $P$ is pre-periodic, then $F$ is connected and simply connected.*

Only Corollary 9.5.3 should require further explanation. Suppose now that every finite critical point of $P$ is pre-periodic so, by Theorem 9.5.1, $F_\infty$ is simply connected. If $F$ is not connected then, by Sullivan's Theorem (Theorem 8.1.2) and the complete invariance of $F_\infty$, there must exist a periodic cycle of components of $F$ other than $F_\infty$. However, as we have seen in §9.3, such a cycle requires a critical point which is either in a super-attracting cycle in $\mathbb{C}$, or which has an infinite forward orbit. As pre-periodicity explicitly excludes both of these two possibilities, we find that $F = F_\infty$, and so is both connected and simply connected.

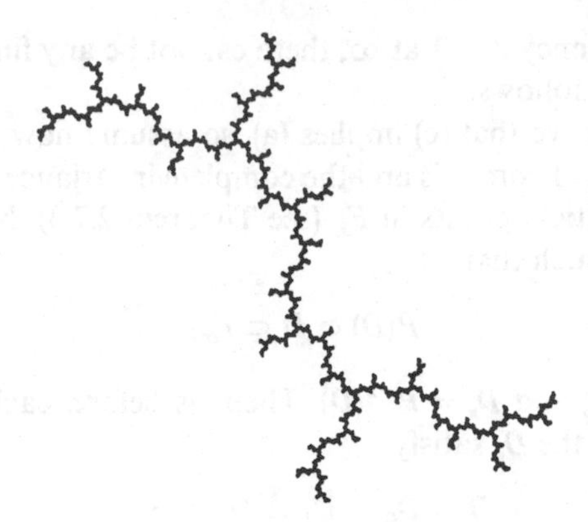

Figure 9.5.1. $z \mapsto z^2 + i$. Fractal image reprinted with permission from *The Beauty of Fractals* by H.-O. Peitgen and P.H. Richter, 1986, Springer-Verlag, Heidelberg, New York.

As examples of the three hypotheses in these corollaries (and in the same order), we find that $F_\infty$ is simply connected in each of the cases

$$z \mapsto z^2 - 2, \qquad z \mapsto z^2 - 1, \qquad z \mapsto z^2 + i.$$

The reader is invited to supply the details. The Julia set of $z^2 + i$ is given above in Figure 9.5.1: the Julia set for $z^2 - 2$ is $[-2, 2]$, and the Julia set for $z^2 - 1$ is illustrated in Figure 1.5.1.

The reader will almost certainly have seen many illustrations of Julia sets for polynomials $z^2 + c$, and it is evident that these Julia sets have rotational symmetry of order two about the origin. We shall now show how to compute the symmetry group of the Julia set of any polynomial $P$ of degree $d$, where $d \geq 2$.

First, by definition, the symmetry group $\Sigma(P)$ of $J(P)$ is the group of Euclidean isometries $\gamma$ (of the form $z \mapsto az + b$, $|a| = 1$) such that $\gamma(J(P)) = J(P)$. Now every element of $\Sigma(P)$ is either a rotation or a translation, and as $J(P)$ is bounded in $\mathbb{C}$, $\Sigma(P)$ can only contain rotations. However, if two rotations $\sigma$ and $\gamma$ have distinct fixed points (in $\mathbb{C}$), then their commutator $\sigma\gamma\sigma^{-1}\gamma^{-1}$ is a non-trivial translation, so we deduce that $\Sigma(P)$ is some group of Euclidean rotations about a point $\zeta$ in $\mathbb{C}$.

There are two cases to consider, depending on whether $\Sigma(P)$ is finite or infinite. If $\Sigma(P)$ is infinite (in fact, it must then be the group of *all* rotations about $\zeta$), then $J(P)$ must consist of a union of circles centred at $\zeta$ (see Exercise 9.5.4). However, $J(P)$ is also the boundary of $F_\infty$; thus $J(P)$ must consist of exactly one circle. By conjugation we may assume that $J(P)$ is the unit circle, then $P$ is of the form $az^d$, where $|a| = 1$, and $\Sigma(P)$ is the group of all rotation about the origin. Clearly, this is an exceptional case.

Now suppose that $\Sigma(P)$ is finite (and not trivial). Our task is to find the common centre of the rotations and the order of $\Sigma(P)$. Let

$$P(z) = a_0 + \cdots + a_{d-1}z^{d-1} + a_d z^d,$$

where $a_d \neq 0$. Now take any $w$ in $\mathbb{C}$, and let $z_1, \ldots, z_d$ be the solutions of $P(z) = w$; thus

$$P(z) = w + a_d(z - z_1)\cdots(z - z_d)$$

and so (because $d \geq 2$)

$$(z_1 + \cdots + z_d)/d = -a_{d-1}/(da_d).$$

This says that the centre of gravity of the $z_j$ is independent of the choice of $w$, and clearly this extends to the fact that the centre of gravity of $P^{-n}(w)$ is independent of both $n$ and $w$. As (roughly speaking) $P^{-n}(w) \to J(P)$, it is reasonable to suppose that this centre of gravity is indeed the centre of symmetry of $J(P)$. It is important to note that this centre of gravity is defined even when $J(P)$ has no symmetries, so we define $\xi$ by

$$\xi = -a_{d-1}/(da_d)$$

and call it the *centroid* of $P$.

Before stating our result, observe that if $\gamma$ is a Euclidean similarity, then (either by computing coefficients, or by examining inverse images) the centroid of $\gamma P \gamma^{-1}$ is the $\gamma$-image of the centroid of $P$. Clearly,

$$\Sigma(\gamma P \gamma^{-1}) = \gamma \Sigma(P) \gamma^{-1},$$

so in attempting to identify the symmetry group $\Sigma(P)$, it is sufficient to replace $P$ by any polynomial conjugate (by a similarity) to $P$. Conjugating $P$ by a translation, we may shift the centroid to the origin (this means that the coefficient of $z^{d-1}$ vanishes) and by a further conjugation of the form $z \mapsto \lambda z$ (which will not move the centroid), we may assume that $P$ is monic. Thus (up to conjugacy) we may assume that $P$ is in the form

$$P(z) = a_0 + \cdots + 0z^{d-1} + z^d, \tag{9.5.1}$$

and we now rewrite this as

$$P(z) = z^a P_0(z^b), \tag{9.5.2}$$

where $a$ and $b$ are maximal for this form, and $P_0$ is a monic polynomial. If $a = d$, then $P(z) = z^d$ and $J(P)$ is a circle. Excluding this exceptional case, we find that $0 \leq a < d$ and $P_0(0) \neq 0$. We can now state our basic result.

**Theorem 9.5.4.** *The elements of the symmetry group $\Sigma(P)$ are rotations about the centroid of $P$. Further, if $P$ is given by* (9.5.1) *and* (9.5.2) and $a < d$, *then $\Sigma(P)$ has order $b$.*

As an example, consider

$$P(z) = z^3 - 9z^2 + 29z - 3.$$

The centroid of $P$ is 3, and writing $\gamma(z) = z - 3$ we have

$$\gamma P \gamma^{-1}(z) = z(2 + z^2).$$

We find, then, that the only non-trivial symmetry of $J(P)$ has order two and $\Sigma(P) = \{I, \gamma\}$, where $\gamma(z) = 6 - z$.

The proof of Theorem 9.5.4 depends on a knowledge of Green's functions, so we shall discuss this first. The unbounded component $F_\infty$ of $F(P)$ supports a Green's function $g(z)$ with pole at $\infty$, and this is the unique function with the properties:

(i) $g$ is harmonic and positive in $F_\infty \cap \mathbb{C}$;
(ii) $g(z) - \log|z|$ is bounded in a neighbourhood of $\infty$; and
(iii) $g(z) \to 0$ as $z \to \partial F_\infty$.

We shall not enter into the theory of Green's functions here, but the interested reader can consult one of many texts on complex analysis or potential theory. Our next task is to identify $g$ in terms of the function which conjugates $P$ to $z \mapsto z^d$ at $\infty$. As $\infty$ is a super-attracting fixed point there is some function $\varphi$ that fixes $\infty$ and is analytic nearby, and which is such that

$$\varphi P \varphi^{-1}(z) = z^d$$

in some neighbourhood of $\infty$ (Theorem 6.10.1). Explicitly, we can find a neighbourhood $D$ of $\infty$ with the properties:

(iv) the closure of $P(D)$ is a compact subset of $D$;
(v) near $\infty$, $\varphi$ is analytic and of the form

$$\varphi(z) = z + b_0 + b_1/z + \cdots,$$

(vi) $\varphi$ maps $D$ onto some set $\{|z| > r\}$, where $r > 1$; and
(vii) $\varphi P(z) = [\varphi(z)]^d$ on $D$.

For later use, observe that if $P$ satisfies (9.5.1) and (9.5.2) then, by equating coefficients of $z^{d-1}$ in (vii),

$$b_0 = 0. \tag{9.5.3}$$

We prove

**Lemma 9.5.5.** *$g(z) = \log|\varphi(z)|$ in $D$.*

PROOF. As $\varphi P^n$ is defined on $P^{-n}(D)$, we can define a function $g_n$ there by

$$g_n(z) = d^{-n} \log|\varphi(P^n z)|,$$

and from (v) and (vi), $g_n$ is positive and harmonic on $P^{-n}(D)$ (except at $\infty$). Next, $P^{-n}(D) \subset P^{-(n+1)}(D)$, so if $z$ is in $P^{-n}(D)$, then both $g_n$ and $g_{n+1}$ are defined at $z$. As

$$\varphi P^{n+1}(z) = \varphi P(P^n z) = [\varphi(P^n z)]^d,$$

we see that $g_{n+1}(z) = g_n(z)$ on $P^{-n}(D)$. Finally as $F_\infty$ is the union of the domains $P^{-n}(D)$, we can now define a positive harmonic function $g$ on $F_\infty$ by putting $g = g_n$ on $P^{-n}(D)$. Of course, near $\infty$,

$$g(z) = \log|\varphi(z)| = \log|z| + O(1),$$

so in order to complete the proof, we must show that $g(z) \to 0$ as $z \to \partial F_\infty$.

For any $z$ in $F_\infty$, $z$ lies in some $P^{-(m+1)}(D)$ and

$$dg_{m+1}(z) = d^{-m} \log|\varphi P^m(Pz)| = |g_n(Pz)|;$$

thus we have the functional relation

$$d^{-1}g(Pz) = g(z),$$

and hence also

$$d^{-n}g(P^n z) = g(z).$$

Now let

$$M = \sup\{g(w): w \in P^{-1}(D), w \notin D\}.$$

If $z$ lies outside $P^{-n}(D)$, then for some $m$ satisfying $m \geq n$, $z$ lies in $P^{-(m+1)}(D)$ but not in $P^{-m}(D)$; thus $g(P^m z) \leq M$. It follows that

$$d^n g(z) \leq d^m g(z) = g(P^m z) \leq M,$$

and so finally,

$$0 < g(z) < M/d^n$$

on $F_\infty - P^{-n}(D)$. The proof is complete. □

We can now give the

Proof of Theorem 9.5.4. The first step is to identify the elements of $\Sigma(P)$ in terms of the function $\varphi$, so take any isometry $\gamma$, say $\gamma(z) = az + b$. Now clearly, $\gamma$ is a symmetry of $J(P)$ if and only if it is a symmetry of $F_\infty$, and hence if and only if $g$ is invariant under $\gamma$; thus $\gamma \in \Sigma(P)$ if and only if

$$g(\gamma z) = g(z)$$

for all $z$ near $\infty$. Recalling that $g(z) = \log|\varphi(z)|$, we now find that $\gamma \in \Sigma(P)$ if and only if for some $\lambda$ with $|\lambda| = 1$,

$$\varphi(az + b) = \lambda\varphi(z)$$

near $\infty$. If this is so, then from (9.5.3), $a = \lambda$ and $b = 0$, and we have now proved that the elements of $\Sigma(P)$ are rotations about the centroid of $P$.

Note that this argument also shows that if $\gamma \in \Sigma(P)$, then $\varphi\gamma = \gamma\varphi$. Conversely, if $\gamma$ is a rotation about the origin, and if it commutes with $\varphi$, then

$$\begin{aligned} g(\gamma z) &= \log|\varphi\gamma(z)| \\ &= \log|\gamma\varphi(z)| \\ &= \log|\varphi(z)| \\ &= g(z); \end{aligned}$$

thus we have proved that an isometry $\gamma$ is in $\Sigma(P)$ if and only if $\gamma(0) = 0$ and $\gamma\varphi = \varphi\gamma$. □

Next, we identify the elements of $\Sigma(P)$ in terms of $P$. We have

**Lemma 9.5.6.** *An isometry $\gamma$ is in $\Sigma(P)$ if and only if $\gamma(0) = 0$ and $P\gamma = \gamma^d P$.*

PROOF. Suppose first that $\gamma(0) = 0$ and $P\gamma = \gamma^d P$. Then for each $n$ there is some $m$ with

$$P^n\gamma = \gamma^m P^n,$$

so

$$|P^n(\gamma z)| = |P^n(z)|.$$

Letting $n \to \infty$, this shows that $\gamma(z)$ is in $F_\infty$ if and only if $z$ is, thus $\gamma$ must be a symmetry of $J(P)$ (the boundary of $F_\infty$).

Now suppose that $\gamma$ is in $\Sigma(P)$; then (from above) $\gamma$ fixes the origin and commutes with $\varphi$. Writing $\gamma(z) = \lambda z$, this shows that

$$\begin{aligned}
\varphi(P\gamma z) &= \varphi P(\gamma z)\\
&= [\varphi(\gamma z)]^d\\
&= [(\varphi\gamma)(z)]^d\\
&= [(\gamma\varphi)(z)]^d\\
&= [\lambda\varphi(z)]^d\\
&= \lambda^d[\varphi(z)]^d\\
&= \gamma^d(\varphi(z)^d)\\
&= \gamma^d(\varphi P(z))\\
&= \varphi(\gamma^d P(z)),
\end{aligned}$$

and because $\varphi$ is univalent near $\infty$, this shows that $P\gamma = \gamma^d P$ as required. □

We are now in a position to show that if $P$ is given by (9.5.1) and (9.5.2), then $\Sigma(P)$ has order $b$. First, suppose that $\sigma(z) = \mu z$, where $\mu^b = 1$. Then from (9.5.2), $a \equiv d \pmod{b}$ so $\mu^a = \mu^d$, and so

$$P\sigma(z) = \mu^a z^a P_0(\mu^b z^b) = \mu^d z^a P_0(z^b) = \sigma^d P(z),$$

so by Lemma 9.5.6, $\sigma$ is in $\Sigma(P)$. Thus $\Sigma(P)$ contains the group of rotations about the origin of order $b$.

Next, let $\sigma$ be any element of $\Sigma(P)$, so $\sigma(z) = \mu z$, say, where $|\mu| = 1$. As $P\sigma = \sigma^d P$, we have

$$\mu^a z^a P_0(\mu^b z^b) = \mu^d z^a P_0(z^b).$$

Dividing by $z^a$ and letting $z \to 0$ (and recalling that $P_0(0) \neq 0$), we find that $\mu^a = \mu^d$, and so for all $w$,

$$P_0(\mu^b w) = P_0(w). \tag{9.5.4}$$

Now write $P_0$ in the form

$$P_0(w) = \alpha_0 + \sum \alpha_j w^{m_j}, \tag{9.5.5}$$

where each $\alpha_j$ is non-zero, and observe that because the integer $b$ in (9.5.2) is maximal, the integers $m_j$ are coprime; thus there are integers $v_j$ such that

$$\sum_j v_j m_j = 1.$$

Using (9.5.4), we find that for each of the terms in the sum in (9.5.5), $\mu^{bm_j} = 1$, and so

$$\mu^b = \mu^{b \sum_j v_j m_j} = \prod_j \mu^{bm_j v_j} = 1,$$

as required. The proof of Theorem 9.5.4 is now complete. □

## Exercise 9.5

1. Let $P(z) = z^2 - 1$, and for each $w$ in $F(P)$, let $F_w$ be the component of $F(P)$ containing $w$. Show that the only finite critical point of $P$ is in an attracting two-cycle of $P$, and deduce
   (i) $F(P)$ has infinitely many components;
   (ii) every component of $F(P)$ is simply connected;
   (iii) $P$ has no Siegel discs;
   (iv) $\{F_\infty\}$ and $\{F_0, F_{-1}\}$ are the only periodic cycles of components of $F$;
   (v) every bounded component of $F(P)$ is mapped by some $P^n$ into $F_0$, thus for any bounded component $\Omega$, the sequence $P^n(\Omega)$ is eventually $\ldots, F_0, F_{-1}, F_0, F_{-1}, \ldots$.

   Find the multiplicity of $P$ as a map acting between the given components in each of the following cases:

   $$F_\infty \to F_\infty, \qquad F_0 \to F_{-1}, \qquad F_{-1} \to F_0.$$

   Locate the component of $F$ in Figure 1.5.1 (other than $F_{-1}$) which maps onto $F_0$.

2. Let $P$ be a polynomial of degree $d$, where $d \geq 2$, and suppose that $F_\infty$ is simply connected. Show that $P\colon F_\infty \to F_\infty$ is conjugate to $z \mapsto z^d$ acting on the unit disc.

3. Show that if $\sigma$ and $\gamma$ are rotations about different points, then their commutator is a non-trivial translation.

4. Suppose that the symmetry group $\Sigma(P)$ of $J(P)$ is an infinite group of rotations about $\xi$ (as in the text). Let $w$ be any point of $J(P)$. Show that the set $\{\gamma(w)\colon \gamma \in \Sigma(P)\}$ is dense in the circle centre $\xi$ and radius $|w - \xi|$, and deduce that the entire circle lies in $J(P)$.

5. Let $\gamma$ be a similarity and suppose that $\xi$ is the centroid of $P$. Show that $\gamma P \gamma^{-1}$ has centroid $\gamma(\xi)$.

6. Show that if $P(z) = z^2 + c$, where $c \neq 0$, then $\Sigma(P) = \{I, \sigma\}$, where $\sigma(z) = -z$.
7. Let $P(z) = z^3 + z^2 + Az + B$. Find all values of $A$ and $B$ for which $\Sigma(P)$ has order 1, 2 and 3 respectively.
8. Let $P(z) = z(1 + z^2)$ and $Q(z) = -z(1 + z^2)$. Show that $P$ and $Q$ have the same Julia set, but that $P$ and $Q$ are not conjugate (use Theorem 4.2.9 and Lemma 2.6.1)

## §9.6. The Number of Non-Repelling Cycles

In this section we show that a rational map $R$ of degree $d$ has at most $2d - 2$ non-repelling cycles. Theorem 9.3.1 shows that the number $N_0$ of (super)-attracting cycles is at most $2d - 2$, and Fatou showed that by perturbing the coefficients of $R$ in an appropriate way, one can ensure that at least half of the number $N_1$ of indifferent cycles become attracting. If the perturbation is small enough, the (super)attracting cycles remain super(attracting), so Fatou proved that $N_0 + N_1/2 \leq 2d - 2$: in addition to Fatou's paper ([45], p. 66), we refer the reader to [22] and [67] for a discussion of this method. More recently, Shishikura showed that by using quasiconformal maps to provide the perturbation, one can obtain the sharp inequality $N_0 + N_1 \leq 2d - 2$ [88]: we follow his method here and prove

**Theorem 9.6.1.** *A rational map $R$ of degree $d$, $d \geq 2$, has at most $2d - 2$ non-repelling cycles.*

We remark that this (or Fatou's weaker result) provides the missing step in the proof of Theorem 6.9.2, namely that $J(R)$ is the closure of the repelling cycles of $R$.

The idea behind the proof of Theorem 9.6.1 is to compose $R$ with a quasiconformal map in such a way that the non-repelling cycles of $R$ become attracting cycles of the composite map $S$. Now $S$ is not rational, but we can construct a quasiconformal conjugate $\varphi S \varphi^{-1}$ of $S$ which is rational, which leaves the degree invariant, and which preserves the attracting nature of the cycles of $S$. With this, we apply Theorem 9.3.1 to the map $\varphi S \varphi^{-1}$ and so obtain the required bound of $2d - 2$.

The reader is reminded that Chapter 8 contains a brief account of, and references for, quasiconformal maps. In addition to this, we shall say that a map is *quasiregular* if it is a composite map of the form $fg$, where $f$ is rational and $g$ is a quasiconformal map of the sphere onto itself. The rest of this section is devoted to the

PROOF OF THEOREM 9.6.1. Given the rational map $R$, we take any finite collection $C_1, \ldots, C_q$ of non-repelling cycles of $R$, and our objective is to prove that

$q \leq 2d - 2$. By conjugation, we may suppose that the cycles $C_1, \ldots, C_q$ lie in $\mathbb{C}$, and we write

$$C_1 \cup \cdots \cup C_q = \{\zeta_1, \ldots, \zeta_t\}.$$

The proof proceeds by constructing various functions $h$, $R_\varepsilon$, $H_\varepsilon$ and $S_\varepsilon$ which depend on a parameter $\varepsilon$ in $(0, 1)$ and then, eventually, choosing a value of $\varepsilon$ that is small enough to satisfy certain conditions that have arisen during the proof.

We begin by constructing a polynomial $h$ with the properties

$$h(\zeta_j) = 0, \qquad h'(\zeta_j) = -1, \qquad j = 1, \ldots, t$$

(see Exercise 9.6.1), and throughout this proof we shall make frequent use of $k$, where $k = \deg(h)$. We now choose any $\varepsilon$ in $(0, 1)$ and use $h$ (which does not depend on $\varepsilon$) to construct the perturbation $R_\varepsilon$ of $R$ given by

$$R_\varepsilon(z) = R(z + \varepsilon h(z)).$$

Of course, $R_\varepsilon$ is rational, and each $C_j$ is a (super)attracting cycle of $R_\varepsilon$: however, $\deg(R_\varepsilon) > \deg(R)$, so we need to modify this perturbation of $R$ to return its degree back to $\deg(R)$.

In some sense, the deviation of $R_\varepsilon$ from $R$ is most when $\varepsilon|z|^k$ is large, so we want to suppress the perturbation in this region. To do this, we take any decreasing $C^\infty$-function

$$\rho\colon [0, +\infty) \to [0, 1]$$

such that $\rho = 1$ on $[0, 1]$ and $\rho = 0$ on $[2, +\infty)$, and replace $z + \varepsilon h(z)$ by

$$H_\varepsilon(z) = z + \varepsilon\rho(\varepsilon^{1/k}|z|)h(z).$$

With this, we put

$$S_\varepsilon(z) = RH_\varepsilon(z) = R(z + \varepsilon\rho(\varepsilon^{1/k}|z|)h(z)),$$

and observe that

$$S_\varepsilon(z) = \begin{cases} R_\varepsilon(z) & \text{if } \varepsilon^{1/k}|z| < 1, \\ R(z) & \text{if } \varepsilon^{1/k}|z| \geq 2 \end{cases}$$

(see Figure 9.6.1). In particular, if

$$\varepsilon < \min\{|\zeta_1|^{-k}, \ldots, |\zeta_t|^{-k}\},$$

then $S_\varepsilon = R_\varepsilon$ near each $\zeta_j$ and so each cycle $C_j$ is a (super)attracting cycle for $S_\varepsilon$.

Next, we find a number $M$, $M > 1$, such that:

(1) $|h(z)| \leq M$ when $|z| \leq 1$;
(2) $|h(z)| \leq M|z|^k$ and $|h'(z)| \leq M|z|^{k-1}$ when $|z| \geq 1$;
(3) for all real $x$, $|\rho'(x)| \leq M$; and
(4) $\sigma(Rz, Rw) \leq M\sigma(z, w)$ (see Theorem 2.3.1).

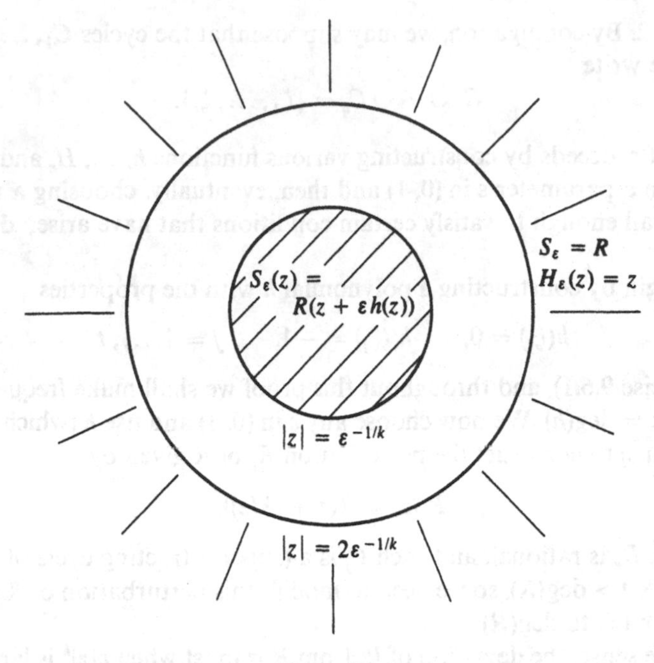


Figure 9.6.1

With these, we can easily establish two basic properties of the functions $H_\varepsilon$ and $S_\varepsilon$, namely

(5) $$\sigma(S_\varepsilon(z), R(z)) \leq 2^k M^2 \varepsilon^{1/k}$$

on $\mathbb{C}_\infty$ (so $S_\varepsilon \to R$ uniformly on $\mathbb{C}_\infty$ as $\varepsilon \to 0$), and

(6) for all sufficiently small $\varepsilon$, $H_\varepsilon$ is a quasiconformal homeomorphism of $\mathbb{C}_\infty$ onto itself.

To prove (5), note that $S_\varepsilon(z) = R(z)$ when $|z| > 2\varepsilon^{-1/k}$ so we need only concern ourselves with those $z$ satisfying $|z| \leq 2\varepsilon^{-1/k}$. For these $z$, we can use (1), (2) and (4) to obtain

$$\begin{aligned}
\sigma(S_\varepsilon(z), R(z)) &\leq M\sigma(H_\varepsilon z, z) \\
&\leq \frac{2M\varepsilon|h(z)|}{(1 + |z|^2)^{1/2}} \\
&\leq \frac{2M^2\varepsilon \max\{1, |z|^k\}}{\max\{1, |z|\}} \\
&\leq 2M^2\varepsilon \max\{1, |z|^{k-1}\} \\
&\leq 2^k M^2 \varepsilon^{1/k},
\end{aligned}$$

which proves (5).

To prove (6), we must estimate $\|\mu_\varepsilon\|_\infty$, where

$$\frac{\partial H_\varepsilon(w)}{\partial \bar{z}} = \mu_\varepsilon(w)\frac{\partial H_\varepsilon(w)}{\partial z},$$

and it is clear that we may assume that

$$\varepsilon^{-1/k} \leq |z| \leq 2\varepsilon^{-1/k},$$

because outside of this region $H_\varepsilon$ is analytic and $\mu_\varepsilon = 0$ there. When $z$ satisfies these inequalities, we have

$$\begin{aligned}\left|\frac{\partial H_\varepsilon}{\partial \bar{z}}\right| &= \left|\varepsilon h(z)\frac{\partial \rho(\varepsilon^{1/k}|z|)}{\partial \bar{z}}\right| \\ &= \left|\varepsilon h(z)\varepsilon^{1/k}\rho'(\varepsilon^{1/k}|z|)\frac{\partial |z|}{\partial \bar{z}}\right| \\ &\leq \varepsilon^{1+1/k}M^2|z|^k \\ &\leq 2^k M^2 \varepsilon^{1/k}.\end{aligned}$$

Similarly,

$$\frac{\partial H_\varepsilon}{\partial z} = 1 + \varepsilon\rho(\varepsilon^{1/k}|z|)h'(z) + \varepsilon h(z)\varepsilon^{1/k}\rho'(\varepsilon^{1/k}|z|)\frac{\partial |z|}{\partial z},$$

and this time simple estimates show that

$$\left|\frac{\partial H_\varepsilon}{\partial z} - 1\right| = O(\varepsilon^{1/k})$$

as $\varepsilon \to 0$. These inequalities imply that for all sufficiently small $\varepsilon$, $\|\mu_\varepsilon\|_\infty < 1$, and in this case, the Jacobian $J$ of $H_\varepsilon$ is positive for (in general)

$$J = \left|\frac{\partial H_\varepsilon}{\partial z}\right|^2 - \left|\frac{\partial H_\varepsilon}{\partial \bar{z}}\right|^2,$$

and so $H_\varepsilon$ (which is $C^\infty$) is locally a homeomorphism. This means that $H_\varepsilon$ is a smooth covering map of the sphere onto itself, and hence (by the Monodromy Theorem) it is a homeomorphism of the sphere onto itself. Finally, as $\|\mu_\varepsilon\|_\infty < 1$, (6) holds.

Observe now that as $H_\varepsilon$ is a homeomorphism of the sphere onto itself, we have $\deg(S_\varepsilon) = \deg(R)$, where here, $\deg(S_\varepsilon)$ is the cardinality of the set $S_\varepsilon^{-1}\{w\}$ for all but a finite number of exceptional points $w$. Suppose now that for any small $\varepsilon$ we can find a quasiconformal map $\varphi$ of the sphere onto itself such that $\varphi S_\varepsilon \varphi^{-1}$ is rational. Clearly, this conjugation leaves the degree unchanged, so

$$\deg(\varphi S_\varepsilon \varphi^{-1}) = \deg(R),$$

and moreover, it preserves the (super)attracting nature of the cycles $C_j$ of $S_\varepsilon$, for such a cycle is characterized by the convergence of the iterates of points near the cycle to the cycle, and this is invariant under any topological conjugacy. Applying Theorem 9.3.1 to this $\varphi S_\varepsilon \varphi^{-1}$, we obtain Theorem 9.6.1, so it only remains to show that such a map $\varphi$ exists. The existence proof is based on the following

**Lemma 9.6.2.** *Let $g$ be a quasiregular map of $\mathbb{C}_\infty$ onto itself and suppose that $g$ maps the open set $E$ into itself. Suppose also that $\partial g/\partial \bar{z} = 0$ on both $E$ and $\mathbb{C}_\infty - g^{-1}(E)$. Then there exists a quasiconformal map $\varphi\colon \mathbb{C}_\infty \to \mathbb{C}_\infty$ such that $\varphi g \varphi^{-1}$ is rational.*

Shishikura calls this *quasiconformal surgery* for it shows that by a quasiconformal conjugation, we can merge the two analytic maps $g$ of $E$ into itself, and $g$ of $\mathbb{C}_\infty - \bigcup_{n=0}^{\infty} g^{-n}(E)$ into itself, into one analytic map of the sphere onto itself. We shall continue with the proof of Theorem 9.6.1 until it is complete, and then return to prove Lemma 9.6.2.

The next result is merely a restatement of Lemma 9.6.2 in the context of our proof, but it does serve to focus attention on the next important step, namely the construction of the set $E$. For each positive $\varepsilon$, let

$$V_\varepsilon = \{z\colon |z| \geq \varepsilon^{-1/k}\}.$$

**Lemma 9.6.3.** *Suppose that $E$ is an open set such that:*

(a) *$E$ and $V_\varepsilon$ are disjoint; and*
(b) *$S_\varepsilon$ maps $E \cup V_\varepsilon$ into $E$.*

*Then for all sufficiently small $\varepsilon$, $S_\varepsilon$ is quasiconformally conjugate to a rational map.*

PROOF. We verify the hypotheses of Lemma 9.6.2 with $g = S_\varepsilon$. By assumption, $S_\varepsilon$ maps the open set $E$ into itself, and we have seen that for sufficiently small $\varepsilon$, $S_\varepsilon$ is quasiregular on the sphere and analytic outside of $V_\varepsilon$. As (a) and (b) imply that both $E$ and $\mathbb{C}_\infty - S_\varepsilon^{-1}(E)$ are disjoint from $V_\varepsilon$, the desired conjugacy follows.

We must now construct a set $E$ satisfying the conditions in Lemma 9.6.3, and to do this, we construct an open set $E_\varepsilon$ depending on the positive parameter $\varepsilon$, and then show that we can take $E$ to be $E_\varepsilon$ for some suitably small $\varepsilon$. We recall that each $\zeta_j$ lies in $\mathbb{C}$, and for each $\varepsilon$, we shall construct a neighbourhood of $N_j$ of $\zeta_j$ (for brevity, we omit the dependence of $N_j$ on $\varepsilon$ from our notation) satisfying

$$N_j \subset \{z\colon |z - \zeta_j| < 1\},$$

and the set $E_\varepsilon$ will be the union of the $N_j$. We shall assume that $\varepsilon$ is chosen so that for each $j$,

$$|\zeta_j| + 1 < \varepsilon^{-1/k},$$

and these assumptions imply that Lemma 9.6.3(a) is satisfied, and also that

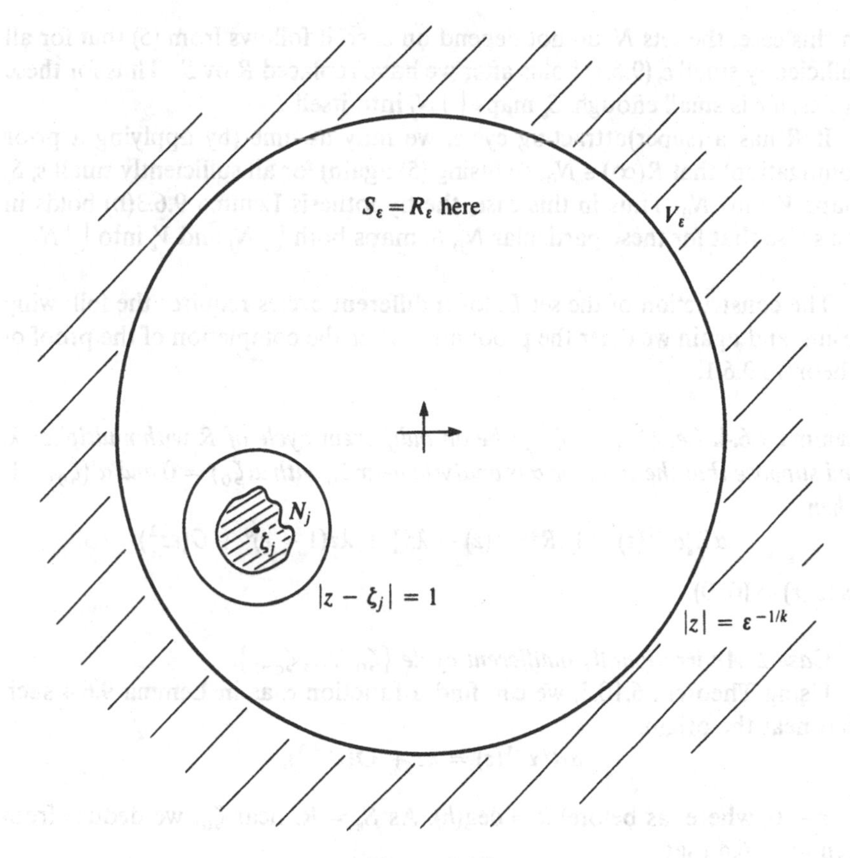


Figure 9.6.2

$S_\varepsilon = R_\varepsilon$ on $E$: see Figure 9.6.2. The verification of Lemma 9.6.3(b) will be part of our construction of the $N_j$.

We now give the detailed construction of the $N_j$, and this splits naturally into the three cases of a (super)attracting, irrationally, or rationally indifferent cycle. Of these, the latter case is the hardest.

*Case* 1: *A (super)attracting cycle* $\{\zeta_0, \ldots, \zeta_{q-1}\}$.

It will be convenient to use the notation $A < B$ for sets $A$ and $B$ to mean that the closure of $A$ is contained in the interior of $B$. We construct a simply connected open neighbourhood $W$ of $\zeta_0$ such that $R^q(W) < W$, and then find open neighbourhoods $W_j$ of $\zeta_0$ such that

$$R^q(W) < W_1 < \cdots < W_{q-1} < W.$$

Now put

$$N_0 = R^q(W), \qquad N_j = R^j(W_j), \qquad j = 1, \ldots, q-1,$$

so

$$R(N_0) < N_1, \quad R(N_1) < N_2, \quad \ldots, \quad R(N_{q-1}) < N_0. \tag{9.6.1}$$

In this case, the sets $N_j$ do not depend on $\varepsilon$, so it follows from (5) that for all sufficiently small $\varepsilon$, (9.6.1) holds after we have replaced $R$ by $S_\varepsilon$. Thus for these cycles, if $\varepsilon$ is small enough, $S_\varepsilon$ maps $\bigcup N_j$ into itself.

If $R$ has a (super)attracting cycle, we may assume (by applying a prior conjugation) that $R(\infty) \in N_0$, so (using (5) again) for all sufficiently small $\varepsilon$, $S_\varepsilon$ maps $V_\varepsilon$ into $N_0$. Thus in this case, the hypothesis Lemma 9.6.3(b) holds in the sense that for these particular $N_j$, $S_\varepsilon$ maps both $\bigcup N_j$ and $V_\varepsilon$ into $\bigcup N_j$.

The construction of the set $E_\varepsilon$ for indifferent cycles requires the following result, and again we defer the proof until after the completion of the proof of Theorem 9.6.1.

**Lemma 9.6.4.** *Let $\{\zeta_0, \dots, \zeta_{q-1}\}$ be an indifferent cycle of $R$ with multiplier $\lambda$, and suppose that the function $\alpha$ is analytic near $\zeta_0$ with $\alpha(\zeta_0) = 0$ and $\alpha'(\zeta_0) = 1$. Then*

$$\alpha R_\varepsilon^q \alpha^{-1}(z) = [\alpha R^q \alpha^{-1}(z) - \lambda z] + \lambda z(1 - \varepsilon)^q + O(\varepsilon z^2)$$

*as $(z, \varepsilon) \to (0, 0)$.*

*Case 2: An irrationally indifferent cycle $\{\zeta_0, \dots, \zeta_{q-1}\}$.*

Using Theorem 6.10.5, we can find a function $\alpha$ as in Lemma 9.6.4 such that near the origin,

$$\alpha R^q \alpha^{-1}(z) = \lambda z + O(z^{k+3}),$$

as $z \to 0$, where (as before) $k = \deg(h)$. As $S_\varepsilon = R_\varepsilon$ near $\zeta_0$, we deduce from Lemma 9.6.4 that

$$\alpha S_\varepsilon^q \alpha^{-1}(z) = \lambda z[(1 - \varepsilon)^q + O(\varepsilon z) + O(z^{k+2})], \tag{9.6.2}$$

where there is some neighbourhood $\mathcal{N}$ of (0, 0), and some constant $A$, such that the terms $O(\varepsilon z)$ and $O(z^{k+2})$ are bounded above by $A\varepsilon|z|$ and $A|z|^{k+2}$ respectively.

Next, for sufficiently small $\varepsilon$, $\alpha^{-1}$ is defined on the disc

$$D = \{z: |z| < \varepsilon^{1/(k+1)}\},$$

so we can write

$$N_0 = \alpha^{-1}(D), \qquad N_1 = S_\varepsilon(N_0), \dots, N_{q-1} = S_\varepsilon^{q-1}(N_0).$$

By definition, then,

$$S_\varepsilon(N_0 \cup \cdots \cup N_{q-1}) \subset N_1 \cup \cdots \cup N_{q-1} \cup S_\varepsilon^q(N_0),$$

so we want to show that for sufficiently small $\varepsilon$, $S_\varepsilon^q(N_0) \subset N_0$ or, equivalently,

$$\alpha S_\varepsilon^q \alpha^{-1}(D) \subset D.$$

Now from (9.6.2), if $z$ is in $D$ then

$$\alpha(S_\varepsilon)^q \alpha^{-1}(z) = \lambda z[1 - q\varepsilon + O(\varepsilon^{1+t})],$$

where $t = (k+1)^{-1}$, and it follows that for sufficiently small $\varepsilon$,

$$|\alpha(S_\varepsilon)^q\alpha^{-1}(z)| < |\lambda z| = |z|,$$

so $\alpha S_\varepsilon^q \alpha^{-1}$ maps $D$ into itself. We deduce that for these cycles, and for all sufficiently small $\varepsilon$, $S_\varepsilon$ maps $\bigcup N_j$ into itself.

If $R$ has at least one (super)attracting cycle, the previous argument (in Case 1) which shows that $S_\varepsilon$ maps $V_\varepsilon$ into $\bigcup N_j$ remains valid. If not, but if $R$ has least one irrationally indifferent cycle, we argue as follows. The set $V_\varepsilon$ lies in a disc with centre $\infty$ and chordal radius $2\varepsilon^{1/k}$. We may assume (by a prior conjugation) that $R(\infty) = \zeta_0$; then, as $R$ is Lipschitz with respect to the chordal metric, $R(V_\varepsilon)$ lies in a disc with centre $\zeta_0$ and chordal radius $O(\varepsilon^{1/k})$. Now this together with (5) implies that $S_\varepsilon(V_\varepsilon)$ lies in some disc of chordal, and therefore Euclidean, radius $O(\varepsilon^{1/k})$. As $N_0$ has centre $\zeta_0$ and Euclidean radius approximately $\varepsilon^{1/(k+1)}$, and as $\varepsilon^{1/k} < \varepsilon^{1/(k+1)}$, we see that for all sufficiently small $\varepsilon$, $S_\varepsilon(V_\varepsilon) \subset N_0$. This completes the discussion for the irrationally indifferent cycles.

*Remark.* We remark that at this point we have already obtained Fatou's inequality. If $N_0$, $N_1$, $N_2$ and $N_3$ denote the number of (super)attracting, indifferent, irrationally indifferent, and rationally indifferent, cycles respectively (so $N_1 = N_2 + N_3$), then the argument above shows that $N_0 + N_2 \leq 2d - 2$. However, as each (super)attracting and rationally indifferent cycle attracts the forward orbit of a critical point, we also have $N_0 + N_3 \leq 2d - 2$ and together, these yield $N_0 + N_1/2 \leq 2d - 2$.

*Case 3: A rationally indifferent cycle* $\{\zeta_0, \ldots, \zeta_{q-1}\}$.

In this case we use Lemma 9.6.4 in conjunction with the Petal Theorem. We shall suppose that $R$ has $m$ petals at each $\zeta_j$, so there is some map $\alpha$ such that near the origin,

$$\alpha R^q \alpha^{-1}(z) = \lambda z(1 - z^m + O(z^{m+1})).$$

Note that $\lambda^m = 1$: this follows from the Petal Theorem as this map has $m$ petals at the origin which are permuted by $z \mapsto \lambda z$. Next, from Lemma 9.6.4 (and the fact that $S_\varepsilon = R_\varepsilon$ near $\zeta_0$),

$$\alpha S_\varepsilon^q \alpha^{-1}(z) = \lambda z[(1-\varepsilon)^q - z^m + O(\varepsilon z) + O(z^{m+1})] \tag{9.6.3}$$

as $z \to 0$ (note that unlike Case 2, the $m$ here is determined by $R$). Also, the terms $O(\varepsilon z)$ and $O(z^{m+1})$ are at most $A\varepsilon|z|$ and $A|z|^{m+1}$, say, respectively providing that $|z| < r_1$ and $\varepsilon < \varepsilon_1$, where these depend only on $R$.

The construction of the $N_j$ is more delicate in this case, and in some sense it is an amalgamation of the ideas in Cases 1 and 2. Roughly speaking, the set $N_j$ will be the union of regions which resemble the petals at $\zeta_j$, together with a small disc $D_j$ at $\zeta_j$, and we begin by discussing these sets.

We choose a positive number $r_0$ (with $r_0 < 1$, but also chosen to satisfy certain inequalities to be given later), and we insist that

$$\varepsilon < r_0^{m-1/2} < r_0.$$

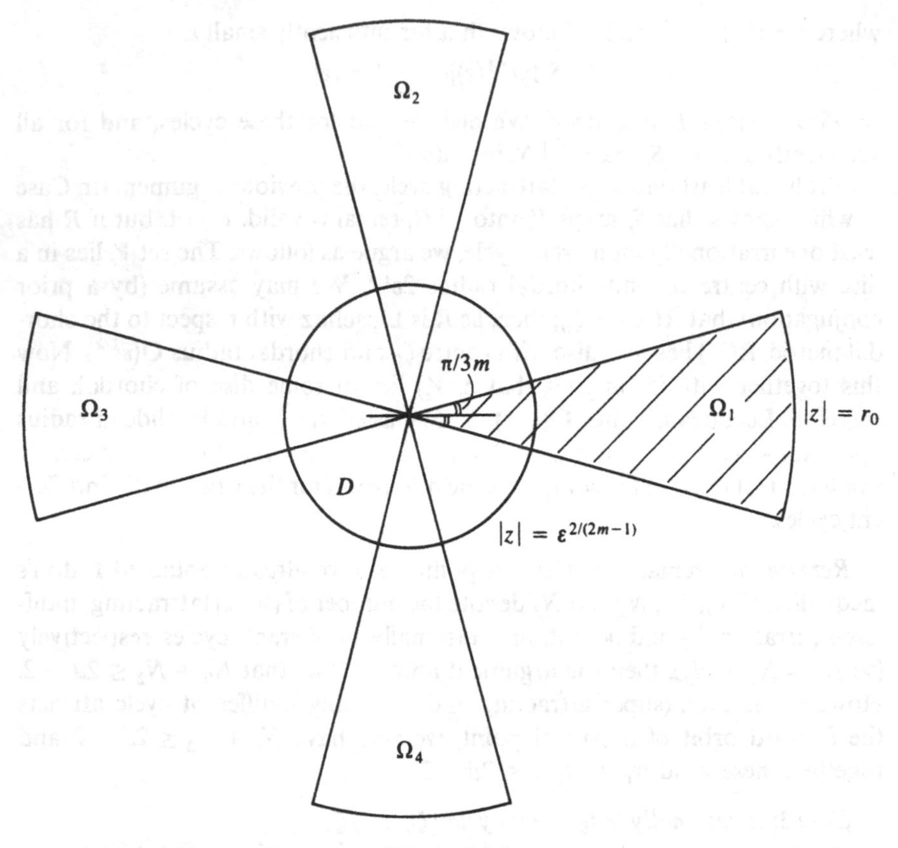


Figure 9.6.3

Now let

$$D = \{z: |z| < \varepsilon^{2/(2m-1)}\},$$

$$\Omega_1 = \{re^{i\theta}: 0 < r < r_0, |\theta| < \pi/3m\},$$

and let $\Omega_2, \ldots, \Omega_m$ be the successive images of $\Omega_1$ under a rotation about the origin of angle $2\pi/m$. Finally, let

$$N = D \cup \Omega_1 \cup \cdots \cup \Omega_m;$$

this is illustrated in Figure 9.6.3 in the case $m = 4$.

If $r_0$ is chosen small enough, the maps $\alpha^{-1}$ and $\alpha S_\varepsilon^q \alpha^{-1}$ for $0 < \varepsilon < \varepsilon_0$, say, are defined on $N$, and our immediate objective is to show that for all sufficiently small $\varepsilon$, $\alpha S_\varepsilon^q \alpha^{-1}$ maps the closure $\overline{N}$ into $N$. As $N$ is invariant under the map $z \mapsto \lambda z$, it suffices to show that $\overline{N}$ is mapped into $N$ by $g_\varepsilon$, where

$$\begin{aligned} g_\varepsilon(z) &= \lambda^{-1}\alpha S_\varepsilon^q \alpha^{-1}(z) \\ &= z[(1-\varepsilon)^q - z^m + O(\varepsilon z) + O(z^{m+1})], \end{aligned}$$

and to do this, it is sufficient to prove the following lemma.

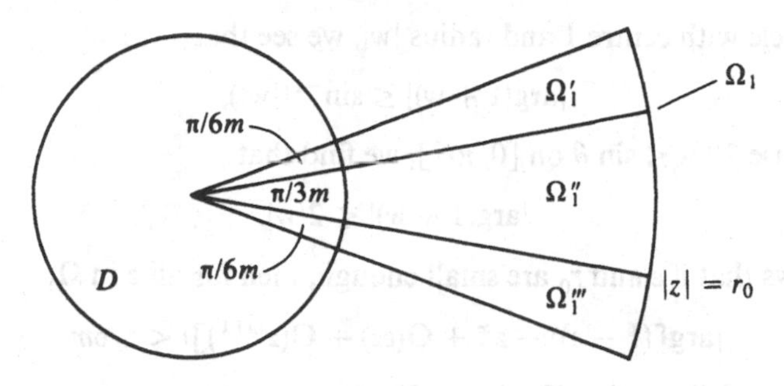


Figure 9.6.4

**Lemma 9.6.5.** *In the notation above*:

(a) $|g_\varepsilon(z)| < |z|$ *on* $\overline{N}$; *and*
(b) $|\arg g_\varepsilon(z)| < \pi/3m$ *on* $\Omega_1 - D$.

*Remark.* Let the sets $\Omega_1'$, $\Omega_1''$ and $\Omega_1'''$ be as illustrated in Figure 9.6.4. It is sufficient to prove (b) on $\Omega_1'$ and $\Omega_1''$ as the argument for $\Omega_1'''$ parallels that for $\Omega_1'$. Further, a similiar statement (and proof) holds for the other "petals" $\Omega_j$, so with these extensions (which we omit), Lemma 9.6.5 does prove that $g_\varepsilon(\overline{N}) \subset N$.

PROOF. First, we prove (a). If $z$ is in $\overline{D}$, then $|\varepsilon z|$, $|z^m|$ and $|z^{m+1}|$ are all at most $\varepsilon^{1+t}$, where $t = 1/(2m-1)$. Thus

$$\begin{aligned}|g_\varepsilon(z)| &\le |z|[(1-\varepsilon)^q + (1+2A)\varepsilon^{1+t}]\\ &= |z|[(1 - q\varepsilon + O(\varepsilon^{1+t})],\end{aligned}$$

and this is strictly less than $|z|$ if $\varepsilon$ is sufficiently small.

Next, take $z$ in $\Omega_1$. Then

$$\begin{aligned}|(1-\varepsilon)^q - z^m|^2 &= (1-\varepsilon)^{2q} + |z|^{2m} - 2\,\mathrm{Re}[(1-\varepsilon)^q z^m]\\ &\le (1-\varepsilon)^{2q} + |z|^{2m} - (1-\varepsilon)^q|z|^m\\ &< [(1-\varepsilon)^q - |z|^m/3]^2,\end{aligned}$$

provided that $8|z|^m < 3(1-\varepsilon)^q$. Taking $\varepsilon$ and $r_0$ small enough, we have

$$\begin{aligned}|g_\varepsilon(z)| &\le |z|[(1-\varepsilon)^q - |z|^m/3 + A\varepsilon|z| + A|z|^{m+1}]\\ &\le |z|[1 - q\varepsilon(1 + O(\varepsilon) + O(z)) - |z|^m(\tfrac{1}{3} + O(z))]\\ &< |z|\end{aligned}$$

as required.

Before starting our proof of (b), take any $w$ with $|w| < 1$. Because $1 + w$ lies on the circle with centre 1 and radius $|w|$, we see that

$$|\arg(1+w)| \leq \sin^{-1}(|w|),$$

and because $2\theta/\pi \leq \sin\theta$ on $[0, \pi/2]$, we find that

$$|\arg(1+w)| \leq 2|w|.$$

This shows that if $\varepsilon$ and $r_0$ are small enough, then for all $z$ in $\Omega_1$,

$$|\arg[(1-\varepsilon)^q - z^m + O(\varepsilon z) + O(z^{m+1})]| < \pi/6m. \tag{9.6.4}$$

From this, it follows that if $z$ lies in $\Omega_1''$, then

$$\begin{aligned}|\arg g_\varepsilon(z)| &\leq |\arg z| + |\arg[(1-\varepsilon)^q - z^m + O(\varepsilon z) + O(z^{m+1})]| \\ &< \pi/6m + \pi/6m,\end{aligned}$$

which proves (b) for these $z$.

Finally, take $z$ in $\Omega_1'$. Then (9.6.4) holds so certainly

$$\begin{aligned}\arg g_\varepsilon(z) &= \arg z + \arg[(1-\varepsilon)^q - z^m + O(\varepsilon z) + O(z^{m+1})] \\ &\geq \pi/6m - \pi/6m \\ &= 0.\end{aligned}$$

Next,

$$\arg g_\varepsilon(z) \leq \pi/3m + \arg[(1-\varepsilon)^q - z^m + O(\varepsilon z) + O(z^{m+1})]$$

so it is sufficient to prove that on $\Omega_1'$,

$$\arg[(1-\varepsilon)^q - z^m + O(\varepsilon z) + O(z^{m+1})] < 0. \tag{9.6.5}$$

Now for $z$ in $\Omega_1'$,

$$\mathrm{Im}[(1-\varepsilon)^q - z^m] = -\mathrm{Im}(z^m) \leq -|z|^m/2.$$

However, as $z$ lies outside of $D$,

$$\begin{aligned}|O(\varepsilon z) + O(z^{m+1})| &\leq A(|z|^{m+1/2} + |z|^{m+1}) \\ &< 2A|z|^m\sqrt{r_0},\end{aligned}$$

so, providing that $2A\sqrt{r_0} < \frac{1}{2}$,

$$\mathrm{Im}[(1-\varepsilon)^q - z^m + O(\varepsilon z) + O(z^{m+1})] < 0$$

and hence (9.6.5) holds. This completes the proof of the lemma. □

We have proved that for small enough $\varepsilon$, $g_\varepsilon(\overline{N}) \subset N$, and it is now easy to construct the neighbourhoods $N_j$ at the points $\zeta_j$. The map $\alpha^{-1}$ maps $N$ into some region $N_0$ at $\zeta_0$ and it follows that $S_\varepsilon^q$ maps $N_0$ into $N_0$. We now write

$$N_1 = S_\varepsilon(N_0), \ldots, N_{q-1} = S_\varepsilon^{q-1}(N_0),$$

and a similiar argument to that in the earlier cases shows that $\alpha S_\varepsilon \alpha^{-1}$ maps $\bigcup N_j$ into itself.

Finally, if $R$ has none of the cycles discussed in Cases 1 and 2, we may assume (by a prior conjugation) that $R(\infty)$ lies in the interior of the image of $\Omega_1$, and as this does not depend on $\varepsilon$, we see that if $\varepsilon$ is small enough, $S_\varepsilon(V_\varepsilon)$ will also lie in the image. This completes the discussion of Case 3, and with it, the proof of Theorem 9.6.1 subject to proving Lemmas 9.6.2 and 9.6.4.

We end by giving the proofs of these lemmas.

Proof of Lemma 9.6.4. The function

$$F(z, \varepsilon) = \alpha(S_\varepsilon)^q\alpha^{-1}(z)$$

is an analytic function of $z$ (for fixed $\varepsilon$), and an analytic function of $\varepsilon$ (for fixed $z$), in each case in some neighbourhood of the origin. Thus $F$ is analytic in the variable $(z, \varepsilon)$, and it can be expressed as a convergent power series

$$F(z, \varepsilon) = \sum_{m=0}^{\infty} \sum_{n=0}^{\infty} a_{m,n} z^m \varepsilon^n$$

in some neighbourhood of $(0, 0)$ in $\mathbb{C}^2$. Next (and we shall justify these shortly),

$$\sum_{m=0}^{\infty} a_{m,0} z^m = \alpha R^q \alpha^{-1}(z);$$

$$\sum_{n=0}^{\infty} a_{0,n} \varepsilon^n = F(0, \varepsilon) = 0;$$

$$\sum_{n=0}^{\infty} a_{1,n} z \varepsilon^n = \lambda z(1-\varepsilon)^q;$$

$$\sum_{m=2}^{\infty} \sum_{n=1}^{\infty} a_{m,n} z^m \varepsilon^n = O(\varepsilon z^2).$$

Given that these sums are as claimed, the result follows by summing over all four expressions and taking account of the two terms counted twice, namely $a_{0,0}$ $(= 0)$ and $a_{0,1}z$ $(= \lambda z)$. It only remains to justify these four sums.

The first sum is $F(z, 0)$, and

$$F(z, 0) = \alpha S_0^q \alpha^{-1}(z) = \alpha R^q \alpha^{-1}(z).$$

The second sum is $F(0, \varepsilon)$, and

$$\begin{aligned} F(0, \varepsilon) &= \alpha S_\varepsilon^q \alpha^{-1}(0) \\ &= \alpha S_\varepsilon^q(\zeta_0) \\ &= \alpha(\zeta_0) \\ &= 0. \end{aligned}$$

The fourth sum is obviously $O(\varepsilon z^2)$. The third sum is $[\partial F(0, \varepsilon)/\partial z]z$. Now

$$\frac{\partial F(z, \varepsilon)}{\partial z} = [\alpha'(R_\varepsilon^q(\alpha^{-1}z))].[(R_\varepsilon q)'(\alpha^{-1}z)].[(\alpha^{-1})'(z)]$$

and putting $z = 0$ we obtain

$$\begin{aligned}\frac{\partial F(0, \varepsilon)}{\partial z} &= \alpha'(\zeta_0).(R_\varepsilon^q)'(\zeta_0).(\alpha^{-1})'(0)\\ &= (R_\varepsilon^q)'(\zeta_0)\\ &= \prod_{j=0}^{q-1} R_\varepsilon'(\zeta_j)\\ &= \prod_{j=0}^{q-1} [(1-\varepsilon)R'(\zeta_j)]\\ &= \lambda(1-\varepsilon)^q\end{aligned}$$

as required, and this completes the proof of Lemma 9.6.4. □

THE PROOF OF LEMMA 9.6.2. The idea is to construct a Beltrami coefficient $\mu$ on $\mathbb{C}_\infty$ such that

$$g\colon \mathbb{C}_\infty[\mu] \to \mathbb{C}_\infty[\mu]$$

is analytic so, if $\varphi$ is a solution of the Beltrami equation

$$\partial\varphi/\partial\bar{z} = \mu\partial\varphi/\partial z, \tag{9.6.6}$$

then the composite map

$$\mathbb{C}_\infty[0] \xrightarrow{\varphi^{-1}} \mathbb{C}_\infty[\mu] \xrightarrow{g} \mathbb{C}_\infty[\mu] \xrightarrow{\varphi} \mathbb{C}_\infty[0]$$

is analytic and so is rational.

As $g$ maps the open set $E$ into itself, we have

$$E \subset g^{-1}(E) \subset g^{-2}(E) \subset \cdots,$$

so $\mathbb{C}_\infty$ is the disjoint union of the measurable sets

$$E_0\,(=E), \qquad E_{n+1} = g^{-(n+1)}(E) - g^{-n}(E) \qquad (n \geq 0),$$

and

$$K = \mathbb{C}_\infty - \bigcup_{n=0}^{\infty} g^{-n}(E) = \mathbb{C}_\infty - \bigcup_{n=0}^{\infty} E_n.$$

Now define $\mu$ on $\mathbb{C}_\infty$ by putting $\mu = 0$ on $K \cup E_0$, and defining $\mu$ inductively on $E_1, E_2, \ldots$ by

$$\mu(w) = \frac{\partial g(w)/\partial\bar{z} + \mu(gw)\partial g(w)/\partial z}{\partial g(w)/\partial z + \mu(gw)\partial g(w)/\partial\bar{z}}. \tag{9.6.7}$$

As quasiconformal maps (and their inverses) preserve zero area, $\mu$ is defined a.e. on $\mathbb{C}_\infty$. Further, by definition, (9.6.7) holds throughout $\bigcup_{n=0}^{\infty} E_n$, and it holds in $K$ for if $w \in K$, then $g(w) \in K$, $g(w) \notin g^{-1}(E)$ and so

$$\mu(w) = 0 = \mu(gw), \qquad \partial g(w)/\partial\bar{z} = 0:$$

thus (9.6.7) holds throughout $\mathbb{C}_\infty$.

We shall now show that $\mu$ is a Beltrami coefficient on $\mathbb{C}_\infty$. First, if $w \in E_n$, where $n \geq 2$, then $g(w)$ is not in $g^{-1}(E)$ so $\partial g(w)/\partial \bar{z} = 0$, and hence $|\mu(w)| = |\mu(gw)|$. This shows that the essential supremum of $\mu$ over $\mathbb{C}_\infty$ is the same as the essential supremum of $\mu$ over $E_1$. However, if $w \in E_1$, then $\mu(gw) = 0$; thus the essential supremum of $\mu$ over $E_1$ is at most $\|\mu_g\|_\infty$, where $\mu_g$ is the complex dilatation of $g$, and so $\|\mu\|_\infty \leq \|\mu_g\|_\infty < 1$. A standard argument shows that $\mu$ is measurable (in our application, $g$ is $S_\varepsilon$, a smooth function), so $\mu$ is a Beltrami coefficient on $\mathbb{C}_\infty$.

Now let $\varphi$ be any solution of (9.6.6) on $\mathbb{C}_\infty$. If we compute the complex dilatation of the composition $\varphi g$ we find (see [1], or [63]) that it is precisely the right-hand side of (9.6.7): thus $\varphi g$ is also a solution of (9.6.6) throughout $\mathbb{C}_\infty$. It follows from Theorem 8.3.1 that the composite function $(\varphi g)\varphi^{-1}$ is analytic except, possibly, at the finite set of points where the valency of $g$ exceeds one. As these points are removable singularities of $\varphi g \varphi^{-1}$, we deduce that $\varphi g \varphi^{-1}$ is analytic on the sphere and hence a rational map. The proof is complete. □

EXERCISE 9.6

1. Given distinct points $z_1, \ldots, z_n$ in $\mathbb{C}$, define the polynomials $\Pi$ and $Q_j$ by

$$\Pi(z) = (z - z_1) \cdots (z - z_n), \qquad Q_j(z) = \Pi(z)/(z - z_j).$$

Use Lagrange's Interpolation Formula to find a polynomial $P$ with $P(z_j) = -1/Q_j(z_j)$, and show that if $h(z) = \Pi(z)P(z)$, then for all $j$, $h(z_j) = 0$ and $h'(z_j) = -1$. Find $h$ explicitly when $n = 2$.

## §9.7. Expanding Maps

An important class of rational maps $R$ are those which are expanding on their Julia set. Roughly speaking, this means that if we take a small open neighbourhood $W$ of $J$, then the inverse images $W, R^{-1}(W), R^{-2}(W), \ldots$ of $W$ contract at some specified rate towards $J$. For example, this happens for $P: z \mapsto z^2$, where $|P'(z)| = 2$ on $J$, and it fails to happen for $Q: z \mapsto z^2 - 2$, where $Q'(0) = 0$ and $0 \in J$. This section is concerned with obtaining information of this kind which can then be used to develop metric properties of the Julia set as, for example, in §9.8 and §9.9. Our discussion will focus on two different points of view; the first is the more elementary (and follows from Theorem 9.2.1), while the second is in terms of general Riemannian metrics and uses the hyperbolic metric and the Uniformization Theorem.

We begin with the more elementary discussion and we prove

**Theorem 9.7.1.** *Let $R$ be a rational map with* $\deg(R) \geq 2$. *Suppose that* $\infty \in F(R)$, *and that the closure of the forward orbit* $C^+$ *of the critical points of*

*R is disjoint from J. Then*

$$\liminf_{n\to\infty} \inf_{z\in J} |(R^n)'(z)| = +\infty.$$

PROOF. We begin by explaining how this follows from certain properties of the branches $S_n$ of $(R^n)^{-1}$, and after this, we use Theorem 9.2.1 to derive these properties. The hypothesis on $C^+$ ensures that we can find local branches $S_n$ of $(R^n)^{-1}$ at each point $\zeta$ in $J$, and by Theorem 9.2.1, the derivatives of these branches converge to zero on a neighbourhood of $\zeta$. As $J$ is compact, we can convert this local information into a global statement which says (roughly speaking) that $S_n' \to 0$ uniformly on $J$ (the problem, of course, is that we do not know that the $S_n$ are unambiguously defined throughout $J$), and this shows that $|(R^n)'| \to +\infty$ uniformly on $J$. The following lemma contains a precise statement of the properties of the $S_n$ that we require. □

**Lemma 9.7.2.** *Given any positive $\varepsilon$, there is an integer $n_0$ such that for all $n \geq n_0$, and all $z$ in $J$, and each branch $S_n$ of $(R^n)^{-1}$ at $z$, $|S_n'(z)| < \varepsilon$.*

Assuming this, take any $w$ in $J$. Let $z_n = R^n(w)$, and let $S_n$ be the branch of $(R^n)^{-1}$ which maps $z_n$ to $w$. Now for $z$ near $z_n$, $R^n S_n(z) = z$ so

$$|(R^n)'(S_n z)| \,.\, |S_n'(z)| = 1.$$

Given any positive $\varepsilon$, choose $n_0$ as in Lemma 9.7.2 and put $z = z_n$. Then for $n \geq n_0$,

$$1 = |(R^n)'(w)| \,.\, |S_n'(z_n)| < \varepsilon |(R^n)'(w)|$$

which yields the desired result. It remains to give the

PROOF OF LEMMA 9.7.2. We suppose the contrary so there is a positive $\varepsilon$, an infinite set $N$ of positive integers, points $z_n$ in $J$ (for $n$ in $N$), and branches $B_n$ of $(R^n)^{-1}$ at $z_n$ such that for all $n$ in $N$,

$$|B_n'(z_n)| \geq \varepsilon. \tag{9.7.1}$$

By passing to a subsequence we may assume that $z_n \to \zeta$, where $\zeta$ is in $J$. Now find a disc $D$ at $\zeta$ not meeting $C^+$: then $D$ does not contain any critical value of any $R^n$ and so for every $n$, all branches of $(R^n)^{-1}$ exist in $D$. For $n \geq m$, say, $z_n$ lies in $D$ and so we can define branches $S_n$ of $(R^n)^{-1}$ in $D$ with $S = B_n$ at $z_n$ and nearby. Now by Theorem 9.2.1, $S_n' \to 0$ locally uniformly in $D$, whence

$$B_n'(z_n) = S_n'(z_n) \to 0$$

contrary to (9.7.1). □

We turn now to the second part of our discussion. Let $R$ be a rational map with Julia set $J$, and let $\omega(z)|dz|$ be a Riemannian metric defined on some neighbourhood $W$ of $J$ (so $\omega$ is positive and continuous on $W$). We do *not* insist that $\omega$ is defined throughout the sphere. If $z$ is in $W \cap R^{-1}(W)$, then the change of scale of $R$ at $z$ (measured in the $\omega$-metric) is

$$\|R'(z)\| = \frac{\omega(Rz)|R'(z)|}{\omega(z)},$$

and this is defined on a neighbourhood of $J$.

**Definition 9.7.3.** We say that $R$ is *expanding on $J$* (with respect to $\omega$) if there are positive numbers $c$ and $\lambda$, with $\lambda > 1$ and

$$\|(R^n)'(z)\| \geq c\lambda^n, \quad n \geq 1,$$

on $J$. We then call $\lambda$ a *dilatation constant* of $R$ on $J$.

Our first task is to show that the property of $R$ being expanding on $J$ is independent of the choice of the metric $\omega(z)|dz|$; thus if there is one metric with respect to which $R$ is expanding, then, automatically, $R$ will be expanding with respect to both the Euclidean and the spherical metrics.

**Theorem 9.7.4.** *If $R$ is expanding with respect to one metric, then it is expanding with respect to all such metrics. Moreover, the values of the dilatation constants are independent of the metric.*

Proof. Let $\rho(z)|dz|$ and $\omega(z)|dz|$ be two metrics, each defined on some neighbourhood of $J$. Then, from the continuity of the map $z \mapsto \omega(z)/\rho(z)$ and the compactness of $J$, there are positive numbers $m$ and $M$ such that on $J$,

$$m \leq \omega(z)/\rho(z) \leq M.$$

Using subscripts to denote the metric concerned, we then have

$$\|(R^n)'(z)\|_\omega \leq (M/m)\|(R^n)'(z)\|_\rho, \tag{9.7.2}$$

and also a similiar inequality with $\omega$ and $\rho$ interchanged, and this completes the proof. □

Next, we establish a criterion for $R$ to be expanding on $J$.

**Theorem 9.7.5.** *Suppose that each critical point of $R$ has a forward orbit that accumulates at a (super)attracting cycle of $R$. Then $R$ is expanding on $J$.*

Proof. First, we construct disjoint open (topological) discs, one at each point of each (super)attracting cycle, such that the union $V$ of these discs is forward invariant under $R$. By taking the discs to be small enough, the complement of $V$ is connected.

Let $W$ be the complement of $V \cup C^+$, so $W$ supports a hyperbolic metric which we denote by $\rho(z)|dz|$. Now by assumption, only finitely many points of $C^+$ lie outside $V$, and $R$ maps $V \cup C^+$ into itself. It follows that $R^{-1}(W) \subset W$, so if we take any branch $S$ of $R^{-1}$ at a point in $W$, its values, and all values obtained by analytic continuation of $S$ in $W$, will lie in $W$. Note that such continuations exist as no critical points of $R$ lie in $W$.

Now choose $w$ in $W$ and some branch $S$ of $R^{-1}$ at $w$ such that $S(w) \neq w$. Let $\pi: \Delta \to W$ be the universal cover map; we may assume that $\pi(0) = w$, and we choose $\zeta$ in $\Delta$ such that $\pi(\zeta) = S(w)$. The branch $S$ at $w$ provides an analytic branch $\varphi$ of $\pi^{-1} S\pi$ with $\varphi(0) = \zeta$, and we can continue $\varphi$ analytically throughout $\Delta$ to obtain a single-valued analytic map $\varphi$ of $\Delta$ into itself.

Now Schwarz's Lemma implies that $\varphi$ is either an isometry or a contraction on $\Delta$ (with respect to the hyperbolic metric on $\Delta$). If $\varphi$ is a contraction, its scaling factor is strictly less than one at each point of $\Delta$ (again, this is Schwarz's Lemma), and as $\pi: \Delta \to W$ is a local isometry (with respect to the two hyperbolic metrics), we deduce that any local branch of an analytic continuation of $S$ is a local contraction in the metric $\rho$. This means that $\|R'(z)\|_\rho > 1$ at each point of $W$, so writing

$$\lambda = \inf\{\|R'(z)\|_\rho : z \in J\},$$

we have $\lambda > 1$. Further, the Chain Rule now implies that, for all $n \geq 1$, and all $z$ in $J$,

$$\|(R^n)'(z)\|_\rho \geq \lambda^n > 1 \tag{9.7.3}$$

on $J$.

If $\varphi$ were an isometry, the same argument would show that the branches of $S$, and so $R$ also, would be a local isometry. This cannot be so, however, for if $w$ is in a repelling cycle of length $m$, $R^m$ is expanding at $w$ in the Euclidean metric, and hence in the hyperbolic metric also. The proof is now complete. □

We end with the remark that although we have shown that

$$\|R'(z)\|_\rho \geq \lambda > 1$$

on $J$, where $\rho$ is the hyperbolic metric, it does not necessarily follow from this that $|R'(z)| \geq \lambda > 1$ on $J$. However, from (9.7.2) there is some constant $c$, and some integer $n$, such that on $J$,

$$|(R^n)'(z)| \geq c\lambda^n > 1.$$

These observations add weight to the view that the hyperbolic metric is the instrinsic metric here, and constantly changing between $R$ and $R^n$ is the price that we pay for using the Euclidean (or spherical) metric.

## Exercise 9.7

1. Let $P(z) = (3z - z^3)/2$. Show that all of the critical points of $P$ are super-attracting fixed points (so $P$ is expanding on $J$). Show also that all components of $F(P)$ are simply connected.

## §9.8. Julia Sets as Cantor Sets

Let $P$ be a polynomial of degree $d$, where $d \geq 2$. We know that the unbounded component $F$ has infinite connectivity if it contains some finite critical point of $P$ (Theorems 5.2.3 and 9.5.1), but we can say much more if it contains *all* finite critical points of $P$. This section is devoted to a discussion of this situation, but in a form that is applicable to general rational maps.

We need the notion of a Cantor set, the prototype being the famous example of the Cantor "middle-third" set (see Exercise 9.8.1). A subset $E$ of the complex sphere is said to be a *Cantor set* if it is non-empty, closed, perfect (there are no isolated points), and totally disconnected (each component of $E$ is a single point). This is a purely topological description, but later in this section we shall consider metric properties. First, though, we prove

**Theorem 9.8.1.** *Let $R$ be a rational map of degree $d$, where $d \geq 2$, and let $\zeta$ be a (super)attracting fixed point of $R$. If all of the critical points of $R$ lie in the immediate attracting basin of $\zeta$, then $J(R)$ is a Cantor set.*

Theorem 9.8.1 is applicable to some rational maps that are not polynomials (see Exercise 9.8.2). For quadratic polynomials, the converse of Theorem 9.8.1 is true (but only because there is a single finite critical point): in this case, if $J$ is a Cantor set, then $F_\infty$ is of infinite connectivity, and so by Theorem 9.5.1, $F_\infty$ contains the unique finite critical point of $P$. On the other hand, the converse is false for cubic polynomials (which have two finite critical points): see §11.6. It is natural to ask whether $J$ is a Cantor set whenever it has infinitely many components, and again the answer is "no"; see §11.5. Finally, we note that there is a rational map $R$ for which all critical points are attracted towards a super-attracting fixed point of $R$, yet for which $J$ is not a Cantor set: see §11.8.

We now have the

Proof of Theorem 9.8.1. By conjugation, we may assume that $\zeta = \infty$, and we let $F_\infty$ be the component of $F$ that contains $\infty$ (so $F_\infty$ is the immediate basin in Theorem 9.8.1). We denote the distinct critical values of $R$ by $c_1, c_2, \ldots$ (there are only finitely many of these): by assumption, each $c_j$ lies in $F_\infty$, hence so does each $R^n(c_j)$ and $R^n(c_j) \to \infty$ as $n \to \infty$.

We are not yet in a position to say that $F$ is connected (that is, $F = F_\infty$), but we can show now that

$$R^n \to \infty \quad \text{on } F. \tag{9.8.1}$$

To see this, recall that by the No Wandering Domains Theorem (Chapter 8) each component of $F$ is eventually mapped by $R$ into some periodic cycle of components $F_1, \ldots, F_t$. Now this cycle of components must attract critical points (in the precise sense described in Theorems 9.3.1, 9.3.2 and 9.3.3), and our hypothesis about critical points implies that this cycle can only be $F_\infty$. It follows that any component of $F$ is eventually mapped into $F_\infty$, and (9.8.1) follows. We now begin our construction of $J$ as a Cantor set.

As $J$ is a compact subset of $\mathbb{C}$, and as $C^+(R)$ accumulates only at $\infty$, we can find a Jordan curve $\Gamma$ that separates $J$ from $C^+(R)$. [For example, we can take a compact disc at $\infty$ and in $F_\infty$ and join $Q$ to each of the finite number of points $R^n(c_j)$ not in $Q$ by a simple arc. We may assume that these arcs $\tau_1, \ldots, \tau_q$ are pairwise disjoint and that the connected compact set $Q \cup \tau_1 \cup \cdots \cup \tau_q$ lies in $F_\infty$. It is now easy to separate this set from $J$ by a Jordan curve.] Let $V$ and $W$ be the interior and exterior of $\Gamma$ respectively and let $K = V \cup \Gamma$. Clearly, $w \cup \Gamma$ is a compact subset of $F_\infty$, so for some integer $N$,

$$R^N(W \cup \Gamma) \subset W,$$

and therefore also

$$(R^N)^{-1}(V \cup \Gamma) \subset V.$$

Observe also that as the critical values of $R^N$ are in $C^+(R)$, $R^N$ has no critical values in $K$.

As $J(R) = J(R^N)$, it is sufficient to work with $R^N$ so, relabelling $R^N$ as $R$, we may now assume that

$$R(W \cup \Gamma) \subset W, \qquad R^{-1}(V \cup \Gamma) \subset V, \qquad C^+(R) \subset W, \qquad J \subset V,$$

and we let $d = \deg(R)$. Now $R$ is a smooth covering map of each component of $R^{-1}(V)$ onto $V$, and as $V$ is simply connected, the restriction of $R$ to each such component is a homeomorphism of that component onto $V$. Thus we can define branches $S_1, \ldots, S_d$ of $R^{-1}$ on $K$ $(= V \cup \Gamma)$ and it is clear that the sets $S_1(K), \ldots, S_d(K)$ are pairwise disjoint compact subsets of $K$. It is these sets which form the basis of the construction of $J$ as a Cantor set: see Figure 9.8.1.

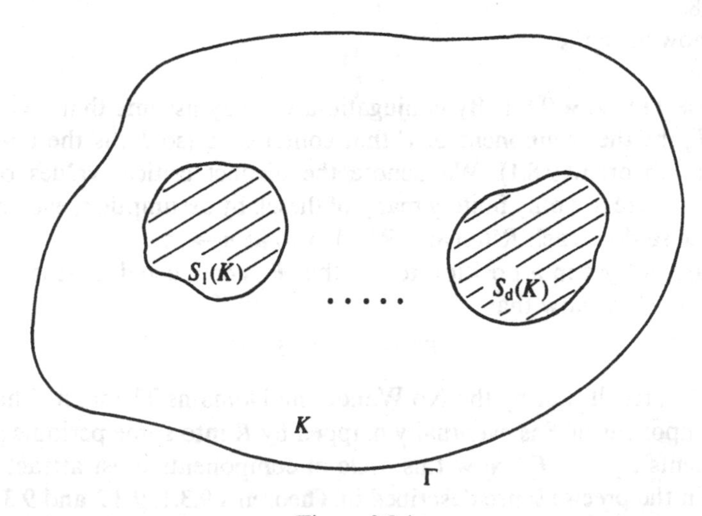


Figure 9.8.1

To realize $J$ as a Cantor set, we consider the images of $K$ under the elements of the semi-group generated by $S_1, \ldots, S_d$; thus for each sequence of integers $i_j$ in $\{1, \ldots, d\}$ we define

$$K(i_1, \ldots, i_n) = S_{i_1} \cdots S_{i_n}(K). \tag{9.8.2}$$

It is clear that

$$K(i_1, \ldots, i_n, i_{n+1}) \subset K(i_1, \ldots, i_n), \tag{9.8.3}$$

and that the $d^n$ sets $K(i_1, \ldots, i_n)$ are pairwise disjoint compact sets with

$$\bigcup_{i_1, \ldots, i_n = 1}^{d} K(i_1, \ldots, i_n) = R^{-n}(K).$$

We write

$$K_\infty = \bigcap_{n=0}^{\infty} R^{-n}(K),$$

so certainly, $K_\infty$ is non-empty, compact and perfect. We now show that $K_\infty = J$, and that $K_\infty$ is totally disconnected.

It is easy to see that $J = K_\infty$. By construction, $J \subset K$, and this with the complete invariance of $J$ shows that $J \subset K_\infty$. Next, if $z$ is in $K_\infty$, then for all $n$, $R^n(z) \in K$, and so by (9.8.1), $z$ is not in $F$. This proves that $K_\infty \subset J$, and hence that $K_\infty = J$.

Now let $K'$ be any component of $K_\infty$: then $K_0$ can lie in at most one of the sets $K(j_1, \ldots, j_n)$ for a given $n$, so there is a sequence $(j_n)$ such that

$$K' \subset \bigcap_{n=0}^{\infty} K(j_1, \ldots, j_n)$$

(in fact, equality holds here for the intersection of a decreasing sequence of compact connected sets is connected, but we do not need to know this). In order to prove that $K$ is totally disconnected, it is enough to prove that for any given sequence $(j_n)$,

$$\text{diameter}[K(j_1, \ldots, j_n)] \to 0 \tag{9.8.4}$$

as $n \to \infty$, and this can be derived from our earlier results on the family of branches of the inverses of the $(R^n)^{-1}$. As each of the functions $S_{j_1} \cdots S_{j_n}$ map of $V$ into itself, together they form a normal family in $V$; this also follows from Theorem 9.2.1, for each of these maps are branches of some $(R^m)^{-1}$. Now by Lemma 9.2.2, any locally uniform limit of a subsequence of

$$S_{j_1}, S_{j_1}S_{j_2}, S_{j_1}S_{j_2}S_{j_3}, \ldots$$

is constant (and such limits exist by normality). As (9.8.2) and (9.8.3) hold, it is clear that this sequence converges locally uniformly to some number $\zeta$ on $K$, and that for all sufficiently large $n$, the sets $K(j_1, \ldots, j_n)$ lie in some pre-assigned arbitrarily small neighbourhood of $\zeta$. This verifies (9.8.4) and completes the proof of Theorem 9.8.1. □

We can say more, both topologically and metrically, about the situation described in the proof of Theorem 9.8.1. First let $\Pi$ be the class of sequences $x = (x_1, x_2, \ldots)$, where each $x_j$ is in $\{1, \ldots, d\}$, and define the *shift map* $\sigma: \Pi \to \Pi$ by

$$\sigma(x) = (x_2, x_3, \ldots).$$

It is easy to see that $\rho$, defined by

$$\rho(x, y) = \sum_{n=1}^{\infty} \frac{|x_n - y_n|}{3^n},$$

is a metric on $\Pi$, and that $\rho(x, y) < 3^{-n}$ if and only if

$$(x_1, \ldots, x_n) = (y_1, \ldots, y_n).$$

Further, for all $x$ and $y$,

$$\rho(\sigma x, \sigma y) \leq 3\rho(x, y)$$

so $\sigma$ is continuous.

We shall prove that if $R$ satisfies the hypotheses of Theorem 9.8.1, then $R: J \to J$ is topologically conjugate to the shift map $\sigma: \Pi \to \Pi$, and together with the elementary properties of $\sigma$ described in Exercise 9.8.4, this will yield a proof of the following result.

**Theorem 9.8.2.** *Suppose that R satisfies the hypotheses of Theorem* 9.8.1. *Then*:

(a) *for a dense set of $\zeta$ in J, the forward orbit $\{R^n(\zeta): n \geq 1\}$ is dense in J; and*
(b) *the periodic points are dense in J.*

We know that in general, the periodic points of $R$ are dense in $J$ (see Theorem 6.9.2 and §9.6), but here we have more explicit information and, in any event, this provides a good illustration of the use of the topological conjugacy between $R$ and $\sigma$. For further details about the shift map $\sigma$ in this context, we refer the reader to [35] and [67].

Proof of Theorem 9.8.2. We take $R$ satisfying the hypotheses of Theorem 9.8.1, and we use the same notation as in the proof of the theorem. Because of (9.8.4), each $x$ in $\Pi$ gives rise to a point

$$\zeta(x) = \bigcap_{n=1}^{\infty} K(x_1, \ldots, x_n),$$

and each point in $J$ arises in this way; thus this defines a bijection $\zeta: x \mapsto \zeta(x)$ of $\Pi$ onto $J$. The proof that this map is a homeomorphism is straightforward and is omitted.

Next,

$$\begin{aligned} R(K(x_1, \ldots, x_n)) &= RS_{x_1} \cdots S_{x_n}(K_1) \\ &= S_{x_2} \cdots S_{x_n}(K_1) \\ &= K(x_2, \ldots, x_n), \end{aligned}$$

so, by intersecting these sets over all values of $n$, we find that $R(\zeta(x)) = \zeta(\sigma(x))$: thus $R\zeta = \zeta\sigma$ on $\Pi$ and as $\zeta$ is a bijection,

$$R = \zeta\sigma\zeta^{-1},$$

so $R$ is conjugate to $\sigma$.

By constructing a sequence $x$ in $\Pi$ such that every finite sequence $(y_1, \ldots, y_n)$ with $n \geq 1$ and $y_j \in \{1, \ldots, d\}$, occurs somewhere in $x$, and considering the point $\zeta(x)$ of $J$, we obtain (a). Similarly, by constructing periodic sequences in $\Pi$, we obtain (b). □

We end this section with a brief discussion of the metric properties of the construction of $J$ from the sets $K(j_1, \ldots, j_n)$. Consider the family

$$\mathscr{F} = \{S_{j_1} \cdots S_{j_n}: j_1, \ldots, j_n = 1, \ldots, d, n \geq 1\}$$

as used in the proof of Theorem 9.8.1. This family is normal in $V_2$, and as each member of the family maps $V_2$ (which is bounded in $\mathbb{C}$) into itself, the family of derivatives, namely

$$\mathscr{F}' = \{(S_{j_1} \cdots S_{j_n})': j_1, \ldots, j_n = 1, \ldots, d, n \geq 1\},$$

is locally uniformly bounded, and hence normal, in $V_2$. By Lemma 9.2.2, the locally uniform limits of sequences from $\mathscr{F}$ are constant, thus any locally uniform limit of any sequence from $\mathscr{F}'$ is the zero map. It follows from this that, for example,

$$|(S_{j_1} \cdots S_{j_n})'(z)| \leq \tfrac{1}{2}$$

on the compact subset $K_1$ of $V_2$ for all but a finite set of $n$ (for if not, there is some sequence of these maps whose maximum modulus on $K_2$ is at least $\frac{1}{4}$). Taking some $n$ for which all $d^n$ maps of the form $S_{j_1} \cdots S_{j_n}$ have this property, we relabel these maps as $T_1, \ldots, T_k$, where $k = d^n$, and then use these maps to generate the Cantor set. Of course, we again obtain $J$ as the Cantor set, but in this construction the sets in the $n$-th generation decrease in size at a uniform rate $O(1/2^n)$. With a little more effort, we can show that a similar statement is true (perhaps with a different constant) for the original construction, but in any case, the point is that $J$ is realized as the limit of a fairly regular process.

## EXERCISE 9.8

1. Let $E_0 = [0, 1]$, and construct $E_n$ inductively as follows. Given that $E_n$ is a finite union of disjoint closed intervals, $E_{n+1}$ is obtained from $E_n$ by removing the open middle third of each of these intervals (that is, we remove $(a + \frac{1}{3}t, a + \frac{2}{3}t)$, where $t = b - a$, from $[a, b]$). For example, $E_1 = [0, \frac{1}{3}] \cup [\frac{2}{3}, 1]$. Prove that $E$, $E = \bigcap E_n$, is a Cantor set (this is the famous "middle-third" Cantor set). Prove that the complement of $E$ in $[0, 1]$ has unit length; thus $E$ has zero length.

2. Show that Theorem 9.6.1 is applicable to the map $z \mapsto 2z - 1/z$. [$J$ is a Cantor set (see §1.8) but the results here yield more information about $J$.]

3. Let $P(z) = z^2 + c$. Show that if $|z| \geq |c|$, then

$$|P(z)| \geq (|c| - 1)|z|.$$

Deduce that if $|c| > 2$, then $J$ is a Cantor set. [When $c = -2$, $J$ is not a Cantor set.]

4. Let $\Pi$ be the class of sequences $x = (x_1, x_2, \ldots)$, where each $x_j$ is in $\{1, \ldots, d\}$, and define the *shift map* $\sigma\colon \Pi \to \Pi$ by

$$\sigma(x) = (x_2, x_3, \ldots).$$

(i) Show that $\rho$ defined by

$$\rho(x, y) = \sum_{n \geq 1} \frac{|x_n - y_n|}{3^n}$$

is a metric on $\Pi$, and that $\rho(x, y) < 3^{-n}$ if and only if

$$(x_1, \ldots, x_n) = (y_1, \ldots, y_n).$$

(ii) Show that $\sigma$ is surjective, but not injective, and that

$$\rho(\sigma x, \sigma y) \leq 3\rho(x, y),$$

so $\sigma$ is continuous.

(iii) The set of all *finite* sequences with elements in $\{1, \ldots, d\}$ is countable, and so can be written in a list, say $s_1, s_2, \ldots$. Making the obvious interpretation of

$$\zeta = (s_1, s_2, \ldots)$$

as a point in $\Pi$, show that the forward orbit of $\zeta$ under iterates of $\sigma$ is dense in $\Pi$.

(iv) By constructing periodic sequences in $\Pi$, show that the set of periodic points of $\sigma$ is dense in $\Pi$.

## §9.9. Julia Sets as Jordan Curves

In this section our principal objective is to obtain sufficient conditions for a Julia set to be a Jordan curve. Suppose that a (super)attracting fixed point $\zeta$ of a polynomial $P$ lies in a simply connected component $F_0$ of $F(P)$ (and recall that every bounded component of $F(P)$ is simply connected). We shall show that *the boundary $\partial F_0$ of $F_0$ is a closed curve, and each point of $\partial F_0$ is accessible from $F_0$* (in fact, $\partial F_0$ is a quasicircle, [105]). We remark, in passing, that not every simply connected domain has a closed curve as its boundary (for example, the complement of a compact, connected, but not arcwise connected, set is a simply connected domain whose boundary is not even arcwise connected).

Now suppose that a given polynomial $P$ has two completely invariant components, say $F_0$ and $F_1$, each containing a (super)attracting fixed point. Then these are the only components of the Fatou set of $P$, they are both simply connected, and $J$ is their common boundary (Theorems 5.6.1 and 9.4.3). It follows from the remarks above that each point of $J$ is accessible from both $F_0$ and $F_1$, and we are now in a position to apply the *converse* of the Jordan Curve Theorem ([77], p. 170) which asserts that in these circumstances, *$J$ is a Jordan curve.*

It is possible to avoid the use of the converse of the Jordan curve theorem here in the following way. First, one shows (by the method given below) that the boundary of $F_0$ is a continuous image, say $e^{i\theta} \mapsto \Gamma(e^{i\theta})$ of the unit circle $\partial\Delta$. To show that $\partial F_0$ is a Jordan curve, we must realize it as a homeomorphic image of $\partial\Delta$. Now in general, $\Gamma$ is not a homeomorphism, but one can show that for each $\zeta$ in $\partial F_0$, $\Gamma^{-1}\{\zeta\}$ is a closed arc on the unit circle (see, for example, [35], p. 285). This induces an equivalence relation $\sim$ on $\partial\Delta$ whose equivalence classes are the closed arcs $\Gamma^{-1}\{\zeta\}$, and $\partial F_0$ is homeomorphic to the quotient space $\partial\Delta/\sim$. Strictly speaking, it is still necessary to prove that $\partial\Delta/\sim$ is homeomorphic to $\partial\Delta$, and this can be done, for example, by using the topological characterization of Jordan curves, [77].

We begin our formal discussion with the special case in which the fixed point is at $\infty$.

**Lemma 9.9.1.** *Let $P$ be a polynomial of degree $d$, where $d \geq 2$. If $F_\infty$ is simply connected, and if $P$ is expanding on its Julia set $J$, then $J$ is a closed curve and every point of $J$ is accessible from $F_\infty$.*

*Remark.* With these hypotheses, $J(P)$ $(= \partial F_\infty)$ need not be a Jordan curve; for example, the map $z \mapsto z^2 - 1$ is expanding on its Julia set (see Theorem 9.7.5 and Figure 1.5.1).

THE PROOF OF LEMMA 9.9.1. Applying the Riemann–Hurwitz formula to $P: F_\infty \to F_\infty$, we find that $F_\infty$ has exactly $d - 1$ critical points of $P$, and all of these lie at $\infty$. Now take any disc $D$ centred at $\infty$, which is such that $P(\overline{D}) \subset D \subset F_\infty$. For each $n$, let $D_n = P^{-n}(D)$: then $D_n$ is open and connected,

$$D = D_0 \subset D_1 \subset D_2 \subset \cdots,$$

and as

$$\chi(D_{n+1}) + (d - 1) = d\chi(D_n),$$

we see that each $D_n$ is simply connected. Let $\gamma_n$ be the boundary of $D_n$; then $\gamma_n$ is a Jordan curve and $P^n$ is a $d^n$-fold map of $\gamma_n$ onto $\gamma_0$. Roughly speaking, we shall show that $\gamma_n$ converges to $\partial F_\infty$.

Let $A = \{z: 1 < |z| < r\}$. The topological annulus $\mathscr{A}$ bounded by $\gamma_0$ and $\gamma_1$ is conformally equivalent to $A$ (for some $r$), with $\gamma_0$ corresponding to $\{|z| = 1\}$, and we can use this conformal equivalence to transfer the radial cross-cuts

$$t \mapsto (1 + t(r - 1))e^{i\theta}, \qquad 0 \leq t \leq 1,$$

of $A$ to $\mathscr{A}$. We call the image curves the *radial cross-cuts of $\mathscr{A}$*, and we write the radial cross-cut starting at $z$ on $\gamma_0$ as the map

$$t \mapsto \sigma(z, t), \qquad t \in [0, 1].$$

It is convenient to give a label to this curve as a set (rather than as a map), and we denote this by $\sigma_0(z)$. Note also that the conformal equivalence between $A$ and $\mathscr{A}$ extends to a homeomorphism between the closed domains, so the map $(z, t) \mapsto \sigma(z, t)$ is cntinuous on $\gamma_0 \times [0, 1]$. Further, as $\partial\mathscr{A}$ is an analytic curve, the conformal equivalence extends analytically across the boundaries (by the Reflection Principle) so each $\sigma_0(z)$ has finite length.

Now consider a point $z$ on $\gamma_0$ and suppose that the radial cross-cut $\sigma_0(z)$ from $z$ ends at $z_1$ on $\gamma_1$; thus $\sigma(z, 0) = z$ and $\sigma(z, 1) = z_1$. Then $P(z_1)$ lies on $\gamma_0$ and so we can lift the radial cross-cut $\sigma_0(Pz_1)$ under $P^{-1}$ to produce a curve

$$t \mapsto \sigma(z, t), \qquad t \in [1, 2],$$

starting at $z_1$ and joining $\gamma_1$ to $\gamma_2$ (see Figure 9.9.1); we denote this curve as a set by $\sigma_1(z)$. This process can be continued in the obvious way to construct curves

$$t \mapsto \sigma(z, t), \qquad t \in [n, n + 1]$$

$(n = 1, 2, \ldots)$ which join $\gamma_n$ to $\gamma_{n+1}$. We denote these curves as sets by $\sigma_n(z)$, and (for the same $z$) these are joined end to end and are otherwise disjoint. It is clear that each point of each $\gamma_n$ has a unique curve of this type passing through it.

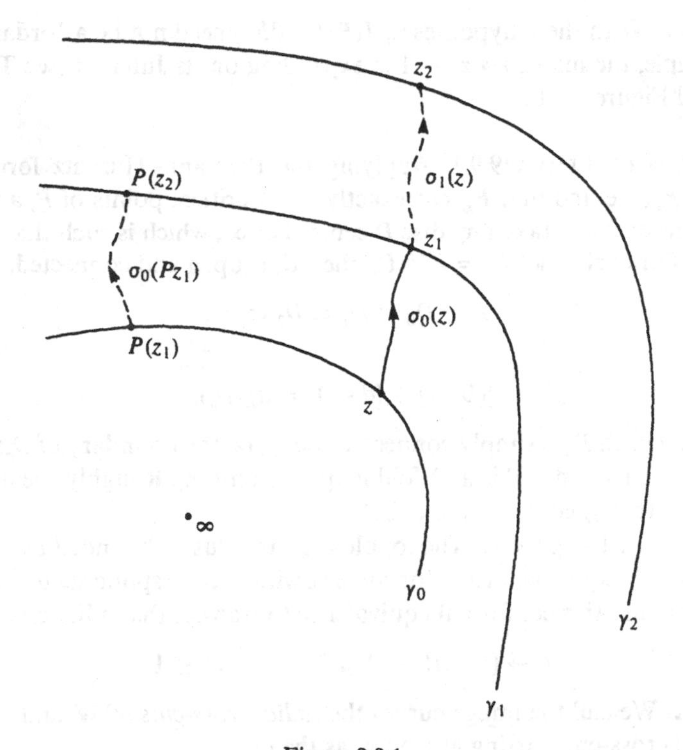


Figure 9.9.1

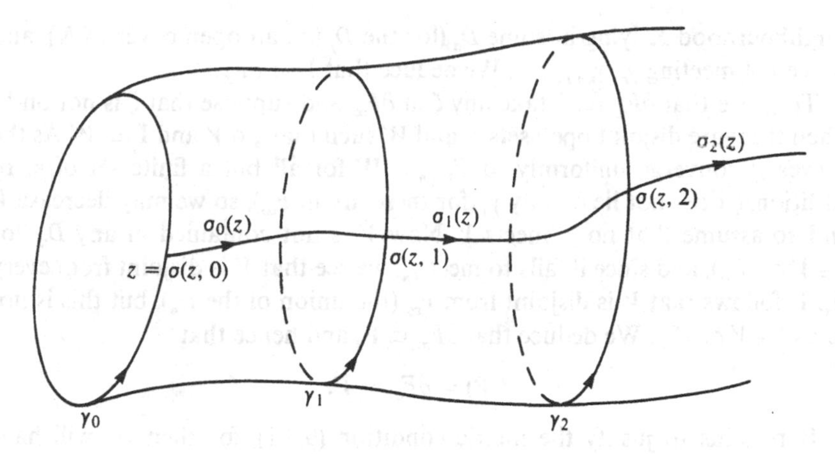


Figure 9.9.2

Let us now examine this in detail. We have constructed a map

$$(z, t) \mapsto \sigma(z, t)$$

of the product space $\gamma_0 \times [0, +\infty)$ (a cylinder) into $F_\infty$ such that:

(a) $\sigma(z, t)$ is continuous in $(z, t)$;
(b) the end-points of $\sigma_n(z)$ are $\sigma(z, n)$ and $\sigma(z, n + 1)$;
(c) $z \mapsto \sigma(z, n)$ maps $\gamma_0$ onto $\gamma_n$.

This is illustrated in Figure 9.9.2.

We come now to a metric condition which is closely related to $P$ being expanding on $J$ and which, for the moment, we regard as an assumption. We shall assume that there is some constant $k$, $0 < k < 1$, and some integer $n_0$, such that for $n \geq n_0$, and all $z$ on $\gamma_0$,

$$\text{length}[\sigma_{n+1}(z)] \leq k\,.\,\text{length}[\sigma_n(z)]. \tag{9.9.1}$$

Using this we see that

$$|\sigma(z, n) - \sigma(z, n + m)| \leq \sum_{j=n}^{n+m-1} \text{length}[\sigma_j(z)] \leq Mk^{-n},$$

for some $M$, and combining this with the General Principle of Uniform Convergence, we find that the sequence of maps

$$z \mapsto \sigma(z, n) \tag{9.9.2}$$

converges uniformly on $\gamma_0$ to some function $\Phi$. We denote the image $\Phi(\gamma_0)$ by $\Gamma$, say, and as the maps (9.9.2) are continuous in $z$, so is the map $\Phi$. It follows that $\Gamma$ is a closed curve, and we shall now show that $\Gamma = \partial F_\infty$.

If $\zeta \in \Gamma$ then there are points $\zeta_n$ on $\gamma_n$ which converge to $\zeta$, so, in particular, $\zeta$ is in the closure of $F_\infty$. However, $\zeta$ cannot lie in $F_\infty$ else it has a compact neighbourhood $K$ lying in some $D_n$ (for the $D_j$ are an open cover of $K$), and hence not meeting $\gamma_n, \gamma_{n+1}, \ldots$. We deduce that $\Gamma \subset \partial F_\infty$.

To prove that $\partial F_\infty \subset \Gamma$, take any $\zeta$ in $\partial F_\infty$ and suppose that $\zeta$ is not on $\Gamma$. Then there are disjoint open sets $V$ and $W$ such that $\zeta \in V$ and $\Gamma \subset W$. As the curves $\gamma_n$ converge uniformly to $\Gamma$, $\gamma_n \subset W$ for all but a finite set of $n$. In addition, $\zeta$ does not lie on any $\gamma_n$ (for these are in $F_\infty$), so we may decrease $V$ and so assume that no $\gamma_n$ meets $V$. Now $V$ is not contained in any $D_n$ (for $\zeta \in V \cap \partial F_\infty$), and since it fails to meet $\gamma_n$, we see that $V$ is disjoint from every $D_n$. It follows that $V$ is disjoint from $F_\infty$ (the union of the $D_n$), but this is not so as $\zeta \in V \cap \partial F_\infty$. We deduce that $\partial F_\infty \subset \Gamma$, and hence that

$$J(P) = \partial F_\infty = \Gamma.$$

It remains to justify the metric condition (9.9.1), for then we will have shown that $J$ is a closed curve, every point of which is, by construction, accessible from $F_\infty$. Now (9.9.1) is a almost consequence of the fact that $P$ is expanding on its Julia set $J$. Suppose for the moment that

$$|P'(z)| \geq k > 1 \tag{9.9.3}$$

on $J$. Then (as $J$ is compact) a similar inequality holds (with a smaller value of $k$) on some open neighbourhood $W$ of $J$, and so we may assume that (9.9.3) holds on $W$. The set $F_\infty - W$ is compact and so lies in some $D_n$; thus for $n \geq n_0$, say, $F_\infty - D_n$ lies in $W$. With this, we deduce that for these $n$,

$$\text{length}[\sigma_{n-1}(Pz)] = \text{length}[P(\sigma_n(z))] \geq k \,\text{length}[\sigma_n(z)].$$

Writing

$$L_n = \sup_{z \in \gamma_0} \text{length}[\sigma_n(z)]$$

(which is finite as $\gamma_0$ is compact) we find that

$$kL_n \leq L_{n-1},$$

and hence that

$$\text{length}[\sigma_n(z)] = O(k^{-n})$$

as $n \to \infty$. Clearly, it is sufficiently to replace (9.9.1) by this in the argument above, so we have now completed the proof under the assumption (9.9.3).

In general, however, (9.9.3) will not hold, but there will be some $m$ with

$$|(P^m)'(z)| \geq k > 1$$

holding on some neighborhood $W$ of $J$ (see §9.7). In this case, we must make a further modification by replacing $\sigma_n(z)$ (joining $\gamma_n$ to $\gamma_{n+1}$) by the curve

$$\tau_n(z) = \sigma_n(z) \cup \cdots \cup \sigma_{n+m-1}(z)$$

which joins $\gamma_n$ to $\gamma_{n+m}$. As

$$\tau_{n-1}(P^m z) = P^m(\tau_n(z)),$$

the proof remains valid with only minor alterations. □

Lemma 9.9.1 readily extends to the case of a general (super)attracting fixed point, and we have

**Theorem 9.9.2.** *Let $P$ be a polynomial which is expanding on its Julia set, and suppose that the immediate attracting basin $F_0$ of a (super)attracting fixed point $\zeta$ of $P$ is simply connected. Then $J$ is a closed curve, and every point of $J$ is accessible from $F_0$.*

Proof. The case when $\zeta = \infty$ follows from Lemma 9.9.1, so we may assume that $\zeta \in \mathbb{C}$, and by applying a conjugation (with respect to a translation) we may now suppose that $\zeta = 0$. It follows that $F_0$ is simply connected and bounded (in $\mathbb{C}$). Now construct a disc $D$, say $\{|z| < \varepsilon\}$, in $F_0$ which is such that $P(\bar{D}) \subset D$. We denote the component of $P^{-n}(D)$ which contains $0\ (= \zeta)$ by $D_n$; thus

$$D = D_0 \subset D_1 \subset D_2 \subset \cdots,$$

and $F_0$ is the union of the $D_n$.

We shall now show that each $D_n$ is simply connected, and to do this we take any simple closed curve $\sigma$ in $D_n$, and let $W$ be the interior of $\sigma$. Our aim is to prove that $W \subset D_n$. Now as $\sigma$ lies in $D_n$, $P^n(\sigma)$ lies in $D$ and so $|P^n| \leq \varepsilon$ on $\sigma$. The Maximum Modulus Theorem now shows that $|P^n| \leq \varepsilon$ throughout $W$, hence $W$ is a connected subset of $P^{-n}(D)$. As $\sigma\ (= \partial W)$ is in $D_n$, we conclude that $W \subset D_n$ and so each $D_n$ is simply connected.

Now for a suitably large $N$, $D_N$ contains all of the critical points of $P$ which lie in $F_0$, and we now relabel $D_N, D_{N+1}, \ldots$ as $D_0, D_1, \ldots$. In this way, we have constructed an increasing sequence of Jordan subdomains $D_n$ of $F_0$, with $D_n$ bounded by a Jordan curve $\gamma_n$, such that:

(1) $D_0 \subset \bar{D}_0 \subset D_1 \subset \bar{D}_1 \subset \cdots \subset \bigcup_n D_n = F_0$;
(2) $D_0$ contains all of the critical points of $P$ in $F_0$;
(3) $P(\gamma_{n+1}) = \gamma_n$, $P(D_{n+1}) = D_n$.

With these, the proof proceeds exactly as for Lemma 9.9.1 and we omit the details. □

We can now argue as follows. If $F(P)$ has exactly two components, then both are simply connected (because each is completely invariant under $P^2$). Thus, finally, and as described above, we have

**Theorem 9.9.3.** *Let $P$ be a polynomial that is expanding on its Julia set, and suppose that $F(P)$ is the union of exactly two components, each of which contains a (super)attracting fixed point. Then $J(P)$ is their common boundary and is a Jordan curve.*

For examples of this, and an example where this is not applicable, see Exercises 9.9.3 and 9.9.4. Roughly speaking, when $J$ is a Jordan curve, it is either very smooth or highly irregular: for further information, we refer the reader to [67] (pp. 80–85), [28] (pp. 116–117) and [86].

EXERCISE 9.9

1. Let
$$K = \{iy: |y| \leq 1\} \cup \{x + i\sin(1/x): 0 < x \leq 1\}.$$
Show that the complement of $K$ in $\mathbb{C}_\infty$ is a simply connected domain whose boundary is not a closed curve.

2. Construct an increasing sequence of domains $D_n$, each bounded by a Jordan curve, whose union $D$ is a simply connected domain with the property that not every point of $\partial D$ is accessible from within $D$.

3. Let $P_c(z) = z^2 + c$, and recall that if $c$ lies in a certain cardioid (see §1.6) then $P_c$ has two attracting fixed points, say $\zeta$ and $\infty$. Let $F_0$ be the component of $F(P)$ which contains $\zeta$. Using Theorem 9.3.1, deduce that $P$ is a two-fold map of $F_0$ onto itself, and hence that $F_0$ is a completely invariant component of $F(P)$. This shows that $F_\infty$ and $F_0$ are the only components of $F(P)$ so $J(P_c)$ is a Jordan curve.

4. Verify that for the map $R$ given in Example 9.4.2, $F(R)$ has exactly two components, neither of which contain an attracting fixed point.

## §9.10. The Mandelbrot Set

Given a family of rational functions $R(z, c_1, \ldots, c_m)$ parametrized by the complex parameters $c_1, \ldots, c_m$, we can partition the parameter space into disjoint regions $\Omega_j$ in such a way that as $(c_1, \ldots, c_m)$ moves through each $\Omega_j$ the dynamics of the iterates of $R$ display essentially the same features, while as $(c_1, \ldots, c_m)$ passes from one $\Omega_j$ to another, some significant change in the dynamics take place. For the family of polynomials
$$P_c(z) = z^2 + c, \qquad c \in \mathbb{C},$$
the partitioning of $\mathbb{C}$ leads to the Mandelbrot set $\mathcal{M}$: for example, we saw in §1.6 that as $c$ moves from the cardioid into the disc, the unique attracting fixed point of $z^2 + c$ is converted into an attracting two-cycle. Much has been written about the Mandelbrot set, and the reader may consult [38], [39], [59] and [80] for more information. Here, we shall confine ourselves to giving the details of the proof that $\mathcal{M}$ is connected [38].

For each $c$ in $\mathbb{C}$ let $J_c$ be the Julia set of $P_c$, and let $F_c$ be the unbounded component of the Fatou set $F(P_c)$. We write $c_n = P_c^n(0)$, so $(c_n)$ is the forward orbit of the finite critical point of $P_c$. Because $P_c^n \to \infty$ in some neighbourhood of $\infty$, either the sequence $(c_n)$ converges to $\infty$ (and $J_c$ is a Cantor set), or it is bounded (and $J_c$ is connected); thus (see §1.6)
$$\mathcal{M} = \{c \in \mathbb{C}: (c_n) \text{ is bounded}\}. \tag{9.10.1}$$

We need a sharper result than this, and as $-2 \in \mathcal{M}$ the following result is the best of its type.

**Theorem 9.10.1.** $\mathcal{M} = \{c \in \mathbb{C}: \text{for } n \geq 1, |c_n| \leq 2\}$.

PROOF. Given any $c$, let

$$W = \{z: |z| \geq |c|, |z| > 2\}.$$

Now for each $z$ in $W$, there is a positive $\varepsilon$ such that $|z| \geq 2 + \varepsilon$, and then

$$|P_c(z)| \geq |z|^2 - |c| \geq |z|^2 - |z| \geq (1 + \varepsilon)|z|,$$

so $P_c(W) \subset W$. Replacing $z$ by $P_c(z)$ (and using the same $\varepsilon$) we see that $P_c^m \to \infty$ on $W$ as $m \to \infty$.

As (9.10.1) holds, it suffices to show that if $c \in \mathcal{M}$, then $|c_n| \leq 2$ for $n \geq 1$, so suppose now that $c \in \mathcal{M}$. Then for all $n$, $c_n \notin W$ (else $c_{m+n} \to \infty$ as $m \to \infty$) so

$$|c_n| < |c| \qquad \text{or} \qquad |c_n| \leq 2.$$

As $c_1 = c$, this shows first that $|c| \leq 2$, and then for all $n$, $|c_n| \leq 2$. □

It is evident that $c_n = Q_n(c)$ for some polynomial $Q_n$, and as $c_{n+1} = P_c(c_n)$ for each $c$, the $Q_n$ can be defined inductively by $Q_1(c) = c$ and

$$Q_{n+1}(c) = [Q_n(c)]^2 + c.$$

In terms of the $Q_n$, Theorem 9.10.1 becomes

$$\mathcal{M} = \bigcap_{n=1}^{\infty} Q_n^{-1}(K),$$

where $K = \{|z| \leq 2\}$, so *$\mathcal{M}$ is compact*, and by taking complements we obtain

$$\mathscr{C}(\mathcal{M}) = \mathbb{C}_\infty - \mathcal{M} = \bigcup_{n=1}^{\infty} Q_n^{-1}(D),$$

where $D = \mathscr{C}(K)$. Now for any non-constant polynomial $Q$, $Q^{-1}(D)$ is open, connected, and contains $\infty$, and as any union of such sets also has these properties, we see that *$\mathscr{C}(\mathcal{M})$ is open and connected.* Our objective here is to prove

**Theorem 9.10.2.** *There is a conformal map of $\mathscr{C}(\mathcal{M})$ onto $\{|z| > 1\}$, thus $\mathscr{C}(\mathcal{M})$ is simply connected.*

We can motivate this result as follows. First, we have

$$Q_1^{-1}(K) \supset Q_2^{-1}(K) \supset \cdots, \tag{9.10.2}$$

and this shows that $\mathcal{M}$ is the intersection of the decreasing sequence of compact sets $Q_n^{-1}(K)$. If these sets are connected (as computer experiments suggest), then so is their intersection $\mathcal{M}$ and hence by Proposition 5.1.3, $\mathscr{C}(\mathcal{M})$ is simply connected.

We pause to verify (9.10.2) even though this will not be needed. First, we prove (by induction) that if $|z| > 2$, then

$$2 < |z| = |Q_1(z)| < |Q_2(z)| < |Q_3(z)| < \cdots, \tag{9.10.3}$$

the inductive step being

$$\begin{aligned} |Q_{n+1}(z)| &\geq |Q_n(z)|^2 - |z| \\ &\geq |Q_n(z)|(|Q_n(z)| - 1) \\ &> |Q_n(z)|. \end{aligned}$$

Now suppose that $z \in Q_{n+1}^{-1}(K)$; then (9.10.3) fails so $|z| \leq 2$, and as this implies that

$$|Q_n(z)|^2 = |Q_{n+1}(z) - z| \leq |Q_{n+1}(z)| + |z| \leq 4,$$

(9.10.2) follows.

The remainder of this section is devoted to the

Proof of Theorem 9.10.2. We begin by recalling the properties of the Green's function $g_c(z)$ of the unbounded component $F_c$ of the Fatou set $F(P_c)$: see §9.5. For each $c$, there is a function $g_c$: $F_c \to (0, +\infty]$ that is positive and harmonic in $F_c$ except at $\infty$ where $g_c(z) - \log|z|$ is bounded. Further, $g_c(z) \to 0$ as $z \to \partial F_c$ $(= J_c)$, and for sufficiently large $|z|$,

$$g_c(z) = \log|\varphi_c(z)|,$$

where $\varphi_c$ is the unique function which is analytic and satisfies

$$\varphi_c(z) \sim z \quad \text{as } z \to \infty, \qquad \varphi_c(P_c(z)) = \varphi_c(z)^2,$$

near $\infty$. Finally, in $F_c$,

$$g_c(P_c(z)) = 2g_c(z). \tag{9.10.4}$$

Roughly speaking, to prove the theorem we construct an analytic continuation of $(z, c) \mapsto \varphi_c(z)$ and then show that the desired conformal map is $c \mapsto \varphi_c(c)$, and to do this we require a little of the theory of functions of two complex variables. For any open subset $D$ of $\mathbb{C} \times \mathbb{C}$, a function $f$: $D \to \mathbb{C}$ is *analytic* if it can be expressed as a power series (in two variables) in some neighbourhood of each point of $D$, and this is so if $f$ is analytic in each variable when the other variable is kept fixed. Most of our maps are defined on subsets of $\mathbb{C}_\infty \times \mathbb{C}$, and analyticity at $\infty$ (in the first variable) is defined by the usual application of $z \mapsto 1/z$. As an example, $P_c(z)$ is analytic in $\mathbb{C}_\infty \times \mathbb{C}$.

It seems clear that we should confine our attention to the set

$$\Omega = \{(z, c) \in \mathbb{C}_\infty \times \mathbb{C} \colon c \in \mathscr{C}(\mathcal{M}), z \in F_c\},$$

and our first task is to show that this is open. Let

$$\Omega_0 = \{(z, c): c \in \mathscr{C}(\mathscr{M}), |z| > 2, |z|^2 > 2|c|\},$$

clearly, this is open in $\mathbb{C}_\infty \times \mathbb{C}$. Moreover, if $(z, c) \in \Omega_0$ then

$$|P_c(z)| \geq |z|^2 - |c| > |z|^2/2,$$

and from this, we obtain

$$2 < |z| < |P_c(z)| < |P_c^2(z)| < \cdots. \tag{9.10.5}$$

A simple induction argument now yields

$$|P_c^m(z)| \geq 2(|z|/2)^{2^m} \to \infty \tag{9.10.6}$$

and we deduce that $P_c^n(z) \to \infty$ locally uniformly on $\Omega_0$.

Now define $F$ by

$$F: (z, c) \mapsto (P_c(z), c).$$

Then

$$(z, c) \in \bigcup_{n=0}^{\infty} F^{-n}(\Omega_0)$$

if and only if for some $n \geq 0$,

$$c \in \mathscr{C}(\mathscr{M}), \qquad |P_c^n(z)| > 2, \qquad |P_c^n(z)|^2 > 2|c|,$$

that is, if and only if $c \in \mathscr{C}(\mathscr{M})$ and $P_c^n(z) \in \Omega_0$. In view of the remarks above, it is now clear that

$$\Omega = \bigcup_{n=0}^{\infty} F^{-n}(\Omega_0),$$

and as $F$ is continuous and $\Omega_0$ is open, we find that $\Omega$ is open. Moreover, as $P_c^n \to \infty$ locally uniformly in $\Omega_0$, we now see that the same holds locally uniformly in $\Omega$.

For any set $D$ in $\mathbb{C}_\infty \times \mathbb{C}_\infty$, the *c-section of D* is

$$D(c) = \{z: (z, c) \in D\}.$$

Now for each $c$ in $\mathscr{C}(\mathscr{M})$, $\Omega(c) = F_c$, and as this is infinitely connected (it is the complement of the Cantor set $J_c$), it is easier to work with an open subset $\Omega_1$ of $\Omega$ which has simply connected $c$-sections, but which still contains $(c, c)$ for all $c$ in $\mathscr{C}(\mathscr{M})$. We now construct such an $\Omega_1$. If $c \in \mathscr{C}(\mathscr{M})$, then $0 \in F_c$ so we can define

$$\Omega_1 = \{(z, c) \in \Omega: g_c(z) > g_c(0)\}.$$

**Lemma 9.10.3.**

(i) $(\infty, c)$ *and* $(c, c)$ *lie in* $\Omega_1(c)$;
(ii) *each c-section of* $\Omega_1$ *is simply connected*;
(iii) $\Omega_1$ *is an open subset of* $\mathbb{C}_\infty \times \mathbb{C}$.

**Proof.** Clearly $(\infty, c) \in \Omega_1(c)$, and putting $z = 0$ in (9.10.4) we see that $(c, c) \in \Omega_1(c)$.

To prove (ii), we fix $c$ and write

$$\Delta_n = \{z \in F_c \colon g_c(z) > 2^n g_c(0)\},$$

so

$$\Omega_1(c) = \Delta_0 \supset \Delta_1 \supset \Delta_2 \supset \cdots.$$

As the $\Delta_n$ form a sequence of neighbourhoods of $\infty$ which shrink uniformly to $\infty$, for a sufficiently large $N$, $z \in \Delta_N$ if and only if

$$|\varphi_c(z)| = \exp[g_c(z)] > \exp[2^N g_c(O)],$$

and as $\varphi_c$ is conformal near $\infty$, this inequality defines a simply connected domain. Thus for some $N$, $\Delta_N$ is a simply connected domain.

Now from (9.10.4), for all $n$,

$$z \in \Delta_n \quad \textit{if and only if} \quad P_c(z) \in \Delta_{n+1};$$

thus $P_c$ is a two-fold branched covering map of $\Delta_n$ onto $\Delta_{n+1}$ which is branched only at $\infty$ (for the only other branch point of $P_c$ is at 0 and $0 \notin \Delta_0$). This implies that each $\Delta_n$ is a domain, and the Riemann–Hurwitz relation

$$\chi(\Delta_n) + 1 = 2\chi(\Delta_{n+1})$$

shows that if one of the $\Delta_n$ is simply connected, then they all are. We deduce that $\Delta_0$ is simply connected and as $\Delta_0 = \Omega_1(c)$, this is (ii).

As $\Omega$ is open in $\mathbb{C}_\infty \times \mathbb{C}$, to prove (iii) it is enough to prove that the map

$$g \colon (z, c) \mapsto g_c(z)$$

is continuous on $\Omega$. Moreover, this will be so if $g$ is continuous on $\Omega_0$: indeed, on $\Omega$,

$$g(z, c) = g_c(z) = 2^{-n} g_c(P_c^n(z)) = 2^{-n}(gF^n)(z, c),$$

and this is continuous on $F^{-n}(\Omega_0)$ (whose union is $\Omega$) if $g$ is continuous on $\Omega_0$. The proof that $g$ is continuous on $\Omega_0$ will come later, and we emphasize that until we have proved this, we do not know that $\Omega_1$ is open.

Our next task is to fix a value of $n$ and define a single-valued $2^n$-th root of $P_c^n(z)$ throughout $\Omega_1$. We shall denote this by $\Phi_n(z, c)$, so we claim that there are functions $\Phi_n(z, c)$ on $\Omega_1$ such that

$$(z, c) \mapsto \Phi_n(z, c) \text{ is analytic in } \Omega_1;$$

$$\text{for each } c \text{ in } \mathscr{C}(\mathscr{M}),\ \Phi_n(z, c) \sim z \quad \text{as} \quad z \to \infty; \tag{9.10.7}$$

$$\Phi_n(P_c(z), c) = [\Phi_{n+1}(z, c)]^2 \quad \text{in } \Omega_1. \tag{9.10.8}$$

To construct the $\Phi_n$, note that the function

$$(z, c) \mapsto 1/P_c^n(z^{-1}, c) = z^{2^n} H_n(z, c)$$

is defined on

$$\Sigma = \{(z, c): (z^{-1}, c) \in \Omega_1\},$$

and $H_n$ is never zero there. For each $c$, $\Sigma(c)$ is simply connected (see Lemma 9.10.3(ii)) so we can define an analytic $2^n$-th root of $z \mapsto H_n(z, c)$ in $\Sigma(c)$ by

$$z \mapsto \exp\left\{\frac{1}{2^n}\int_\gamma \frac{\partial H_n/\partial z(w, c)}{H_n(w, c)}\,dw\right\}, \tag{9.10.9}$$

where the integral is over any curve $\gamma$ joining 0 to $z$ in $\Sigma(c)$. Conjugating by $z \mapsto 1/z$, this provides us with maps $\Phi_n$ which are defined and which satisfy (9.10.7) and (9.10.8) in $\Omega_1$. It is clear from (9.10.9) that each $\Phi_n$ is analytic in the interior of $\Omega_1$, and we recall that do not yet know that $\Omega_1$ is open. □

We proceed by analysing the functions $\Phi_n$ on $\Omega_0$, and we prove

**Lemma 9.10.4.** *$\Phi_n$ converges uniformly on $\Omega_0$ to some $\Phi$, where:*
(1) *$\Phi$ is analytic in $\Omega_0$;*
(2) *$\Phi(c, c) \sim c$ as $c \to \infty$;*
(3) *for each $c$, $\Phi(z, c) \sim z$ as $z \to \infty$;*
(4) *$|\Phi(z, c)| > 1$ in $\Omega_0$;*
(5) *$\Phi(P_c(z), c) = \Phi(z, c)^2$.*

PROOF. Write

$$\Phi_n(z, c) = \Phi_1(z, c) \prod_{k=1}^{n-1} \frac{\Phi_{k+1}(z, c)}{\Phi_k(z, c)},$$

and as $P_c^{k+1}(z) = P_c(P_c^k(z))$, we have

$$\frac{\Phi_{k+1}(z, c)}{\Phi_k(z, c)} = \left[1 + \frac{c}{[P_c^k(z)]^2}\right]^{1/2^{k+1}} = 1 + \theta_k(z, c),$$

say. To estimate $\theta_k$, we integrate the derivative of $(1 + w)^{1/k}$ along the radial segment from 0 to $w$, $|w| < \frac{1}{2}$, to obtain

$$|(1 + w)^{1/k} - 1| \le 2|w|/k,$$

and using this and (9.10.5) we have

$$|\theta_k(z, c)| \le \frac{2|c|}{2^{k+1}|P_c^k(z)|^2} \le \frac{|c|}{2^k|z|^2} \le \frac{1}{2^{k+1}}. \tag{9.10.10}$$

This shows that the $\Phi_n$ converge uniformly on $\Omega_0$ to some $\Phi$ and (1) follows as the $\Phi_n$ are analytic there.

To prove (2), we recall that $\mathcal{M} \subset \{|z| \le 2\}$. Thus if $|c| > 2$, then $(c, c) \in \Omega_0$ and so from (9.10.10),

$$|\theta_k(c, c)| \le \frac{1}{2^k|c|} \to 0 \qquad \text{as} \quad c \to \infty.$$

This shows first that $\Phi_n \to \Phi$ uniformly on $\{(c, c): |c| > 2\}$, and second (by taking the limit term by term),

$$\lim_{c\to\infty} c^{-1}\Phi(c, c) = \lim_{c\to\infty}\left(\frac{(c^2 + c)^{1/2}}{c}\right)\prod_{n=1}^{\infty}[1 + \theta_n(c, c)]$$

$$= 1.$$

Next, take a fixed $c$ in $\mathscr{C}(\mathscr{M})$. The functions $z \mapsto \Phi_n(z, c)$ converge uniformly to $z \mapsto \Phi(z, c)$ on $\Omega_0(c)$, and (after conjugating by $z \mapsto 1/z$) the derivative of $\Phi_n$ at $\infty$ converges to that of $\Phi$ and this gives (3).

Next, from (9.10.5), $|\Phi_n(z, c)| > 1$ on $\Omega_0$, so $|\Phi(z, c)| \geq 1$ there. Now for each $c$, $z \mapsto \Phi(z, c)$ is analytic in the disc $\Omega_0(c)$ and, by (3), it is not constant there: thus (4) follows from the Minimum Modulus Theorem on $\Omega_0(c)$. Finally, (5) follows by letting $n \to \infty$ in (9.10.8) and this completes the proof of Lemma 9.10.4. □

At this point, we can complete our proof that $\Omega_1$ is open by showing that $g$ is continuous on $\Omega_0$. For each $c$, the functions

$$g_c(z), \qquad \log|\varphi_c)|, \qquad \log|\Phi(z, c)|,$$

of $z$ are defined in some neighbourhood of $\infty$. The first two are equal there, and the second two are equal there (by Lemma 9.10.4 and the uniqueness of $\varphi_c$). It follows that for each $c$,

$$g_c(z) = \log|\Phi(z, c)| \tag{9.10.11}$$

as functions of $z$ throughout $\Omega_0(c)$ (for they are both harmonic there), thus this holds throughout $\Omega_0$ and as $\Phi$ is analytic in $\Omega_0$, $g$ is continuous there. It now follows that $\Omega_1$ is open and each $\Phi_n$ is analytic there.

We have now assembled all of the preliminary material for the proof. The functions $\Phi_n$ are defined in $\Omega_1$ and they converge uniformly to $\Phi$ in the open subset $\Omega_0 \cap \Omega_1$ of $\Omega_1$. We have also seen that $P_c^n(z) \to \infty$ locally uniformly on $\Omega_1$, and this shows that given any compact subset $K$ of $\Omega_1$, there is some $n(K)$ such that if $n \geq n(K)$, then $|\Phi_n| > 1$ on $K$. Now Cauchy's Integral formula holds in several variables and in the usual way (when combined with the map $z \mapsto 1/z$) this shows that the family $\{\Phi_n(z, c)\}$ is locally equicontinuous, and hence normal, in $\Omega_1$. The usual proof of Vitali's Theorem is valid (see Theorem 3.3.3) and using this to guarantee the analytic continuation, we deduce that $\Phi$ extends to an analytic map in $\Omega_1$, where $\Phi_n \to \Phi$ locally uniformly in $\Omega_1$.

As $(c, c) \in \Omega_1$ for every $c$ in $\mathscr{C}(\mathscr{M})$, we can now define an analytic map

$$c \mapsto \varphi(c) = \Phi(c, c)$$

on $\mathscr{C}(\mathscr{M})$. Further, as (9.10.11) holds throughout $\Omega_0$, and as $g_c(z) > 0$, we find that $|\varphi(c)| > 1$, thus $\varphi$ maps $\mathscr{C}(\mathscr{M})$ into $\{|z| > 1\}$.

Next, we need to know that

$$|\varphi(c)| \to 1 \quad \text{as} \quad c \to \partial\mathscr{M}. \tag{9.10.12}$$

Assuming this for the moment, we see first that if $|w| > 1$, then $\varphi^{-1}(w)$ is finite, and second that every curve in $\{|w| > 1\}$ lifts under $\varphi^{-1}$ to some curve in $\mathscr{C}(\mathscr{M})$. It follows that for some $k$, $\varphi$ is a $k$-fold covering map of $\mathscr{C}(\mathscr{M})$ onto $\{|w| > 1\}$ (see the discussion of complete covering surfaces in [5]), and $k = 1$ for the only pole of $\varphi$ is at $\infty$, and as $\Phi(c, c) \sim c$ when $c \to \infty$, this is a simple pole.

It only remains to prove (9.10.12) and to do this note that

$$\begin{aligned} 1 < |\varphi(c)| &= |\Phi(c, c)| \\ &= \lim_{n\to\infty} |P_c^n(c)|^{1/2^n} \\ &= \left( \lim_{n\to\infty} |Q_n(c)|^{1/2^n} \right)^2 \end{aligned}$$

as

$$Q_{n+1}(c) = c_{n+1} = P_c^{n+1}(0) = P_c^n(c).$$

Now take any positive $\varepsilon$, choose a positive integer $k$ such that

$$4 < (1 + \varepsilon)^{2^{k-1}},$$

and let

$$E = \{z: |z| < 3, |Q_k(z)| < 3\}.$$

By Theorem 9.10.1, $E$ is an open neighbourhood of $\mathscr{M}$, and we shall verify (9.10.12) by showing that if $c \in E$, then $|\varphi(c)| \leq 1 + \varepsilon$.

Let $T(x) = x^2 + 3$; then (by induction) for $n \geq 1$,

$$T^n(3) \leq (\tfrac{3}{4})4^{2^n}.$$

Now suppose that $c \in E$. As $|c| \leq 3$, for each $n$ we have

$$|Q_{n+1}(c)| \leq T(|Q_n(c)|),$$

and as $T$ is increasing, repeated applications of this yield

$$\begin{aligned} |Q_{n+k}(c)| &\leq T^n(|Q_k(c)|) \\ &\leq T^n(3) \\ &< 4^{2^n}. \end{aligned}$$

We deduce that

$$\begin{aligned} 1 < |\varphi(c)| &\equiv \left( \lim_{n\to\infty} |Q_n(c)|^{1/2^n} \right)^2 \\ &\leq 1 + \varepsilon \end{aligned}$$

as required. □

CHAPTER 10

# Hausdorff Dimension

We define the Hausdorff dimension of a set, and after giving some examples we show that the Julia set of a rational map has positive dimension.

## §10.1. Hausdorff Dimension

In 1918, Hausdorff introduced the $t$-dimensional measure $m_t(E)$ of a set $E$ and used this to create the dimension of $E$. The idea is that if the set $E$ has dimension $d(E)$, and if $t > d(E)$, then $m_t(E) = 0$ (the measure is too crude to recognize $E$), while if $t < d(E)$, then $m_t(E) = +\infty$ (the measure $m_t$ is too delicate to measure $E$). For example, a disc in $\mathbb{C}$ has dimension 2, infinite length and zero volume, and Hausdorff extended this familiar idea to the whole spectrum of positive numbers.

We shall consider subsets of $\mathbb{C}$ (and later of $\mathbb{C}_\infty$ with the chordal and spherical metrics), but the ideas are equally applicable to any metric space. The open disc in $\mathbb{C}$ with center $w$ and radius $r$ is denoted by $D(w, r)$, and for any set $A$, $|A|$ denotes the diameter of $A$, so

$$|A| = \sup\{|z - w|: z, w \in A\}.$$

Let $E$ be any set and $t$ a positive number. For each positive $\delta$, we consider the possible coverings $\{A_j\}$ of $E$ by sets of diameter less than $\delta$, and in an attempt to minimize the sum $\sum_j |A_j|^t$, we define

$$m_t^\delta(E) = \inf\left\{\sum_j |A_j|^t: |A_j| < \delta, E \subset \bigcup_j A_j\right\}.$$

As $\delta$ decreases, the class of such coverings of $E$ diminishes, so $m_t^\delta(E)$ increases and we define the $t$-dimensional measure $m_t(E)$ of $E$ by

$$m_t(E) = \lim_{\delta \to 0} m_t^\delta(E) = \sup_{\delta > 0} m_t^\delta(E).$$

This limit always exist (we allow it to take the value $+\infty$), and $m_t(E)$ is called the *t-dimensional Hausdorff measure* of $E$. In fact, $m_t$ is an outer measure on the class of all subsets of $\mathbb{C}$.

The following simple result enables us to go on to define the Hausdorff dimension of $E$.

**Lemma 10.1.1.** *If $m_t(E) < +\infty$ and $t < T$, then $m_T(E) = 0$.*

PROOF. Suppose that the family $\{A_j\}$ satisfies $E \subset \bigcup_j A_j$ and $|A_j| < \delta$; then

$$m_T^\delta(E) \le \sum_j |A_j|^T \le \delta^{T-t} \sum_j |A_j|^t.$$

Varying $\{A_j\}$ over all such coverings of $E$, we obtain

$$m_T^\delta(E) \le \delta^{T-t} m_t^\delta(E),$$

and letting $\delta \to 0$, we obtain $m_T(E) = 0$. □

An immediate consequence of Lemma 10.1.1 is the existence of a non-negative number $d(E)$ such that

$$m_t(E) = \begin{cases} +\infty & \text{if } t < d(E); \\ 0 & \text{if } t > d(E), \end{cases}$$

and we call $d(E)$ the *Hausdorff dimension* of $E$. Notice that we make no claims about the size of $m_t(E)$ when $t = d(E)$ as this can take any value in the range $[0, +\infty]$.

There are many other ways to measure the size of a set $E$, for example, the more recently introduced box dimension, and also the notion of capacity from potential theory. We shall concern ourselves only with Hausdorff's approach, but we refer the reader to [42], [43], [60], [84], [86], [97] and [98] for more information on these and related topics.

We end this section with some useful remarks concerning the computation of the dimension. First, any set $A$ can be enclosed in an open disc of radius $2|A|$, so if each set $A_j$ in a covering is replaced by such a disc $D_j$, then

$$\sum_j |A_j|^t < \sum_j |D_j|^t \le 4^t \sum_j |A_j|^t,$$

and this shows that we may confine ourselves to coverings of $E$ by open discs in our definition of $d(E)$.

Next, suppose that $E$ is a subset of $\mathbb{C}_\infty$, and define $d_0(E)$ to be the dimension of $E$ as above, except that for this, we use the chordal metric to measure diameters (we could equally well use the spherical metric here, and the corresponding value of $d_0(E)$ would be the same). As any Möbius map $g$ is Lipschitz with respect to the chordal metric, it is easy to see that $d_0(g(E)) = d_0(E)$. In addition, if $E$ is bounded in $\mathbb{C}$, then the ratio of the chordal and Euclidean metrics near $E$ is bounded (above and below) and so $d(E) = d_0(E)$. These observations lead us to the conclusion that we should really use the chordal (or spherical metric) to define $d(E)$, and that when $E \subset \mathbb{C}$, we may, if we wish, use the Euclidean metric. Henceforth, we shall use $d(E)$ (and not $d_0(E)$) for the dimension computed in this way.

Finally, given a rational map $R$ and a Möbius map $g$, the remarks above show that

$$d(J(gRg^{-1})) = d(g(J(R))) = d(J(R)),$$

so when estimating the dimension of $J(R)$, we may freely replace $R$ by any of its conjugates.

EXERCISE 10.1

1. Show:
   (i) if $A$ is countable, then $d(A) = 0$;
   (ii) if $A \subset \mathbb{R}$, then $d(A) \leq 1$;
   (iii) if $A \subset \mathbb{C}$, then $d(A) \leq 2$.

2. (i) Show that $d(\mathbb{R}) = 1$, and that $m_1(\mathbb{R}) = +\infty$.
   (ii) Suppose that for the sets $E_n$, $d(E_n) = \alpha - n^{-1}$, and let $E$ be the union of the $E_n$. Show that $d(E) = \alpha$ and $m_\alpha(E) = 0$.

## §10.2. Computing Dimensions

This section contains a brief discussion of how one can estimate (or, in rare circumstances, compute) the Hausdorff dimension of a set. Generally speaking, it is not difficult to establish an upper bound for $d(E)$ (for any covering will suffice), but it is much harder to obtain a lower bound.

To obtain an upper bound of $\dim(E)$, it is only necessary to find an efficient covering of $E$ by sets of small diameters and then compute the appropriate sum. For example, suppose that for each $\delta$ we can find a covering $\{A_j\}$ of $E$ with $|A_j| < \delta$ and

$$\sum_j |A_j|^t \leq 1.$$

Then for all $\delta$, $m_t^\delta(E) \leq 1$; thus $m_t(E) \leq 1$ and hence $d(E) \leq t$.

To illustrate this, consider the classical Cantor middle-third set $E$ (see Exercise 9.8.1). In this case, the construction of $E$ provides us with an efficient covering for, at the $n$-th stage, there is a covering $\{I_j\}$ of $E$ by $2^n$ intervals, each of length $3^{-n}$. Given any $\delta$, choose $n$ sufficiently large so that $3^{-n} < \delta$, and define

$$t = \frac{\log 2}{\log 3}. \tag{10.2.1}$$

Then

$$\sum_{j=1}^{2^n} |I_j|^t = 2^n(3^{-n})^t = 1,$$

and so by our remarks above, $d(E) \leq t$. Later, we shall see that $d(E) = t$.

Next, we describe a general method for obtaining a lower bound on $d(E)$. Recalling that $D(w, r)$ is a disc, we prove

**Lemma 10.2.1.** *Suppose that $\mu$ is a probability measure on the subset $E$ of $\mathbb{C}$, and that for some positive constants $c$, $t$ and $r_0$,*

$$\mu(D(z, r)) \leq cr^t$$

*whenever $z \in E$ and $0 < r < r_0$. Then $\dim(E) \geq t$.*

PROOF. Take any positive $\delta$, and as we let $\delta \to 0$ when we compute the Hausdorff dimension, we may assume that $\delta < r_0/2$. Now consider any covering of $E$ by discs $D(w_k, r_k)$, $k = 1, 2, \ldots$, with $r_k < \delta$. For each disc $D(w_j, r_j)$ that meets $E$, we select a point $\zeta_j$ in this intersection and create the disc $D(\zeta_j, \rho_j)$, where $\rho_j = 2r_j \leq r_0$. Clearly, $E$ lies in the union of the newly created discs, and as

$$\begin{aligned} 1 = \mu(E) &\leq \sum_j \mu(D(\zeta_j, \rho_j)) \\ &\leq c \sum_j \rho_j^t \\ &\leq c \sum_k (2r_k)^t \\ &\leq c \sum_k [\operatorname{diam} D(w_k, r_k)]^t, \end{aligned}$$

we find that $m_t^\delta(E) \geq c^{-1}$. Letting $\delta \to 0$, this gives $m_t(E) \geq c^{-1}$ and hence $\dim(E) \geq t$. □

Let us now attempt to complete the proof that the classical Cantor set $E$ has dimension $\log 2/\log 3$ by using Lemma 10.2.1. At the $n$-th stage of our construction of $E$ we have $2^n$ intervals of length $3^{-n}$, and we denote by $\mu_n$ the probability measure which has a constant probability density of $(3/2)^n$ on the union of these intervals. Intuitively, as $n \to \infty$, the probability measures $\mu_n$ converge to a probability measure $\mu$ on $E$, and providing that

$$\mu_n(A) \to \mu(A) \tag{10.2.2}$$

for any set $A$, we can apply Lemma 10.2.1 to $\mu$ and estimate $d(E)$. We apply Lemma 10.2.1 next, then we discuss the existence of $\mu$.

First, (10.2.2) implies that $\mu$ is a probability measure on $E$. Now select any $x$ in $E$. Then $x$ lies in exactly one of the intervals used in each stage of the construction of $E$ and denoting these intervals by $I_1, I_2, \ldots$, we have

$$|I_n| = 3^{-n}, \qquad I_1 \cap I_2 \cap \cdots = \{x\}.$$

Now take any $r$ in $(0, 1)$, and let $N$ be the largest integer such that $(2r)3^N < 1$. This means that $(6r)3^N \geq 1$, and with $t$ given by (10.2.1), we have

$$(6r)^t \geq 3^{-Nt} = 2^{-N}.$$

However, as $|D(x, r)| < 3^{-N}$, $D(x, r)$ can meet at most one of the intervals used at the $N$-th stage of the construction (for two such intervals are separated by a distance of $3^{-N}$), and by definition, this interval must be $I_N$. It follows that for all $n \geq N$,

$$\mu_n(D(x, r)) \leq \mu_n(I_N) = 2^{-N} \leq (6r)^t,$$

so by (10.2.2),

$$\mu(D(x, r)) \leq (6r)^t.$$

Lemma 10.2.1 now shows that $d(E) \geq t$, and so $d(E) = t$ as required.

Now this argument does not need the full strength of (10.2.2), and a full justification of what is actually needed here and in the next section can be found in [60], pp. 1–12. Given a sequence of probability measures $\mu_n$ on $\mathbb{C}$, one can always find a subsequence $\mu_{n(j)}$ such that

$$\mu(D) \leq \liminf_{j\to\infty} \mu_{n(j)}(D)$$

for every open set $D$, and

$$\limsup_{j\to\infty} \mu_{n(j)}(D) \leq \mu(K)$$

for every compact set $K$. These show that in our example above, a subsequence of $(\mu_n)$ converges in this sense to a probability measure $\mu$ on $E$ and, moreover, that the inequality

$$\mu_n(D(z, r)) \leq cr^t$$

for $\mu_n$ is inherited by $\mu$, and this is all that is required. A similar (but slightly different) argument will be used in §10.3.

We end this section with a brief discussion of other examples of sets whose dimension can be estimated. First, given any compact connected set $E$, suppose that there are Euclidean similarities $\varphi_1, \dots, \varphi_k$ such that each $\varphi_j$ maps $E$ into itself, say with $\varphi_j(E) = E_j$, and such that the sets $E_j$ are pairwise disjoint. We can use this pattern to generate a Cantor set, the $k^n$ sets at the $n$-th stage of the construction being of the form $\varphi(E)$, where $\varphi$ runs over all possible words of length $n$ in the semi-group generated by $\varphi_1, \dots, \varphi_k$. Such sets are said to be *self-similiar* and exactly the same arguments as those given above for the classical Cantor set show that $d(E)$ is the unique positive solution $d$ of the equation

$$|E|^d = |E_1|^d + \cdots + |E_k|^d.$$

In the case of the classical Cantor set, $k = 2$, $E = [0, 1]$, and $\varphi_1(x) = x/3$, $\varphi_2(2) = (x + 2)/3$.

In fact, self-similiar sets rarely arise out of other interesting situations, but the methods above allow for some flexibility. For example, there are situations where this type of construction arises, but where the maps $\varphi_j$ are given by analytic functions, and in such circumstances Koebe's Distortion Theorem ([46]) allows one to conclude that there is only a bounded deviation from self-similarity. This is enough to obtain estimates on the dimension of the set and, in particular, to show that it has positive dimension. Such sets are examples of *general Cantor sets*, and for examples of these situations, see [17], [42], [43] or [98]. In §9.7 we described conditions under which a Julia set is a general Cantor set of this type, and the methods outlined above enable us to conclude that these Julia sets have positive dimension (and one can obtain explicit bounds on their dimension).

More information about the dimension of Julia sets is known, and we mention a result of Ruelle, [86], in which $J$ is the Julia set of the polynomial $z^2 + c$. If $c$ is small, then $J$ is a Jordan curve (see §9.8), and Ruelle proved that as $c \to 0$,

$$d(J) = 1 + |c|^2/(4 \log 2) + O(c^3).$$

For other results on the dimension of the Julia set, and deeper problems about the ergodicity of the action of $R$, the reader can consult, for example, [28], [65], [71], [75], [78], and [86].

## §10.3. The Dimension of Julia Sets

Let $R$ be a rational map of degree $d$, where $d \geq 2$, and suppose that $\infty \in F(R)$. Then $R$ has no poles in $J$ and so $|R'(z)|$ attains its (finite) maximum on $J$. As $J$ contains some repelling cycle $\{z_1, \ldots, z_m\}$, and as

$$|R'(z_1) \cdots R'(z_m)| = |(R^m)'(z_1)| > 1,$$

we have

$$K_0 = \max\{|R'(z)|: z \in J\} > 1.$$

With this definition, we now have

**Theorem 10.3.1.** *Let $R$ be a rational map of degree $d$, where $d \geq 2$. If $\infty \in F(R)$, then*

$$\dim(J) \geq \frac{\log d}{\log K_0}, \tag{10.3.1}$$

*and this lower bound is best possible.*

*Remark.* The lower bound is attained when $R(z) = z^d$. Also, for any $R$, $\dim(J) \leq 2$ and so $K_0 \geq \sqrt{d} \geq \sqrt{2}$.

Given any rational map $R$, we can choose a Möbius map $g$ such that $\infty \in F(gRg^{-1})$, and because $J$ and $g(J)$ have the same dimension (see §10.1), we have

**Theorem 10.3.2.** *If* $\deg(R) \geq 2$, *then* $\dim(J) > 0$.

The idea in the proof of Theorem 10.3.1 is to find a point $\zeta$ in $F(R)$ such that for every $n$, $R^{-n}(\zeta)$ has precisely $d^n$ elements, and also such that as $n \to \infty$, $R^{-n}(\zeta)$ accumulates precisely at $J$. We then let $\mu_n$ be the probability measure distributed uniformly over $R^{-n}(\zeta)$, and by letting $n \to \infty$ through a suitable sequence, the measures $\mu_n$ converge to a probability distribution $\mu$ on $J$. By analysing the distribution of the points in $R^{-n}(\zeta)$ and their distances from $J$ we obtain inequalities on $\mu_n$, and hence on $\mu$, which enable us to apply Lemma 10.2.1 and so obtain a lower bound of $\dim(J)$. This proof is essentially that given in [47].

Before giving the proof, let $\mathrm{dist}[U, V]$ be the Euclidean distance between the sets $U$ and $V$, and let $\mathrm{card}(E)$ be the cardinality of the set $E$.

Proof of Theorem 10.3.1. As $\infty \in F$, we can select a positive number $\delta$ such that each pole of $R$ is at a distance at least $2\delta$ from $J$. Now let

$$J(\delta) = \{z: \mathrm{dist}[z, J] < \delta\},$$

and

$$K(\delta) = \sup\{|R'(z)|: z \in J(\delta)\},$$

so obviously, $1 < K_0 < K(\delta) < +\infty$, and

$$K(\delta) \to K_0 \quad \text{as} \quad \delta \to 0.$$

We shall prove that

$$\dim(J) \geq \frac{\log d}{\log K(\delta)}, \tag{10.3.2}$$

and we obtain (10.3.1) from this by letting $\delta \to 0$. As $\delta$ will not vary in proving (10.3.2), we can now write $K$ for $K(\delta)$.

To select $\zeta$, we choose a compact disc $E$ in $F(R)$ to which Theorem 4.2.8. is applicable (if $R$ has (super)attracting or rationally indifferent cycles, we take $E$ to be a disc in the immediate basin of the cycle; if no such cycles exist, then $F(R)$ contains a Siegel disc or a Herman ring, say $F_0$, and we take $E$ to be a disc in $R^{-1}(F_0)$). We have used Sullivan's Theorem here, but this is not really necessary for if $R$ had a wandering domain $\Omega$, we could take $E$ to be in disc in $\Omega$.

We now take $\zeta$ to be any point of $E$ whose orbit does not meet the orbit of any critical point; then for every $n$, $R^{-n}(\zeta)$ has precisely $d^n$ elements. From Theorem 4.2.8, there is an integer $m$ such that $R^{-n}(z) \subset J(\delta)$ whenever $n \geq m$, and by replacing $\zeta$ by some point in $R^{-m}(\zeta)$ and relabelling, we may assume that

$$\bigcup_{n=0}^{\infty} R^{-n}(\zeta) \subset J(\delta). \tag{10.3.3}$$

We now introduce various quantities that we shall need for the proof; these are:

(a) $R^{-n}(\zeta) = \{\zeta_n(i): i = 1, \ldots, d^n\}$;
(b) $d_n = \text{dist}[J, R^{-n}(\zeta)]$, so $d_0 = \text{dist}[J, \zeta]$; and
(c) the uniform probability distribution $\mu_n$ on $R^{-n}(\zeta)$.

Note that $\mu_n$ places a mass of $d^{-n}$ on each $\zeta_n(i)$, and for any $E$,

$$\mu_n(E) = d^{-n}.\text{card}\{i: \zeta_n(i) \in E\}.$$

Next, from (10.3.3),

$$d_n < \delta. \tag{10.3.4}$$

Also, Theorem 4.2.8 implies that $d_n \to 0$ as $n \to \infty$, and we shall now show that this convergence is not too rapid (that is, the sets $R^{-n}(\zeta)$ do not approach $J$ too quickly).

**Lemma 10.3.3.** *In the notation above, $d_n \leq Kd_{n+1}$, and so*

$$d_0 \leq d_1 K \leq d_2 K^2 \leq \cdots.$$

PROOF. Take any $z$ in $J$ and any $\zeta_{n+1}(i)$, so either:

(i) $\delta \leq |z - \zeta_{n+1}(i)|$; or
(ii) $|z - \zeta_{n+1}(i)| < \delta$.

If (i) holds, then

$$d_n < \delta < K\delta \leq K|z - \zeta_{n+1}(i)|.$$

If (ii) holds, then the linear segment from $z$ to $\zeta_{n+1}(i)$ lies entirely in $J(\delta)$ and so $|R'| \leq K$ on this segment. As $R(z)$ and $R(\zeta_{n+1}(i))$ are in $J$ and $R^{-n}(\zeta)$ respectively, we have

$$d_n \leq |R(z) - R(\zeta_{n+1}(i))| \leq K|z - \zeta_{n+1}(i)|.$$

In both cases, then, we have

$$d_n \leq K|z - \zeta_{n+1}(i)|,$$

and by letting $z$ and $\zeta_{n+1}(i)$ vary, we obtain $d_n \leq Kd_{n+1}$. □

The proof of Theorem 10.3.1 is easiest if there are no critical points in $J$, so *we shall proceed under this assumption*, and subsequently consider the changes necessary to prove the result in the general case. It is important to note, however, that *the part of the proof given so far remains valid without this assumption.*

This assumption means that for each $z$ in $J$, there is a positive $r_z$ such that $R$ is univalent on $D(z, r_z)$. Now the discs $D(z, r_z)$, $z \in J$, form an open cover of the compact set $J$, so by Lebesgue's Covering Theorem, there is a positive number $\tau$ such that $R$ is univalent on each disc $D(z, \tau)$, $z \in J$. Decreasing $\tau$ if necessary, we may assume that $\tau < \delta$, and we now choose a positive integer $q$ such that

$$d_0/K^q < \tau < \delta.$$

The next result says that in some sense, $R^{-n}(\zeta)$ must be uniformly distributed near $J$.

**Lemma 10.3.4.** *For all $z$ in $J$, and all integers $n$ and $m$ satisfying $n \geq 1$, $m \geq q$, there are at most $d^{n+q-m}$ points of $R^{-n}(\zeta)$ in $D(z, d_0/K^m)$; thus*

$$\operatorname{card}(D(z, d_0/K^m) \cap R^{-n}(\zeta)) \leq d^{n+q-m}. \tag{10.3.5}$$

PROOF. We shall prove that (10.3.5) holds for all $z$ in $J$ and all $n \geq 1$, by induction on $m$ for $m \geq q$. Now (10.3.5) holds for all $z$ in $J$ and all $n \geq 1$ when $m = q$, for in this case the upper bound is $d^n$. We now assume that the assertion is true when $q \leq m \leq M$ and prove that for all $z$ in $J$ and all $n \geq 1$,

$$\operatorname{card}(D(z, d_0/K^{M+1}) \cap R^{-n}(\zeta)) \leq d^{n+q-(M+1)}. \tag{10.3.6}$$

First, suppose that $n = 1$. Then by Lemma 10.3.3 (and as $K > 1$),

$$d_0/K^{M+1} \leq d_0/K \leq d_1 = \operatorname{dist}[J, R^{-1}(\zeta)],$$

and so in this case none of the points $\zeta_1(i)$ are within a distance $d_0/K^{M+1}$ of $J$, and (10.3.6) holds because the left-hand side of the inequality is zero.

Now suppose that $n \geq 2$, so $n - 1 \geq 1$. In this case,

$$d_0/K^{M+1} < d_0/K^q < \tau < \delta,$$

so $R$ is univalent on the disc $D(z, d_0/K^{M+1})$ and $|R'| \leq K$ there. This means that $R$ is a univalent map of $D(z, d_0/K^{M+1})$ into $D(Rz, d_0/K^M)$ and so distinct points of $R^{-n}(\zeta)$ in $D(z, d_0/K^{M+1})$ are mapped into distinct points of $R^{-(n-1)}(\zeta)$ in $D(Rz, d_0/K^M)$. Thus, using the inductive step (and $n - 1 \geq 1$), we have

$$\begin{aligned}\operatorname{card}(D(z, d_0/K^{M+1}) \cap R^{-n}(\zeta)) &\leq \operatorname{card}(D(z, d_0/K^M) \cap R^{-(n-1)}(\zeta)) \\ &\leq d^{(n-1)+q-M} \\ &= d^{n+q-(M+1)}.\end{aligned}$$

This proves (10.3.6) when $n = 1$ and when $n \geq 2$ and so the proof of Lemma 10.3.4 is complete. □

If we rewrite Lemma 10.3.4 in terms of the probability measure $\mu_n$ we obtain

$$\mu_n(D(z, d_0/K^m)) \leq d^q/d^m,$$

and it is easy to modify this to obtain a general result of the type described in Lemma 10.1.1. We show that for all $r < d_0/K$,

$$\mu_n(D(z, r)) \leq (d^{q+1} d_0^{-t}) r^t \tag{10.3.7}$$

where $t = (\log d)/(\log K)$. If $r < d_0/K$, there is some positive integer $m$ such that $d_0/K^{m+1} < r \leq d_0/K^m$, and so as $K^t = d$ we have

$$\begin{aligned}\mu_n(D(z, r)) &\leq \mu_n(D(z, d_0/K^m))\\ &\leq d^q/d^m\\ &= d^q/K^{mt}\\ &\leq d^q(rK/d_0)^t\\ &\leq (d^{q+1}/d_0^t)r^t,\end{aligned}$$

which is (10.3.7).

We now observe that some subsequence of the $\mu_n$ converges weakly to some probability measure $\mu$ on $J$ in such a way that the condition (10.3.7) is inherited by $\mu$ (see the discussion in §10.2). Thus for all $z$ in $J$ and all $r$ in $(0, d_0)$, we have

$$\mu(D(z, r)) \leq (d^{q+1}/d_0^t)r^t, \tag{10.3.8}$$

and this with Lemma 10.2.1 shows that

$$\dim(J) \geq (\log d)/(\log K).$$

This completes the proof of Theorem 10.3.1 in the case when $R$ has no critical points in $J$. □

We must now discuss the modifications necessary to prove the general case. By assumption, $R$ now has critical points in $J$, so let the set of these be

$$C = \{c_1, \ldots, c_k\}.$$

Now $R$, and hence any $R^m$, has derivative zero at each $c_j$ and so no iterate $R^m$ can fix any $c_j$ (else $c_j$ would be a super-attracting fixed point of $R^m$ and so lie in $F$). We deduce that for each $j$, $R^n(c_j)$ lies outside of $C$ whenever $n$ is sufficiently large, and so there is some integer $p$ such that $R^p(C)$ is disjoint from $C$.

Now choose any positive $\rho$ satisfying the conditions:

(a) the discs $D(c_j, 2\rho)$ are pairwise disjoint;
(b) $|(R^p)'(z)| \leq 1$ on each $D(c_j, 2\rho)$; and
(c) $4\rho < \mathrm{dist}[R^p(C), C]$.

Next, define

$$H = J - \bigcup D(c_j, \rho),$$

so $H$ is a compact subset of $J$ and $H \cap C = \varnothing$. It follows (as in the earlier part of the proof) that there is a positive number $\tau$ such that if $z \in H$, then $R$ is univalent on the disc $D(z, \tau)$. Finally, our choice of the integer $q$ is modified to take the presence of $C$ into account and here we choose $q$ such that

$$d_0/K^q < \min\{\tau, \rho, \delta\}, \qquad q > p.$$

We now claim that in this modified situation, Lemma 10.3.4 remains valid and with this, the proof proceeds exactly as before. Thus it only remains to give (in these new circumstances) the

PROOF OF LEMMA 10.3.4. We prove this by induction on $m$ for $m \geq q$ and as before, the inequality holds when $m = q$. Next, we assume that the inequality holds for all $z$ in $J$, all $n \geq 1$, and all $m$ with $q \leq m \leq M$, and proceed to show that

$$\operatorname{card}(D(z, d_0/K^{M+1}) \cap R^{-n}(\zeta)) \leq d^{n+q-(M+1)}. \tag{10.3.9}$$

First, we suppose that $1 \leq n \leq M + 1$. Then by Lemma 10.3.3,

$$d_0/K^{M+1} \leq d_0/K^n \leq d_n$$

so

$$\begin{aligned}\operatorname{card}(D(z, d_0/K^{M+1}) \cap R^{-n}(\zeta)) &\leq \operatorname{card}(D(z, d_n) \cap R^{-n}(\zeta))\\ &= 0\end{aligned}$$

by virtue of the definition of $d_n$.

We may now suppose that $n > M + 1$ so

$$n > M + 1 > q > p.$$

When $z \in H$, (10.3.9) follows from the induction hypothesis exactly as before, so we may assume that $z \in \bigcup D(c_j, \rho)$. From (a), $z$ lies in one of the discs $D(c_j, \rho)$, say in $D(c_1, \rho)$. Noting that

$$d_0/K^{M+1} \leq d_0/K^q < \rho,$$

and using (b), we find that $R^p$ maps $D(z, d_0/K^{M+1})$ into $D(R^p(z), d_0/K^{M+1})$. Clearly, at most $d^p$ points in $D(z, d_0/K^{M+1})$ can map into any given point in $D(R^p(z), d_0/K^{M+1})$ and from this it follows that

$$\operatorname{card}(D(z, d_0/K^{M+1}) \cap R^{-n}(\zeta)) \leq d^p \operatorname{card}(D(R^p(z), d_0/K^{M+1}) \cap R^{-(n-p)}(\zeta)).$$

Now from (c), $R^p(z)$ lies in $H$, and as we have already derived the inequality (10.3.9) for points $z$ in $H$, we deduce that

$$\operatorname{card}(D(z, d_0/K^{M+1}) \cap R^{-n}(\zeta)) \leq d^p d^{(n-p)+q-(M+1)}$$

and the proof is complete. □

# CHAPTER 11

# Examples

We end the book as we began it, with examples of rational maps and their Julia and Fatou sets. Now that the general theory is available, we can consider more interesting examples than those in Chapter 1 and these, in turn, serve to illustrate some of the theorems we have proved since then. In each case, we point out the most interesting features first, and we recommend that the reader try to sketch the Julia and Fatou sets before considering the details.

## §11.1. Smooth Julia Sets

For the sake of completeness, we recall the cases in which $J$ is a segment or a circle. Theorem 1.3.1 shows that for a polynomial $P$, $J(P)$ is the unit circle if and only if $P(z) = az^n$, where $|a| = 1$ and $n \geq 2$, while Theorem 1.4.1 shows that $J(P)$ is the segment $[-1, 1]$ if and only if $P$ is one of the Tchebychev polynomials $T_n$ or $-T_n$, where $T_n(\cos\theta) = \cos(n\theta)$.

The point of this section is to remark that there are results available which imply that in general, $J(P)$ is *not* a smooth curve; see, for example, [28] and [67]. Indeed, the repelling cycles are dense in $J$, and if $\zeta$ is a repelling fixed point of $P^m$, then $P^m$ will be conjugate near $\zeta$ to some map $z \mapsto \lambda z$. In general, $\lambda$ is not real nor of modulus one; in these cases $z \mapsto \lambda z$ has spirals as invariant curves, and points of $J$ (near $\zeta$) will lie on conformal images of these spirals.

EXERCISE 11.1

1. Investigate rational functions $R$ whose Julia set is the unit circle. [*Hint*: $\Delta$ is completely invariant under $R$ or $1/R$.]

## §11.2. Dendrites

The Julia set $J(P)$ of a polynomial $P$ is a *dendrite* if $F(P)$ is connected and simply connected, and $J(P)$ is connected, and a sufficient condition for this to be so is that each finite critical point of $P$ is pre-periodic (see Corollary 9.5.3). The simplest example is $J = [-1, 1]$ (as in §11.1), and another example is $z^2 + i$ (see Figure 9.5.1). A more illuminating example is with $P(z) = z^2 + c$, where $c$ is chosen in $(-2, 0)$ so that

$$(c^2 + c)^2 + c = \tfrac{1}{2}(1 - \sqrt{1 - 4c})$$

(and $c$ is approximately $-1.5436\ldots$). The left-hand side here is $P^3(0)$, while the right-hand side is the negative fixed point $\alpha$ of $P$. By drawing graphs of the two functions, we see that there is a solution in $(-2, 0)$ and with this $c$, the critical point 0 is pre-periodic and $J$ is a dendrite.

To obtain some idea of the structure of $J$ in this case, we note first that as $F = F_\infty$, $z \in F$ if and only if $P^n(z) \to \infty$. Now let $\beta$ be the positive fixed point of $P$: then

$$P([-\beta, \beta]) = P([0, \beta]) = [c, \beta] \subset [-\beta, \beta],$$

so by the previous remark, $[-\beta, \beta] \subset J$. Now $P$ maps $[-\beta, \beta]$ in a two-fold manner onto $[c, \beta]$, and (as an easy calculation shows) it maps some interval $\{iy: |y| \leq q\}$ in a two-fold manner onto $[-\beta, c]$. Thus with the obvious interpretation of $[-iq, iq]$,

$$[-\beta, \beta] \cup P^{-1}([-\beta, \beta]) = [-\beta, \beta] \cup [-iq, iq].$$

By repeating this process we see that

$$\bigcup_{n=0}^{\infty} P^{-n}([-\beta, \beta])$$

has a tree-like structure and this is the dendrite $J(P)$: see [35], p. 290, for further details.

## §11.3. Components of $F$ of Infinite Connectivity

We shall show that if

$$R(z) = z^3\left(\frac{1 - az}{z - a}\right),$$

and $0 < a < 1/13$, then $F(R)$ *contains infinitely many components, each of which has infinite connectivity.*

There is a component $F_0$ of $F(R)$ containing the origin (a super-attracting fixed point), and if $|z| \leq 3a/4$, then

$$|R(z)/z| \leq \left(\frac{3a}{4}\right)^2 \left(\frac{4 + 3a^2}{4a - 3a}\right) < 1,$$

so $R^n(z) \to 0$ as $n \to \infty$. It follows that

$$\{z: |z| \leq 3a/4\} \subset F_0.$$

As $R(1/a) = 0$, there is a component $F_1$ of $F(R)$ containing $1/a$, and as $R(F_1) = F_0$, we see that $R^n \to 0$ on $F_1$. Now $R$ maps the unit circle $\partial\Delta$ into itself, so for $z$ in $\partial\Delta$, $R^n(z)$ does not converge to zero. This shows that $\partial\Delta$ is disjoint from $F_0$ and $F_1$, and as $|1/a| > 1$, we see that $F_0$ lies inside $\partial\Delta$, while $F_1$ lies outside $\partial\Delta$.

Next, because $\deg(R) = 4$, $R$ has only four zeros, and as these are 0, 0, 0 and $1/a$, it follows that

$$R^{-1}(F_0) = F_0 \cup F_1. \tag{11.3.1}$$

Of course, $\infty$ is also a super-attracting fixed point lying in the component $F_\infty$, say, so $F(R)$ has at least three, and hence infinitely many, components. Note also that $\sigma R \sigma^{-1} = R$, where $\sigma(z) = 1/z$, so $\sigma(F_0) = F_\infty$.

Our next task is to examine the critical points of $R$. These are at 0, 0, $\infty$, $\infty$, $\alpha$ and $\beta$, say, where by computing $R'(z)$, we see that $\alpha$ and $\beta$ are the solutions of

$$2z - 3a = az(3z - 4a).$$

Note that $\alpha\beta = 1$, so $\sigma(\alpha) = \beta$. We claim that one critical point lies inside the circle

$$\Gamma = \{|z - 3a/2| = a/4\},$$

and to verify this we use Rouche's Theorem. On $\Gamma$, we have

$$\begin{aligned} |az(3z - 4a)| &< 3a|z|(|z - 3a/2| + |3a/2 - 4a/3|) \\ &< 3a(7a/4)(a/4 + a/6) \\ &< a/2 \\ &= |2z - 3a|, \end{aligned}$$

and so, as $2z - 3a$ has one zero inside $\Gamma$, there is one critical point, say $\alpha$, inside $\Gamma$. Because $\alpha$ lies inside $\Gamma$, we can now obtain the estimate

$$\begin{aligned} |R(\alpha)| &\leq |\alpha|^3 \left(\frac{1 + a|\alpha|}{|\alpha| - a}\right) \\ &\leq \left(\frac{7a}{4}\right)^3 \left(\frac{4 + 7a^2}{a}\right) \\ &< 3a/4, \end{aligned}$$

and this shows that $\alpha$ is in $R^{-1}(F_0)$. However, $|\alpha| < 1$, so from (11.3.1), we find that $\alpha$ lies in $F_0$. This means that $F_0$ contains at least three critical points of $R$ (namely 0, 0 and $\alpha$) and so applying the Riemann–Hurwitz relation to $R: F_0 \to F_0$, we have

$$\chi(F_0) + 3 \leq 3\chi(F_0).$$

This shows that $F_0$ is not simply connected, so it must be of infinite connectivity (Corollary 7.5.5). Because $\sigma(F_0) = F_\infty$, $F_\infty$ is also of infinite connectivity.

Finally, observe that as $\beta = \sigma(\alpha)$, the forward images of all critical points accumulate only at 0 and $\infty$, and this implies that any component of $F(R)$ is a pre-image of either $F_0$ or $F_\infty$, and so has infinite connectivity.

## §11.4. $F$ with Infinitely Connected and Simply Connected Components

We shall show that the polynomial

$$P(z) = z^2 - z^3/9$$

has the properties:

(a) *$F_\infty$ is infinitely connected;*
(b) *$F$ has infinitely many other components and each is simply connected; and*
(c) *$J$ has infinitely many non-degenerate components.*

The reader should sketch the graph of $P$ for real $x$ and confirm that $P$ has:

(1) zeros at 0, 0 and 9;
(2) critical points at 0, 6, $\infty$ and $\infty$;
(3) super-attracting fixed points at 0 and $\infty$ lying in components $F_0$ and $F_\infty$ respectively, and repelling fixed points at

$$\alpha = 1{\cdot}145898\ldots, \qquad \beta = 7{\cdot}854102\ldots.$$

As 0 and $\infty$ are fixed points, $F_0 \neq F_\infty$. We claim that

(4) $F_\infty$ contains the disc $D_1 = \{z: |z| \geq 10\}$;
(5) $F_\infty$ contains the circle $C = \{z: |z| = 6\}$;
(6) $F_0$ contains the disc $D_2 = \{z: |z| \leq 9/10\}$.

First, if $|z| \geq 10$, then

$$|P(z)/z| \geq |z|(|z|/9 - 1) \geq 10/9,$$

so (4) follows. Similarly, (6) holds because if $|z| \leq 9/10$, then

$$|P(z)/z| \leq |z|(|z|/9 + 1) \leq 99/100.$$

Finally, (5) holds because if $|z| = 6$, then

$$|P(z)| \geq |z|^2(1 - |z|/9) = 12,$$

so $P(C)$, and hence $C$, lies in the completely invariant $F_\infty$.

We know that, in general, $F_\infty$ is either simply or infinitely connected: in this case, $F_\infty$ contains $C$ but not the point 9 (which maps to 0), hence it is of infinite connectivity. For another proof, observe that $F_\infty$ contains the three critical points 6, ∞ and ∞, so

$$\chi(F_\infty) + 3 = 3\chi(F_\infty):$$

and hence $\chi(F_\infty) = -\infty$.

Next $P(9) = 0$, so $9 \in F$. Moreover, the components $F_0$ and $F_9$ (containing 9) are distinct because they are separated by the circle $C$ in $F_\infty$: thus $F$ has at least three, and hence infinitely many components. As $P$ is a polynomial, all components of $F$ other than $F_\infty$ are bounded and simply connected, and as the forward images of the critical points accumulate only at 0 and ∞, every component of $F$ other than $F_\infty$ is eventually mapped onto $F_0$. Of course, (c) follows from (b).

EXERCISE 11.4

1. Discuss the polynomial $z(z + 1)(z + 2)$. [$P(x) > x$ if $x > 0$.]

## §11.5. *J* with Infinitely Many Non-Degenerate Components

In the example given in §11.4, $J$ had infinitely many non-degenerate components because the complement of $F_\infty$ contained infinitely many topological discs. Here we give an example in which $J$ has infinitely many non-degenerate components, but $F = F_\infty$. We shall show that the polynomial

$$P(z) = (3\sqrt{3}/2)z(z + 1)(z + 2)$$

has the properties:

(a) *F is connected and of infinite connectivity*; *and*
(b) *J contains infinitely many non-degenerate components.*

First, the reader should draw a graph of $P(x)$ for real $x$, and confirm that:

(i) 0 is a repelling fixed point of $P$;
(ii) the critical points of $P$ are ∞, ∞, $\alpha$ and $\beta$, where

$$\alpha = -1 - 1/\sqrt{3}, \qquad \beta = -1 + 1/\sqrt{3};$$

(iii) $P$ maps the interval $[-1, 0]$ onto itself.

If $x > 0$, then $P(x) > 3x$, so $P^n \to \infty$ on $(0, +\infty)$ and so this interval lies in $F_\infty$. Because $P(\alpha) = 1$, the critical point $\alpha$ lies in $F_\infty$, and so $F_\infty$ has infinite connectivity (Corollary 7.5.5 and Theorem 9.5.1).

Our knowledge of the critical points of $P$ yields more information. If $F$ had a component other than $F_\infty$, then (by Sullivan's Theorem) there would be a cycle of components which would attract a critical point in the manner described in §9.3. However, $P^n(\alpha) \to \infty$, while $P^2(\beta) = 0$, so $\beta$ is in $J$; thus $F$

cannot have a component other than $F_\infty$, so $F = F_\infty$, and $F$ is both connected and of infinite connectivity.

It is now easy to see that $J$ contains the interval $[-1, 0]$. Indeed, as $F = F_\infty$, we have $P^n \to \infty$ on $F$, and as $[-1, 0]$ is forward invariant under $P$, it lies in $J$. It follows that $J$ contains at least one non-degenerate component, and as any such component cannot be completely invariant under $P$ (else it would be $J$, and $F_\infty$ would be simply connected), it has infinitely many pre-images, and these too are non-degenerate components of $J$.

Exercise 11.5

1. Show that $-2$ is in $J$. By considering the graph of $P$, show that there is only one real solution of $P(z) = -2$, and deduce that $J$ does not lie entirely in the real axis.

## §11.6. $F$ of Infinite Connectivity with Critical Points in $J$

We shall show that for the polynomial

$$P(z) = z^3 - 12z^2 + 36z:$$

(a) *$F$ is connected and is of infinite connectivity;*
(b) *$J$ is a Cantor set; and*
(c) *not every finite critical point is attracted towards $\infty$.*

First, note that

$$P(z) = z(z-6)^2 = z + z(z-5)(z-7),$$

so $P$ has:

(1) zeros at 0, 6 and 6;
(2) fixed points at 0, 5, 7 and $\infty$; and
(3) critical points at 2, 6, $\infty$ and $\infty$.

As $P(6) = 0$, $P(0) = 0$ and $P'(0) = 36$, we see that both 0 and 6 are in $J$; thus (c) holds. Next, if $|z| > 8$, then $|P(z)| > 4|z|$. As $P(2) > 8$, we see that $P^n(2) \to \infty$ as $n \to \infty$; thus $2 \in F_\infty$. Now $F_\infty$ contains exactly three critical points, and as

$$\chi(F_\infty) + 3 = 3\chi(F_\infty),$$

we find that $F_\infty$ has infinite connectivity.

Now observe that $P$ is a strictly increasing map of $[0, 2]$ onto $[0, 32]$, a strictly decreasing map of $[2, 6]$ onto $[0, 32]$, and a strictly increasing map of $[6, 8]$ onto $[0, 32]$. As $\deg(P) = 3$, this shows that for any point $x$ in $[0, 32]$, the entire backward orbit of $x$ lies in $[0, 8]$, and hence $J \subset [0, 8]$. It follows that $F = F_\infty$, so $F$ is connected and (a) holds.

Finally, $J$ cannot contain a non-degenerate component for such a component would have to be an interval, say $(w - \varepsilon, w + \varepsilon)$, of positive length in $[0, 8]$. However, all of the inverse images of the point 2 lie in both $F$ and $[0, 8]$, and as these must accumulate at $w$, no such interval can exist. By Theorem 5.7.1, $J$ is a Cantor set.

EXERCISE 11.6

1. Let $P(z) = \alpha z(z - \beta)^2$, where $\alpha$ and $\beta$ are positive. Find conditions on $\alpha$ and $\beta$ which imply that $P$ has the properties (a), (b) and (c) in the text.

   Find conditions on $\alpha$ and $\beta$ such that $P$ is conjugate to a Tchebychev polynomial (see §1.4) and find $J$ explicitly in this case.

## §11.7. A Finitely Connected Component of $F$

If $P$ is a polynomial, then every component of $F(P)$ is simply connected or infinitely connected. If $R$ is rational, then $R$ may have Herman rings and these are doubly connected. Using quasiconformal mappings, I.N. Baker has established (implicitly) the existence of components of a Fatou set of any given connectivity: here, we give an explicit example (suggested by Shishikura) of a Fatou set with a component of finite connectivity greater than two.

Let

$$R(z) = \frac{z^2(1 + t^{12}z^3)}{(1 - t^4z)(1 - tz)^3}, \qquad t > 0.$$

We shall show that if $t$ is sufficiently small, then $F(R)$ *has a component of connectivity* 3 *or* 4 (it seems probable that with a little more work, one could compute the connectivity exactly, but this is not the main point of the example).

First, we observe that $\infty$ is a repelling fixed point of $R$; thus $\infty$ and its pre-images $1/t$ and $1/t^4$ are in $J$.

Next, $R$ has exactly eight critical points, say $0, t^{-1}, t^{-1}, \zeta_1, \ldots, \zeta_5$, and each is in $\mathbb{C}$. We now prove

**Lemma 11.7.1.** *For sufficiently small $t$, the non-zero critical points of $R$ lie outside the circle $\{|z| = 3\}$.*

PROOF. The critical points $\zeta_j$ are those solutions of $R'(z) = 0$ which are distinct from 0 and $1/t$, and a calculation shows that the $\zeta_j$ satisfy an equation of the form

$$2 + tzQ(z, t) = 0,$$

where $Q$ is a polynomial in $z$ and $t$ (which can, but need not, be found explicitly). Now let

$$M = \sup\{|Q(z, t)|: |z| \leq 3, t \leq 1\},$$

and assume that $t < \min\{1/3, 2/3M\}$, so certainly $|t^{-1}| > 3$. Also, $|\zeta_j| > 3$ for otherwise,

$$2 = |t\zeta_j Q(\zeta_j, t)| < 2.$$

This completes the proof. □

Next, we consider the component $F_0$ of $F$ which contains the origin (a super-attracting fixed point). As $R(z) \to z^2$ uniformly on compact subsets of $\mathbb{C}$ when $t \to 0$, we should expect that when $t$ is sufficiently small, $F_0$ looks roughly like the unit disc; this is the content of the next result.

**Lemma 11.7.2.** *Given any $\varepsilon$ in (0, 1), there is some positive number $t_0$ such that if $0 < t < t_0$, then $F_0$ is simply connected, and*

$$(|z| \leq 1 - \varepsilon\} \subset F_0 \subset \{|z| \leq 1 + \varepsilon\}.$$

Proof. First, we prove that $F_0$ lies between the two given discs. Let $K = \{|z| \leq 2\}$ and choose $t_0$ such that if $t < t_0$ and $z \in K$, then

$$(1 + \varepsilon)^{-1} < |R(z)/z^2| < 1 + \varepsilon.$$

If $|z| < 1 - \varepsilon$, then

$$|R(z)| < (1 + \varepsilon)|z|^2 < (1 - \varepsilon^2)|z|,$$

and so $F_0$ contains the disc $\{|z| \leq 1 - \varepsilon\}$.

Now suppose that $F_0$ meets the circle $|z| = 1 + \varepsilon$ at some point $w$, and join $w$ to the origin by a curve $\sigma$ lying in $F_0$. As $\sigma$ is a compact subset of $F_0$, $R^n \to 0$ uniformly on $\sigma$ so there is a unique positive integer $k$ such that $R^k(\sigma)$ meets $\{|z| = 1 + \varepsilon\}$, but $R^n(\sigma)$ does not for any $n$, $n \geq k + 1$. Now let $\zeta$ be a point where $R^k(\sigma)$ meets $\{|z| = 1 + \varepsilon\}$; then

$$|\zeta| > |R(\zeta)| > |\zeta|^2(1 + \varepsilon)^{-1} = |\zeta|,$$

a contradiction. As $F_0$ is arcwise connected and disjoint from $\{|z| = 1 + \varepsilon\}$, $F_0$ lies within $\{|z| \leq 1 + \varepsilon\}$.

It remains to show that $F_0$ is simply connected. We choose a disc $V$ centred at the origin, and such that $R(\overline{V}) \subset V$, and let $V_n$ be that component of $R^{-n}(V)$ which contains the origin with $V = V_0$. It is clear that $V_n \subset V_{n+1}$, $n \geq 0$, and as $R^n \to 0$ on $F_0$, we also have $F_0 = \bigcup_{n=0}^{\infty} V_n$.

Now $R$ has five zeros, namely 0 (twice) and the three cube roots of $-1/t^{12}$ (which lie outside $|z| = 2$, and so outside of $F_0$). It follows that $R$ must be a two-fold map of $F_0$ onto itself, and from Lemma 11.7.1, the only critical points of $R$ in $F_0$ are those at the origin. It follows from this that $R$: $V_{n+1} \to V_n$ is a two-fold covering map with $\delta(V_{n+1}) = 1$, and the Riemann–Hurwitz formula, namely

$$\chi(V_{n+1}) + \delta(V_{n+1}) = 2\chi(V_n),$$

shows (by induction) that every $V_n$ is simply connected. Finally, as $F_0$ is the union of the increasing sequence of simply connected domains $V_n$, $F_0$ itself is simply connected. (The Riemann–Hurwitz formula applied to $F_0$ directly shows that it is simply or infinitely connected, but we must rule out the latter possibility). This completes the proof of Lemma 11.7.2. □

To complete the argument, we define the triply connected domain $\Omega$ by

$$\Omega = \{z: 1/(2t^4) < |z| < 2/t^4, \quad |z - 1/t^4| < 1/(4t^4)\},$$

and prove

**Lemma 11.7.3.** *$R^{-1}(F_0)$ consists of two components, namely $F_0$ and a component $W$ which contains $\Omega$. Further, $W$ has connectivity three or four.*

Proof. We begin by showing that $R(\Omega) \subset F_0$. Assuming that $6t^3 < 1$, we have the following simple estimates on $\Omega$:

$$|z^2(1 + t^{12}z^3)| \leq 36/t^8;$$
$$|1 - t^4 z| \geq \tfrac{1}{4};$$
$$|1 - tz| \geq (1/2t^3) - 1 \geq 1/(3t^3).$$

These show that for some constant $M$,

$$R(\Omega) \subset \{|z| < Mt\},$$

and by taking $\varepsilon = \frac{1}{2}$ in Lemma 11.7.1, and $t < \frac{1}{2}M$, we find that $R(\Omega) \subset F_0$.

It follows that $\Omega$ lies in some component $F_1$ of $R^{-1}(F_0)$, and Lemma 11.7.2 ensures that if $t$ is small, $F_1$ is disjoint from $F_0$. Now of the five zeros of $R$, two are in $F_0$ (at the origin) and three are in $\Omega$ (at the cube roots of $-1/t^{12}$): thus $R^{-1}(F_0)$ is the disjoint union of $F_0$ and $F_1$, $R$ is a two-fold covering map of $F_0$ onto itself, and a three-fold covering map of $F_1$ onto $F_0$.

Now if $t$ is small, each component of the complement of $\Omega$ contains a point of $J$ (namely, $1/t$, $1/t^4$ and $\infty$); thus the connectivity of $F_1$ is at least three. Finally, of the eight critical points of $R$, three (namely $0, 1/t, 1/t$) are not in $F_1$. It follows that $F_1$ contains at most five critical points, and as $R$ is a three-fold map of $F_1$ onto the simply connected $F_0$, we have

$$\chi(F_1) + 5 \geq \chi(F_1) + \delta(F_1) = 3\chi(F_0) = 3,$$

so $\chi(F_1) \geq -2$. As the connectivity of $F_1$ is $2 - \chi(F_1)$, $F_1$ has connectivity 3 or 4 and this completes the discussion. □

## Exercise 11.7

1. Show that $\infty$ is a repelling fixed point of $R$.

## §11.8. $J$ Is a Cantor Set of Circles

Let

$$R(z) = z^2 + \lambda/z^3,$$

where $\lambda > 0$. As $\infty$ is a super-attracting fixed point of $R$, 0 and $\infty$ lie in some components $F_0$ and $F_\infty$ respectively of $F$, and we shall see later that $F_0 \neq F_\infty$. We shall show that when $\lambda$ is sufficiently small:

(a) *$F_0$ and $F_\infty$ are simply connected, while all other components of $F$ are doubly connected;*
(b) *$R^n \to \infty$ on $F$;*
(c) *$\infty$ attracts all critical points of $R$;*
(d) *$J$ is a Cantor set of circles;*
(c) *there is some component of $J$ which does not meet the boundary of any component of $F$;*
(f) *$R$ is expanding on $J$.*

Any component of $J$ with the property (e) is called a *buried component*, [74], and Theorem 5.7.1 shows that these arise whenever $J$ is disconnected and every component of $F$ has finite connectivity. In this example, $J$ is (roughly speaking) like the set

$$\bigcup_\alpha \{|z| = r_\alpha\},$$

where $\{r_\alpha\}$ is some Cantor set on the positive real axis.

The idea behind this example is as follows. There is an annulus $\mathscr{A}$ which separates 0 from $\infty$, and which is such that $R^{-1}(\mathscr{A})$ consists of two annuli, say $\mathscr{A}_1$ and $\mathscr{A}_2$, both lying in $\mathscr{A}$ and separating 0 and $\infty$. It follows that $R^{-2}(\mathscr{A})$ consists of four annuli, say $\mathscr{A}_{11}, \mathscr{A}_{12}, \mathscr{A}_{21}, \mathscr{A}_{22}$, two of which lie in $\mathscr{A}_1$ and two of which lie in $\mathscr{A}_2$. Continuing in this way, at the $n$-th stage we obtain $2^n$ disjoint annuli, each containing two annuli of the next stage. The intersection of any nested sequence of these annuli gives a Jordan curve separating 0 from $\infty$, and the Julia set is the (uncountable) union of these Jordan curves. While this idea is simple enough, the verification is quite long.

In the following discussion, we shall assume implicitly (and without further comment) that $\lambda$ satisfies various inequalities that arise in our discussion; all of these will be valid if $\lambda$ is sufficiently small and we shall not attempt to estimate this range although, clearly, it would be possible to do so. In addition, we shall use positive numbers $a$, $b$, $\alpha$ and $\beta$, which are assumed to satisfy the inequalities

$$\tfrac{1}{5} < \beta < (1 - b)/3 < \tfrac{1}{3} < \alpha < (1 + b)/3 < a < \tfrac{2}{5}$$

and $a + 3\beta > 1$ (for example, take $\beta = \frac{1}{4}$, $b = \frac{1}{10}$ and then choose $a$ and $\alpha$).

We now define

$$\mathscr{A} = \{\lambda^a < |z| < \lambda^{-b}\}$$

and prove

**Lemma 11.8.1.** $R^{-1}(\mathscr{A}) \subset \mathscr{A}$.

PROOF. The components of the complement of $\mathscr{A}$ are

$$A = \{|z| \leq \lambda^a\}, \qquad B = \{|z| \geq \lambda^{-b}\}.$$

On $B$ we have

$$|R(z)/z| > |z| - \lambda > \lambda^{-b} - \lambda > 1,$$

so $R(B) \subset B$. On $A$ we have

$$\begin{aligned} |R(z)| &\geq \lambda|z|^{-3} - |z|^2 \\ &> \lambda^{1-3a} - \lambda^{2a} \\ &> \lambda^{-b} \end{aligned}$$

because $a > 0$ and $3a - 1 > b$; thus $R(A) \subset B$. Together, these imply that $R^{-1}(\mathscr{A}) \subset \mathscr{A}$. □

Next, we describe $J$.

**Lemma 11.8.2.**

$$J = \bigcap_{n=1}^{\infty} R^{-n}(A).$$

PROOF. We are not yet in a position to prove this completely, but the general idea is as follows. From Lemma 11.8.1, the sets $R^{-n}(\mathscr{A})$, $n \geq 0$, are decreasing with $n$ and each lies in $\mathscr{A}$. Now define

$$K = \bigcap_{n=1}^{\infty} R^{-n}(A),$$

and observe that as $R^{-1}$ commutes with the intersection operator, $K$ is completely invariant under $R$. Later, we shall show that $R^n \to \infty$ on $F$; then as $K$ is forward invariant, it must be disjoint from $F$ and so $K \subset J$. We shall also see that we can replace $A$ in the definition of $K$ by its closure, and this then implies that $K$ is closed and non-empty. It will be clear that $K$ has at least three points (it will be a union of Jordan curves) and this with the minimality of $J$ implies that $J \subset K$, and hence $J = K$. □

In order to justify our description of $J$, we must identify $R^{-1}(\mathscr{A})$ and to help us do this we define two annuli by

$$\begin{aligned} V_1 &= \{\tfrac{1}{2} < |z| < \tfrac{3}{2}\}, \\ W_1 &= \{\lambda^\alpha < |z| < \lambda^\beta\}. \end{aligned}$$

With these, we have

**Lemma 11.8.3.** $R^{-1}(\mathscr{A})$ *is the disjoint union of two doubly connected subdomains, say $V$ and $W$ of $\mathscr{A}$, where:*

(1) *$V$ contains $V_1$, and $W$ contains $W_1$;*
(2) *$R$ is a two-fold map of $V$ onto $\mathscr{A}$; and*
(3) *$R$ is a three-fold map of $W$ onto $\mathscr{A}$.*

PROOF. For small $\lambda$, $R(V_1)$ is approximately

$$\{\tfrac{1}{4} < |z| < \tfrac{9}{4}\},$$

and so for all sufficiently small $\lambda$, $R(V_1) \subset \mathscr{A}$ and hence $V_1 \subset R^{-1}(\mathscr{A})$. On $W_1$ we have

$$\lambda|z|^{-3} - |z|^2 \leq |R(z)| \leq |z|^2 + \lambda|z|^{-3} = \varphi(|z|),$$

where $|z| < \lambda^{\beta} < \lambda^{1/5}$. Now $\varphi$ is decreasing on this range so on $W_1$,

$$\lambda^{1-3\beta} - \lambda^{2\beta} \leq |R(z)| \leq \lambda^{2\alpha} + \lambda^{1-3\alpha},$$

and it follows that $R(W_1) \subset \mathscr{A}$ if

$$\lambda^{a} < \lambda^{1-3\beta} - \lambda^{2\beta} \leq \lambda^{2\alpha} + \lambda^{1-3\alpha} < \lambda^{-b}.$$

This is so because $a > 1 - 3\beta$, $\beta > \frac{1}{5}$ and $b > 3\alpha - 1$; thus both $V_1$ and $W_1$ lie in $R^{-1}(\mathscr{A})$.

It follows that there are components $V$ and $W$ of $R^{-1}(\mathscr{A})$ containing $V_1$ and $W_1$ respectively, and by Lemma 11.8.1, $V$ and $W$ lie in $\mathscr{A}$. Assuming for the moment that $V$ and $W$ are disjoint, we can prove (2) and (3) as follows. As $\deg(R) = 5$, there are exactly five solutions of $R(z) = 1$ in $\mathbb{C}_\infty$, and so (2) and (3) will hold if we prove that the equation $R(z) = 1$ has two solutions in $V_1$ and three solutions in $W_1$. Obviously, we use Rouché's Theorem here and with this in mind, we note that for all sufficiently small $\lambda$,

(i) $\quad |z^5| > |z^3 - \lambda| \quad$ on $\quad |z| = \frac{3}{2}$;

(ii) $\quad |z^3| > |z^5 + \lambda| \quad$ on $\quad |z| = \frac{1}{2}$;

(iii) $\quad |z^3| > |z^5 + \lambda| \quad$ on $\quad |z| = \lambda^{\beta}$;

(iv) $\quad \lambda > |z^3 - z^5| \quad$ on $\quad |z| = \lambda^{\alpha}$.

Note that (iii) holds as $\beta < \frac{1}{3}$, and (iv) holds as $\alpha > \frac{1}{3}$. With these inequalities, Rouché's Theorem shows that $R(z) = 1$ does indeed have two solutions in $V_1$ and three solutions in $W_1$, and this verifies (2) and (3). It also shows that $U$ and $V$ are the only components of $R^{-1}(\mathscr{A})$.

To complete the proof of the lemma, we must show that $V$ and $W$ are disjoint, doubly connected domains and to do this we examine the action of $R$ on its critical points. As $\deg(R) = 5$, $R$ has eight critical points, say $0, 0, \infty$ and $\zeta_1, \ldots, \zeta_5$, where all of the $\zeta_j$ lie on the circle

$$\Gamma = \{z\colon |z|^5 = 3\lambda/2\}.$$

Now for $z$ on $\Gamma$ (and sufficiently small $\lambda$), we have

$$|R(z)| \leq |z|^2(1 + \lambda/|z|^5) = (\tfrac{5}{3})|z|^2 < 2\lambda^{2/5} < \lambda^a,$$

so $R(\Gamma)$ is disjoint from $\mathscr{A}$. As $\Gamma$ separates $V_1$ from $W_1$ (because $\beta > \frac{1}{5}$), and as $\Gamma$ is disjoint from $R^{-1}(\mathscr{A})$, we see that $\Gamma$ separates $V$ from $W$ and so $V$ and $W$ are disjoint.

Finally, as $R(\Gamma)$ is disjoint from $\mathscr{A}$, there are no critical values of $R$ in $\mathscr{A}$, and this shows that for every component $\Omega$ of $R^{-1}(\mathscr{A})$, $R: \Omega \to \mathscr{A}$ is a smooth covering map of $\Omega$ onto $\mathscr{A}$. With this, the Riemann–Hurwitz relation shows that each such $\Omega$ is a doubly connected subdomain of $\mathscr{A}$ and the proof is complete. □

By examining this argument, it should be clear that we can indeed replace $\mathscr{A}$ by its closure in the definition of $K$ above so, as asserted, $K$ is both closed and non-empty. Moreover,

$$R^{-2}(\mathscr{A}) = R^{-1}(V) \cup R^{-1}(W),$$

and this has four components, each of which contain a point of $K$ (see the proof of Lemma 11.8.2). To complete the proof of Lemma 11.8.2, we need to prove (b), namely that $R^n \to \infty$ on $F$, and this will follow from a little more knowledge about the dynamics of $R$. We prove

**Lemma 11.8.4.** *$F_0$ contains the disc $\{|z| < (\lambda/2)^{1/3}\}$, and $F_\infty$ contains the disc $\{|z| > \frac{3}{2}\}$.*

PROOF. If $|z| > \frac{3}{2}$, then

$$|R(z)/z| \geq |z| - \lambda|z|^{-4} > \tfrac{3}{2} - \lambda > \tfrac{5}{4},$$

and so $z \in F_\infty$. Next, if $|w|^3 < \lambda/2$, then

$$|R(w)| \geq \lambda|w|^{-3} - |w|^2 \geq 2 - (\lambda/2)^{2/3} > \tfrac{3}{2},$$

so $\{|w| < (\lambda/2)^{1/3}\}$ is a connected subset of $R^{-1}(F_\infty)$. As $R(0) = \infty$, $F_0 \subset R^{-1}(F_\infty)$ and so $w \in F_0$. □

We have seen that $R(\Gamma) \subset \{|z| < 2\lambda^{2/5}\}$ and, from Lemma 11.8.4, this is contained in $F_0$ (because $\frac{1}{3} < \frac{2}{5}$). It follows that $R^2(\Gamma)$ lies in $F_\infty$ and so the forward orbit of each critical point of $R$ converges to $\infty$: this is (c). It follows from this that (b), namely $R^n \to \infty$ on $F$, holds for if not, there would be some cycle of components of $F(R)$ other than $F_\infty$ and this would have to attract the forward orbit of some critical point as described in §9.3. Note that this information about the forward orbits of critical points also yields (f) (see Theorem 9.7.5).

The crucial point now is to show that $F_0 \neq F_\infty$. Exactly as we argued for $\mathscr{A}$, we find that $R^{-1}(V)$ and $R^{-1}(W)$ each consist of two annuli, and so on as we apply $R^{-1}$ repeatedly. The only issue here is to show that these annuli must separate 0 and $\infty$, and we can see that this must be so in the following way. Take any annulus $A$ in $\mathscr{A}$, so $A$ contains no critical values of $R$. Take a simple closed curve $\sigma$ passing once around $A$ and lift this back under $R^{-1}$ to a simple closed curve $\tau$ in a component $B$ of $R^{-1}(A)$. Now for some $m$, $R$ is an $m$-fold map of $B$ onto $A$, and $R(\tau)$ is the curve $\sigma$ traversed $m$ times (and possibly in the opposite direction to $\sigma$). Assume now that $A$ surrounds the origin. Then, using $n(\gamma, w)$ for the winding number of a curve $\gamma$ about $w$, we have

$$n(R(\tau), 0) = \pm mn(\sigma, 0) \neq 0,$$

and so by the Argument Principle, if $R$ has $Z$ zeros and $P$ poles inside $\tau$, then $Z \neq P$. In particular, $B$ must contain some zeros, or some poles, of $R$ in its inner hole. Now the zeros of $R$ do not lie in $R^{-1}(\mathscr{A})$ (for they map to 0), but they do lie between $W_1$ and $V_1$ (for $\beta > \frac{1}{5}$); thus they must lie between the doubly connected domains $V$ and $W$. It follows that if $B \subset W$, then $B$ must contain a pole of $R$, namely 0, in its inner hole, while if $B \subset V$, then $B$ must contain a zero or a pole in its inner hole. In both cases, $B$ must surround to the origin; thus we have proved that if $A$ in $\mathscr{A}$ surrounds the origin, then so does each component of $R^{-1}(A)$. This step is essential if we are to justify our claim that $F_0$ and $F_\infty$ are separated by a closed curve in $J$.

Knowing now that the annuli in our construction all separate 0 from $\infty$, we consider any decreasing sequence of compact annuli, say $A_1, A_2, \ldots$ from our construction. Denoting the two components of the complement of $A_1$ by $C_1$ and $C_2$, it is easy to prove that the intersection $\bigcap A_n$ separates $C_1$ from $C_2$, and hence 0 from $\infty$. This argument shows that $F_0$ and $F_\infty$ are separated by $J$, and hence $F_0 \neq F_\infty$.

Now consider $F_\infty$. The finite critical points of $R$ are not in $F_\infty$ (they map to $F_0$, and $F_\infty$ is forward invariant) and so $F_\infty$ must be simply connected: to see this, take a disc $D$ at $\infty$ with $R(D) \subset D$, and let $D_n$ be the component of $R^{-n}(D)$ that contains $\infty$; then (by induction)

$$\zeta(D_{n+1}) + 1 = 2\zeta(D_n) = 2,$$

and so $F_\infty$ is the union of the exanding sequence of simply connected domains $D_n$. Next, the non-zero critical points of $R$ are not in $F_0$ (if so, then $R(\Gamma)$ would have to lie in $F_\infty$), so $R$ is a three-fold covering map of $F_0$ onto $F_\infty$ which has two critical points at the origin and no others on $F_0$; the Riemann Hurwitz formula

$$\zeta(F_0) + 2 = 3\zeta(F_\infty) = 3$$

shows that $F_0$ must also be simply connected.

Finally, let

$$Q = \{\lambda^{1/5} < |z| < (3\lambda/2)^{1/5}\}.$$

Exactly as for $R(\Gamma)$, we see that $R(Q) \subset F_0$, and so $Q$ lies in some component $F_1$ of $F$. As $Q$ contains all five zeros of $R$, and all five non-zero finite critical points of $R$, $R$ maps $F_1$ in a five-fold manner onto $F_0$, and the Riemann–Hurwitz formula

$$\zeta(F_1) + 5 = 5\zeta(F_0) = 5$$

shows that $F_1$ is doubly connected. We have now shown first that $R^{-1}(F_\infty)$ comprises $F_0$ and $F_\infty$, and second, that $R^{-1}(F_0) = F_1$. All pre-images of $F_1$ are free of critical points and so are also doubly connected.

The proof that $J$ is a union of Jordan curves uses the fact that $R$ is expanding on $J$ (and the argument follows the lines of that used in §9.9), and for the same reason that the classical Cantor set is uncountable, so $J$ includes an uncountable number of Jordan curves. As each component of $F$ has finite connectivity, (e) holds; alternatively, one can argue in the same way as is used to show that not every point of the classical Cantor set is an end-point of one of the intervals used in its construction.

EXERCISE 11.8

1. Given $\varepsilon > 0$, show that if $\lambda < \varepsilon/2$ then $F_\infty$ contains the disc $\{|z| > 1 + \varepsilon\}$.

## §11.9. The Function $(z-2)^2/z^2$

We shall give a direct proof that for the function

$$R(z) = (z-2)^2/z^2,$$

$J(R)$ is the complex sphere. This proof does not use Sullivan's No Wandering Domains Theorem (see §9.4), nor does it depend on the theory of elliptic functions (as does the example in §4.3).

We assume that $F(R)$ is non-empty, and let $F_0$ be any one of its components. Suppose first that there is some sequence $R^n$ converging locally uniformly to a non-constant function $\varphi$ in $F_0$. Then $\varphi(F_0)$ is a domain, and if $\zeta \in \varphi(F_0)$, we can find some closed disc $D$ in $F_0$ with centre $z_0$, where $\varphi(z_0) = \zeta$. As $R^{n_j} \to \varphi$ uniformly on $\partial D$, Rouché's Theorem implies that $\zeta$ lies in $R^{n_j}(F_0)$ for large $j$ and so $\varphi(F_0) \subset F(R)$.

Let $F_1$ be the component of $F(R)$ which contains $\varphi(F_0)$, and now argue as in the proof of Theorem 7.2.4. We may assume that $m_j \to +\infty$, where $m_j = n_j - n_{j-1}$ and on a subsequence, $R^{m_j} \to \psi$, say, locally uniformly on $F_1$. It follows that $\psi\varphi = \psi$ on $F_1$, and so $\psi = I$; thus the identity map $I$ is a limit function in $F_1$. For large $j$, $R^{m_j}(F_1)$ meets, and hence is, $F_1$, and this implies (as in the proof of Theorem 7.2.4) that $R^m$ is injective in $F_1$. Thus some iterate $R^m$ is an automorphism of $F_1$ and so $F_1$ must be either a Siegel disc or a Herman ring of $R^m$. This cannot be so, however, for this requires some critical point of $R^m$ to have an infinite forward orbit, and the critical points of $R$ are 0 and 2 and these are pre-periodic for $R$ acts in the following way:

$$2 \to 0 \to \infty \to 1 \to 1 \to \cdots.$$

We have now shown that any limit function $\varphi$ in $F_0$ is constant there (and such functions must exist). Suppose, then, that $\varphi$ takes the constant value $\lambda$: we claim that $\lambda \in \{0, 1, 2, \infty\}$, To prove this, we suppose that $R^{n_j} \to \lambda$ in $F_0$, where $\lambda$ is not any of these values, and construct an open disc $D$ with centre $\lambda$ not containing 0, 1, 2 or $\infty$, and thus not meeting any critical value of any $R^n$. Choose $z_0$ in $F_0$; then for large $j$, $R^{n_j}(z_0) \in D$ and so we can define a branch $S_{n_j}$ of $(R^{n_j})^{-1}$ which maps $R^{n_j}(z_0)$ to $z_0$. Further, $S_{n_j}$ can be continued analytically over $D$ to obtain a single-valued analytic function there, and $S_{n_j}R^{n_j} = I$ on $F_0$ (as this holds near $z_0$). By Theorem 9.2.1, the family $\{S_{n_j}\}$ is normal in $D$, and so on some subsequence (which we may take to be $n_j$) $S_{n_j} \to \Phi$, say, on $D$. From standard arguments (involving normal families), we now have for *any* $z_1$ in $F_0$, $R^{n_j}(z_1) \to \lambda$, so

$$z_1 = S_{n_j}R^{n_j}(z_1) \to \Phi(\lambda),$$

so $\Phi(\lambda) = z_1$, a contradiction.

We have now shown that if $F(R)$ is not empty, then there is some sequence of iterates, say $R^{n_j}$, converging locally uniformly to some $\lambda$ in $\{0, 1, 2, \infty\}$. By applying $R^3$, it follows that some sequence of iterates converges locally uniformly to 1 on $F_0$, so we may assume that the sequence $R^{n_j}$ does. We shall now obtain a contradiction. Let

$$D = \{z: |z - 1| < \tfrac{1}{8}\}.$$

As

$$R(z) - 1 = 4(1 - z)/z^2$$

we find that on $D$,

$$3|z - 1| < |R(z) - 1| < 6|z - 1|.$$

Now suppose that $R^n(z) \in D$. Then using the lower bound in the preceding inequality, we see that the points $R^{n+1}(z)$, $R^{n+2}(z)$, ... move away from the point 1, at least until one of them leaves $D$. Thus there is some $k$ with $R^{n+k}(z)$ in $D$ and $R^{n+k+1}(z)$ outside $D$. This gives

$$\tfrac{1}{8} < |R^{n+k+1}(z) - 1| < 6|R^{n+k}(z) - 1| < \tfrac{3}{4},$$

and so $R^{n+k+1}(z)$ lies in the annulus

$$A = \{z: \tfrac{1}{8} < |z - 1| < \tfrac{3}{4}\}.$$

It follows that as $n_j \to \infty$, there is a corresponding sequence $k_j$ such that $R^{n_j+k_j}(z)$ lies in $A$. This cannot be so, though, because we have seen that the only limit functions in $F_0$ are constants with one of the values 0, 1, 2 or $\infty$.

# References

[1] Ahlfors, L.V., *Lectures on Quasiconformal Mappings*, Van Nostrand, 1966.
[2] Ahlfors, L.V., *Conformal Invariants, Topics in Geometric Function Theory*, McGraw-Hill, 1973.
[3] Ahlfors, L.V., *Complex Analysis* (third edition), McGraw-Hill, 1979.
[4] Ahlfors, L.V. and Bers, L., Riemann's Mapping Theorem for variable metrics, *Ann. of Math.*, **72** (1960), 385–404.
[5] Ahlfors, L.V. and Sario, L. *Riemann Surfaces*, Princeton University Press, 1960.
[6] Arnold, V., Small denominators I: On the mappings of the circumference onto itself, *Amer. Math. Soc. Translations*, **46** (1965), 213–284.
[7] Baker, I.N., The existence of fixpoints of entire functions, *Math. Z.*, **73** (1960), 280–284.
[8] Baker, I.N. Permutable power series and regular iteration, *J. Austral. Math. Soc.*, **2** (1962), 265–294.
[9] Baker, I.N., Fixpoints of polynomials and rational functions, *J. London Math. Soc.*, **39** (1964), 615–622.
[10] Baker, I.N., Repulsive fixpoints of entire functions, *Math. Z.*, **104** (1968), 252–256.
[11] Baker, I.N., An entire function which has wandering domains, *J. Austral. Math. Soc.* (Ser. A), **22** (1976), 173–176.
[12] Baker, I.N., Wandering domains in the iteration of entire functions, *Proc. London Math. Soc.* (3), **49** (1984), 563–576.
[13] Baker, I.N., Some entire functions with multiply connected wandering domains, *Ergodic Theory Dynamical Systems*, **5** (1985), 163–169.
[14] Baker, I.N. and Eremenko, A., A problem on Julia sets, *Ann. Acad. Sci. Fenn., Series A*, **12** (1987), 229–236.
[15] Baker, I.N., Wandering domains for maps of the punctured plane, *Ann. Acad. Sci. Fenn.*, **12** (1987), 191–198.
[16] Barnsley, M., *Fractals Everywhere*, Academic Press, 1988.
[17] Beardon, A.F., The Hausdorff dimension of singular sets of properly discontinuous groups, *Amer. J. Math.*, **88** (1966), 722–736.
[18] Beardon, A.F., *The Geometry of Discrete Groups*, Springer-Verlag, 1983.

[19] Beardon, A.F., Symmetries of Julia sets, *Bull. London Math. Soc.* **22** (1990), 576–582.
[20] Beardon, A.F. and Pommerenke, Ch., The Poincaré metric of plane domains, *J. London Math. Soc.* (2), **18** (1978), 475–483.
[21] Bers, L., On Sullivan's proof of the Finiteness Theorem and the Eventual Periodicity Theorem, *Amer. J. Math.*, **109** (1987), 833–852.
[22] Blanchard, P., Complex analytic dynamics on the Riemann sphere, *Bull. Amer. Math. Soc.*, **11** (1984), 85–141.
[23] Boothby, W.M., *An Introduction to Riemannian Manifold and Differential Geometry*, Academic Press, 1975.
[24] Branner, B., Iteration of cubic polynomials, characterization of the Julia sets, *Preprint*, 1988.
[25] Branner, B. and Douady, A., Surgery on complex polynomials, *Proceedings of the Symposium on Dynamical Systems*, Mexico, 1986.
[26] Branner, B. and Hubbard, J.H., The iteration of cubic polynomials I, *Acta Math.*, **160** (1988), 143–206.
[27] Branner, B., and Hubbard, J.H., The iteration of cubic polynomials II, *Math. Rep.*, 1989–12.
[28] Brolin, H., Invariant sets under iteration of rational functions, *Ark. Mat.*, **6** (1965), 103–144.
[29] Brooks, R. and Matelski, J.P., The dynamics of 2-generator subgroups of PSL(2, *C*), Riemann surfaces and related topics, Proceedings of 1978 Stony Brook Conference, *Ann. of Math. Stud.*, **97** (1980), 65–72.
[30] Cayley, M.A., Application of the Newton–Fourier method to the imaginary root of an equation, *Quart. J. Math.*, **16** (1879), 179–185.
[31] Cayley, M.A., The Newton–Fourier imaginary problem, *Amer. J. Math.*, **2** (1879), 97.
[32] Cayley, M.A., Sur les racines d'une equation algebrique, *Comptes Rendus Acad. Sci. Paris*, **110** (1890), 215–218.
[33] Comacho, C., On the local structure of conformal mappings and holomorphic vector fields, *Asterisque*, **59–60** (1978), 83–94.
[34] Curry, J.H., Garnett, L. and Sullivan, D.R., On the iteration of a rational function: Computer experiments with Newton's method, *Comm. Math. Phys.*, **91** (1983), 267–277.
[35] Devaney, R.L., *An Introduction to Chaotic Dynamical Systems*, Benjamin/Cummings, 1986.
[36] Douady, A., Systems dynamiques holomorphes, Séminaire Bourbaki, No. 599 (1982); Asterisque, **105–106** (1983), 39–63.
[37] Douady, A., Disques de Siegel et anneaux de Herman, *Asterisque*, **152–153** (1987), 151–172.
[38] Douady, A. and Hubbard, J., Etudes dynamique des polynômes complexes, *Publications Mathematiques d'Orsay*, 1984.
[39] Douady, A. and Hubbard, J., On the dynamics of polynomial-like mappings, *Ann. Sci. Ecole Norm. Sup.*, **18** (1985), 287–344.
[40] du Val, P., *Elliptic Functions and Elliptic Curves*, London Mathematical Society Lecture Notes, No. 9, Cambridge University Press, 1973.
[41] Eremenko, A.E., and Ljubich, M.Ju. Examples of entire functions with pathological dynamics, *J. London Math. Soc.* (2), **36** (1987), 458–468.
[42] Falconer, K.J., *The Geometry of Fractal Sets*, Cambridge University Press, 1985.
[43] Falconer, K.J., *Fractal Geometry: Mathematical Foundations and Applications*, Wiley, 1990.
[44] Fatou, M.P., Sur les equations fonctionelles, *Bull. Soc. Math. France*, **47** (1919), 161–271.

[45] Fatou, M.P., Sur les equations fonctionelles, *Bull. Soc. Math. France*, **48** (1920), 33–94 and 208–314.

[46] Ford, L.R., *Automorphic Functions* (second edition), Chelsea, 1951.

[47] Garber, V.L., On the iteration of rational functions, *Math. Proc. Cambridge Philos. Soc.*, **84** (1978), 497–505.

[48] Gehring, F.W. and Martin, M., Discreteness in Kleinian groups and the iteration theory of quadratic maps, preprint, 1991.

[49] Gehring, F.W. and Martin, G., Iteration theory and inequalities for Kleinian groups, *Bull. Amer. Math. Soc.*, preprint, 1991.

[50] Grzegorczyk, P., Przytycki, F. and Szlenk, W., On iterations of Misiurewicz's rational maps on the Riemann sphere, *Preprint*, 1988.

[51] Guckenheimer, J., Endomorphisms of the Riemann sphere (editors, S.S. Chern and S. Smale), *Proc. Sympos. Pure Math.*, Vol. 14, 95–123, American Mathematical Society, Providence, 1970.

[52] Herman, M.R., Sur la conjugaison differentiable des diffeomorphismes du cerle a des rotations, *Publ. I.H.E.S.*, **49** (1979), 5–233.

[53] Herman, M.R., Exemples de fractions rationelles ayant une orbite dense sur la sphere de Riemann, *Bull. Soc. Math. France*, **112** (1984), 93–142.

[54] Herman, M.R., Are there critical points on the boundaries of singular domains?, *Comm. Math. Phys.*, **99** (1985), 593–612.

[55] Herman, M.R., Recent results and some open questions on Siegel's linearization theorem on germs of complex analytic diffeomorphisms of $C^n$ near a fixed point, *Proc. VIII Int. Cong. Math. Phys.* (1987), 138–184, World Scientific, Singapore.

[56] Hocking, J.G. and Young G.S., *Topology*, Addison-Wesley, 1961.

[57] Holden, A.V., *Chaos*, Manchester University Press, 1986.

[58] Julia, G., Memoire sur l'iteration des fonctions rationelles, *J. Math. Pures Appl.*, **8** (1918), 47–245.

[59] Jungreis, I., The uniformization of the complement of the Mandelbrot set, *Duke Math. J.*, **52** (1985), 935–938.

[60] Landkoff, N.S., *Foundations of Modern Potential Theory*, Springer-Verlag, 1972.

[61] Lang, S. *Elliptic Functions*, Addison-Wesley, 1973.

[62] Lattès, S., Sur l'iteration des substitutions rationelles et les fonctions de Poincaré, *Comptes Rendus Acad. Sci. Paris*, **166** (1918), 26–28.

[63] Lehto, O., *Univalent Functions and Teichmuller Spaces*, Graduate Texts in Mathematics, Vol. 109, Springer-Verlag, 1987.

[64] Lehto, O. and Virtanen, K., *Quasiconformal Mappings in the Plane*, Springer-Verlag, 1973.

[65] Lyubich, M.Ju, Entropy properties of rational endomorphisms of the Riemann sphere, *Ergodic Theory Dynamical Systems*, **3** (1983), 351–385.

[66] Lyubich, M.Y., On typical behaviour of the trajectories of a rational mapping of the sphere, *Soviet Math. Dokl.*, **27** (1983), 22–25.

[67] Lyubich, M.Y., The dynamics of rational transforms: the topological picture, *Uspekhi Mat. Nauk*, **41** (1986), 35–95; *Russian Math. Surveys*, **41** (1986), 43–117.

[68] Mandelbrot, B., Fractal aspects of the iteration of $z \mapsto \lambda z(1 - z)$ for complex $\lambda$ and $z$, *Ann. New York Acad. Sci.*, **357** (1980), 249–259.

[69] Mandelbrot, B., *The fractal Geometry of Nature*, Freeman, 1982.

[70] Mañe, R., On the instability of Herman rings, *Invent. Math.*, **81** (1985), 459–471.

[71] Mañe, R., Sad, P. and Sullivan, D., On the dynamics of rational maps, *Ann. Sci. Ecole Norm. Sup.*, **16** (1983), 193–217.

[72] Manning, A., The dimension of the maximal measure for a polynomial map,

*Ann. of Math.*, **119** (1984), 425–430.
[73] May, R.M., Simple mathematical models with very complicated dynamics, *Nature*, **261** (1976), 459–467.
[74] McMullen, C., Automorphisms of rational maps, *MSRI Preprint 06019-86*, 1986.
[75] McMullen, C., Area and Hausdorff dimension of Julia sets, *Trans. Amer. Math. Soc.* **300** (1987), 329–342.
[76] Nehari, Z., *Conformal Mapping*, McGraw-Hill, 1952.
[77] Newman, M. *Elements of the Topology of Plane Sets*, Cambridge University Press, 1939.
[78] Oba, M.K. and Pitcher, T.S., A new characterization of the *F*-set of a rational function, *Trans. Amer. Math. Soc.*, **166** (1972), 297–308.
[79] Peitgen, H.O., Saupe, D. and v. Haesler, F., Cayley's problem and Julia sets, *Math. Intelligencer*, **6** (1984), 11–20.
[80] Peitgen, H.O. and Richter, P.H., *The Beauty of Fractals*, Springer-Verlag, 1986.
[81] Peitgen, H.O. and Saupe, D., *The Science of Fractal Images*, Springer-Verlag, 1988.
[82] Ritt, J.F., Periodic functions with a multiplication theorem, *Trans. Amer. Math. Soc.*, **23** (1922), 16–25.
[83] Ritt, J.F., On the iteration of rational functions, *Trans. Amer. Math. Soc.*, **21** (1920), 348–356.
[84] Rogers, C.A., *Hausdorff Measures*, Cambridge University Press, 1970.
[85] Rudin, W., *Real and Complex Analysis*, McGraw-Hill, 1966.
[86] Ruelle, D., Repellers for real analytic maps, *Ergodic Theory Dynamical Systems*, **2** (1982), 99–108.
[87] Sad, P., *Introducao a dinamica das funcoes racionais ne esfera de Riemann*, 14th Coloquio Brasiliero be Matematica, 1983.
[88] Shishikura, M., On the quasiconformal surgery of rational functions, *Ann. Sci. Ecole Norm. Sup.*, **20** (1987), 1–29.
[89] Siegel, C.L., Iteration of analytic functions, *Ann. of Math.*, **43** (1942), 607–616.
[90] Siegel, C.L. and Moser, J., *Lectures on Celestial Mechanics*, Springer-Verlag, 1971.
[91] Sternberg, S., Local contractions and a theorem of Poincaré, *Amer. J. Math.*, **79** (1957), 809–824.
[92] Sullivan, D., Conformal dynamical systems, in *Geometric Dynamics*, Springer-Verlag Lecture Notes, 1007 (1983), 725–752.
[93] Sullivan, D., Quasiconformal homeomorphisms and dynamics I: Solution of the Fatou–Julia problem on wandering domains, *Ann. of Math.*, **122** (1985), 401–418.
[94] Sullivan, D., Quasiconformal homeomorphisms and dynamics II, Preprint, 1985.
[95] Sullivan, D., Quasiconformal homeomorphisms and dynamics III: Topological conjugacy classes of analytic endomorphisms, Preprint, 1985.
[96] Targonski, G., *Topics on Iteration Theory*, Studia Mathematica 6, Vandenhoeck & Ruprecht, 1981.
[97] Taylor, S.J., On the connexion between Hausdorff measures and generalized capacities, *Proc. Cambridge Philos. Soc.*, **57** (1964), 524–531.
[98] Tsuji, M., *Potential Theory in Modern Function Theory*, Maruzen, 1959.
[99] Ulam, S.M. and von Neumann, J., On the combination of stochastic and deterministic processes, *Bull. Amer. Math. Soc.*, **53** (1947), 1120.
[100] Ushiki, S., Chaotic phenomena and fractal objects in numerical analysis, *Stud. Math. Appl.*, **18** (1986), 221–258.

[101] Voronin, S.M., Analytic classification of germs of conformal mappings $(C, 0) \to (C, 0)$ with identity linear part, *Functional Anal. Appl.*, **15** (1981), 1–13.
[102] Whyburn, G.T., *Analytic Topology*, Amer. Math. Soc. Coll. Publ., **28** (1942).
[103] Yakobson, M.V., On the problem of classification of polynomial endomorphisms of the plane, *Mat. Sb.*, **80** (1969), 365–387; *Math. USSR-Sb*, **9** (1969).
[104] Yakobson, M.V., On the question of topological classification of rational mappings of the Riemann sphere, *Uspekhi Mat. Nauk*, **28**:2 (1973), 247–248.
[105] Yakobson, M.V., On the boundaries of some normality domains for rational maps, *Uspekhi Mat. Nauk*, **39** (1984), 211–212: *Russian Math. Surveys*, **39** (1984), 229–230.

# Index of Examples

**Maps of Degree One**



**Quadratic Polynomials**



**Cubic Polynomials**



**Rational Maps**

# Index

# Graduate Texts in Mathematics

*(continued from page ii)*

66 WATERHOUSE. Introduction to Affine Group Schemes.
67 SERRE. Local Fields.
68 WEIDMANN. Linear Operators in Hilbert Spaces.
69 LANG. Cyclotomic Fields II.
70 MASSEY. Singular Homology Theory.
71 FARKAS/KRA. Riemann Surfaces. 2nd ed.
72 STILLWELL. Classical Topology and Combinatorial Group Theory. 2nd ed.
73 HUNGERFORD. Algebra.
74 DAVENPORT. Multiplicative Number Theory. 2nd ed.
75 HOCHSCHILD. Basic Theory of Algebraic Groups and Lie Algebras.
76 IITAKA. Algebraic Geometry.
77 HECKE. Lectures on the Theory of Algebraic Numbers.
78 BURRIS/SANKAPPANAVAR. A Course in Universal Algebra.
79 WALTERS. An Introduction to Ergodic Theory.
80 ROBINSON. A Course in the Theory of Groups. 2nd ed.
81 FORSTER. Lectures on Riemann Surfaces.
82 BOTT/TU. Differential Forms in Algebraic Topology.
83 WASHINGTON. Introduction to Cyclotomic Fields. 2nd ed.
84 IRELAND/ROSEN. A Classical Introduction to Modern Number Theory. 2nd ed.
85 EDWARDS. Fourier Series. Vol. II. 2nd ed.
86 VAN LINT. Introduction to Coding Theory. 2nd ed.
87 BROWN. Cohomology of Groups.
88 PIERCE. Associative Algebras.
89 LANG. Introduction to Algebraic and Abelian Functions. 2nd ed.
90 BRØNDSTED. An Introduction to Convex Polytopes.
91 BEARDON. On the Geometry of Discrete Groups.
92 DIESTEL. Sequences and Series in Banach Spaces.
93 DUBROVIN/FOMENKO/NOVIKOV. Modern Geometry—Methods and Applications. Part I. 2nd ed.
94 WARNER. Foundations of Differentiable Manifolds and Lie Groups.
95 SHIRYAEV. Probability. 2nd ed.
96 CONWAY. A Course in Functional Analysis. 2nd ed.
97 KOBLITZ. Introduction to Elliptic Curves and Modular Forms. 2nd ed.
98 BRÖCKER/TOM DIECK. Representations of Compact Lie Groups.
99 GROVE/BENSON. Finite Reflection Groups. 2nd ed.
100 BERG/CHRISTENSEN/RESSEL. Harmonic Analysis on Semigroups: Theory of Positive Definite and Related Functions.
101 EDWARDS. Galois Theory.
102 VARADARAJAN. Lie Groups, Lie Algebras and Their Representations.
103 LANG. Complex Analysis. 3rd ed.
104 DUBROVIN/FOMENKO/NOVIKOV. Modern Geometry—Methods and Applications. Part II.
105 LANG. $SL_2(\mathbf{R})$.
106 SILVERMAN. The Arithmetic of Elliptic Curves.
107 OLVER. Applications of Lie Groups to Differential Equations. 2nd ed.
108 RANGE. Holomorphic Functions and Integral Representations in Several Complex Variables.
109 LEHTO. Univalent Functions and Teichmüller Spaces.
110 LANG. Algebraic Number Theory.
111 HUSEMÖLLER. Elliptic Curves.
112 LANG. Elliptic Functions.
113 KARATZAS/SHREVE. Brownian Motion and Stochastic Calculus. 2nd ed.
114 KOBLITZ. A Course in Number Theory and Cryptography. 2nd ed.
115 BERGER/GOSTIAUX. Differential Geometry: Manifolds, Curves, and Surfaces.
116 KELLEY/SRINIVASAN. Measure and Integral. Vol. I.
117 SERRE. Algebraic Groups and Class Fields.
118 PEDERSEN. Analysis Now.
119 ROTMAN. An Introduction to Algebraic Topology.
120 ZIEMER. Weakly Differentiable Functions: Sobolev Spaces and Functions of Bounded Variation.
121 LANG. Cyclotomic Fields I and II. Combined 2nd ed.
122 REMMERT. Theory of Complex Functions. *Readings in Mathematics*
123 EBBINGHAUS/HERMES et al. Numbers. *Readings in Mathematics*
124 DUBROVIN/FOMENKO/NOVIKOV. Modern Geometry—Methods and Applications. Part III.
125 BERENSTEIN/GAY. Complex Variables: An Introduction.
126 BOREL. Linear Algebraic Groups. 2nd ed.

127 MASSEY. A Basic Course in Algebraic Topology.
128 RAUCH. Partial Differential Equations.
129 FULTON/HARRIS. Representation Theory: A First Course.
*Readings in Mathematics*
130 DODSON/POSTON. Tensor Geometry.
131 LAM. A First Course in Noncommutative Rings.
132 BEARDON. Iteration of Rational Functions.
133 HARRIS. Algebraic Geometry: A First Course.
134 ROMAN. Coding and Information Theory.
135 ROMAN. Advanced Linear Algebra.
136 ADKINS/WEINTRAUB. Algebra: An Approach via Module Theory.
137 AXLER/BOURDON/RAMEY. Harmonic Function Theory.
138 COHEN. A Course in Computational Algebraic Number Theory.
139 BREDON. Topology and Geometry.
140 AUBIN. Optima and Equilibria. An Introduction to Nonlinear Analysis.
141 BECKER/WEISPFENNING/KREDEL. Gröbner Bases. A Computational Approach to Commutative Algebra.
142 LANG. Real and Functional Analysis. 3rd ed.
143 DOOB. Measure Theory.
144 DENNIS/FARB. Noncommutative Algebra.
145 VICK. Homology Theory. An Introduction to Algebraic Topology. 2nd ed.
146 BRIDGES. Computability: A Mathematical Sketchbook.
147 ROSENBERG. Algebraic *K*-Theory and Its Applications.
148 ROTMAN. An Introduction to the Theory of Groups. 4th ed.
149 RATCLIFFE. Foundations of Hyperbolic Manifolds.
150 EISENBUD. Commutative Algebra with a View Toward Algebraic Geometry.
151 SILVERMAN. Advanced Topics in the Arithmetic of Elliptic Curves.
152 ZIEGLER. Lectures on Polytopes.
153 FULTON. Algebraic Topology: A First Course.
154 BROWN/PEARCY. An Introduction to Analysis.
155 KASSEL. Quantum Groups.
156 KECHRIS. Classical Descriptive Set Theory.
157 MALLIAVIN. Integration and Probability.
158 ROMAN. Field Theory.
159 CONWAY. Functions of One Complex Variable II.
160 LANG. Differential and Riemannian Manifolds.
161 BORWEIN/ERDÉLYI. Polynomials and Polynomial Inequalities.
162 ALPERIN/BELL. Groups and Representations.
163 DIXON/MORTIMER. Permutation Groups.
164 NATHANSON. Additive Number Theory: The Classical Bases.
165 NATHANSON. Additive Number Theory: Inverse Problems and the Geometry of Sumsets.
166 SHARPE. Differential Geometry: Cartan's Generalization of Klein's Erlangen Program.
167 MORANDI. Field and Galois Theory.
168 EWALD. Combinatorial Convexity and Algebraic Geometry.
169 BHATIA. Matrix Analysis.
170 BREDON. Sheaf Theory. 2nd ed.
171 PETERSEN. Riemannian Geometry.
172 REMMERT. Classical Topics in Complex Function Theory.
173 DIESTEL. Graph Theory. 2nd ed.
174 BRIDGES. Foundations of Real and Abstract Analysis.
175 LICKORISH. An Introduction to Knot Theory.
176 LEE. Riemannian Manifolds.
177 NEWMAN. Analytic Number Theory.
178 CLARKE/LEDYAEV/STERN/WOLENSKI. Nonsmooth Analysis and Control Theory.
179 DOUGLAS. Banach Algebra Techniques in Operator Theory. 2nd ed.
180 SRIVASTAVA. A Course on Borel Sets.
181 KRESS. Numerical Analysis.
182 WALTER. Ordinary Differential Equations.
183 MEGGINSON. An Introduction to Banach Space Theory.
184 BOLLOBAS. Modern Graph Theory.
185 COX/LITTLE/O'SHEA. Using Algebraic Geometry.
186 RAMAKRISHNAN/VALENZA. Fourier Analysis on Number Fields.
187 HARRIS/MORRISON. Moduli of Curves.
188 GOLDBLATT. Lectures on the Hyperreals: An Introduction to Nonstandard Analysis.
189 LAM. Lectures on Modules and Rings.
190 ESMONDE/MURTY. Problems in Algebraic Number Theory.
191 LANG. Fundamentals of Differential Geometry.

192 HIRSCH/LACOMBE. Elements of Functional Analysis.
193 COHEN. Advanced Topics in Computational Number Theory.
194 ENGEL/NAGEL. One-Parameter Semigroups for Linear Evolution Equations.
195 NATHANSON. Elementary Methods in Number Theory.
196 OSBORNE. Basic Homological Algebra.
197 EISENBUD/HARRIS. The Geometry of Schemes.
198 ROBERT. A Course in $p$-adic Analysis.
199 HEDENMALM/KORENBLUM/ZHU. Theory of Bergman Spaces.
200 BAO/CHERN/SHEN. An Introduction to Riemann–Finsler Geometry.
201 HINDRY/SILVERMAN. Diophantine Geometry: An Introduction.
202 LEE. Introduction to Topological Manifolds.
203 SAGAN. The Symmetric Group: Representations, Combinatorial Algorithms, and Symmetric Function. 2nd ed.
204 ESCOFIER. Galois Theory.
205 FÉLIX/HALPERIN/THOMAS. Rational Homotopy Theory.